An Introduction to Ecological Genomics

An Introduction to Ecological Genomics

Nico M. van Straalen and Dick Roelofs
Vrije Universiteit, Amsterdam

OXFORD
UNIVERSITY PRESS

OXFORD

UNIVERSITY PRESS

Great Clarendon Street, Oxford OX2 6DP

Oxford University Press is a department of the University of Oxford.
It furthers the University's objective of excellence in research, scholarship,
and education by publishing worldwide in

Oxford New York

Auckland Cape Town Dar es Salaam Hong Kong Karachi
Kuala Lumpur Madrid Melbourne Mexico City Nairobi
New Delhi Shanghai Taipei Toronto

With offices in

Argentina Austria Brazil Chile Czech Republic France Greece
Guatemala Hungary Italy Japan Poland Portugal Singapore
South Korea Switzerland Thailand Turkey Ukraine Vietnam

Oxford is a registered trade mark of Oxford University Press
in the UK and in certain other countries

Published in the United States
by Oxford University Press Inc., New York

British Library Cataloguing in Publication Data

Data available

Library of Congress Cataloging in Publication Data
Straalen, N. M. van
  An introduction to ecological genomics / Nico M. van Straalen and Dick Roelofs.
    p. cm.
  ISBN-13: 978-0-19-856671-7 (alk. paper)
  ISBN-10: 0-19-856671-9 (alk. paper)
 1. Molecular microbiology. 2. Microbiology. 3. Ecology. I. Roelofs, Dick. II. Title.
  QR74.S78 2006   576.5′8—dc22   2005029756

Cover design by Janine Mariën

Typeset by Newgen Imaging Systems (P) Ltd., Chennai, India
Printed in Great Britain
on acid-free paper by
Antony Rowe Ltd., Chippenham, Wiltshire

ISBN 0-19-856670-0 978-0-19-856670-0
ISBN 0-19-856671-9 (Pbk.) 978-0-19-856671-7 (Pbk.)

10 9 8 7 6 5 4 3 2 1

Preface

This book is an introduction to the exciting new field of ecological genomics, for use in MSc courses and by those beginning their PhD studies.

When we became involved in a national research programme on ecological genomics, or ecogenomics as it became known, we realized that information on this newly emerging subject needed to be brought together. In order to start up a research programme in such a new discipline, not only the students, but also we as teachers, had to get to grips with the subject. Furthermore, although obtaining a PhD implies mastering a specialized field, the PhD student must be able to place this field in a broader context if he or she is to become a mature scientist. This approach may be called the T-model of education; the horizontal bar of the T representing a broad understanding, and the vertical bar an investigation in depth, going down to the root of the problem. Our book uses this approach.

We assume a basic level of knowledge in the biological sciences to BSc level: ecology, evolutionary biology, microbiology, plant physiology, animal physiology, genetics, and molecular biology. We have tried to link up with the content of the most common textbooks in these fields, at the same time realizing that students of ecological genomics have a variety of backgrounds. However, our main targets are students with subjects closely related to ecology and evolutionary biology, which is why we place the emphasis on aspects that we judge to be particularly new to them.

Evolutionary genomics and bioinformatics are companion disciplines to ecological genomics. In the last 10 years interest in both disciplines has grown enormously. Several textbooks on bioinformatics have already been published and

subjects encompassed by evolutionary genomics, such as comparative genomics, phylogenetic analysis, and molecular evolution, can now be considered as fields in their own right. They are certainly too large to be covered in an introductory book on ecological genomics; indeed, evolutionary genomics deserves a textbook of its own.

We have organized this book around three issues important in modern ecology, choosing questions for which the links to genomics are best developed. At the outset, we perhaps use rather ambitious phrasing to announce the genomics approach to these ecological questions. Maybe our questions cannot be answered at this stage. However, we decided not to suppress unanswered, and thus open, issues. Instead we hope to stimulate discussion as well as provide factual information. We have included an appraisal section at the end of each chapter to emphasize this question-orientated approach. Combined with information given in the introductory section, this allows the reader to grasp the main points of each chapter, even if the detailed treatment of molecular principles and case studies are left aside.

Case studies are taken from literature published since the year 2000. Nevertheless, a book on genomics runs the risk of becoming outdated very quickly: the rate at which knowledge is being accrued and insight developed is unprecedented. However, we hope that our question-orientated set-up will be useful for some years to come, even when new and better case studies are available.

Before this book was written, journal articles comprised the only literature on ecological genomics. These, although very inspiring, were scattered widely. Today, most textbooks on genetics and evolution have a chapter on genomics. Gibson and Muse published a primer on genome science in 2002, but this did not cover ecological questions. So, for us, writing this book was ploughing unknown ground. We have attempted to add structure to the field, and hopefully have put ecological genomics on the map. However, we welcome constructive criticism and suggestions from our readers.

We thank the colleagues who reviewed parts of the book, suggested issues that had escaped us, or helped with correcting the English: Martin Feder, Claire Hengeveld, Jan Kammenga, René Klein Lankhorst, Bas Kooijman, Jan Kooter, Wilfred Röling, and Martijn Timmermans. We thank Desirée Hoonhout and Karin Uyldert for checking the reference list, and Nico Schaefers, for preparation of the figures. Ian Sherman at Oxford University Press provided us with stimulating discussion. We thank members of the Animal Ecology Department at the Vrije Universiteit for your friendship and encouragement. N.M.vS. also thanks the Faculty of Earth and Life Sciences of the Vrije Universiteit for granting the sabbatical leave during which most of this book was written.

Nico M. van Straalen and Dick Roelofs,
Amsterdam, July 2005

Contents

CHAPTER 1

What is ecological genomics?

We define ecological genomics as

a scientific discipline that studies the structure and functioning of a genome with the aim of understanding the relationship between the organism and its biotic and abiotic environments.

With this book we hope to contribute to this new discipline by summarizing the developments over the last 5 years and explaining the general principles of genomics technology and its application to ecology. Using examples drawn from the scattered literature, we indicate where ecological questions can be analysed, reformulated, or solved by means of genomics approaches. This first chapter introduces the main purpose of ecological genomics. We describe its characteristics, its interactions with other disciplines, and its fascination with model species. We also touch on some of its possible applications.

1.1 The genomics revolution invading ecology

The twentieth century has been called the 'century of the gene' (Fox Keller 2000). It began with the rediscovery in 1900 of the laws of inheritance by DeVries, Correns, and Von Tschermak, laws that had been formulated about 40 years earlier by Gregor Mendel. With the appearance of the Royal Horticultural Society's English translation of Mendel's papers, William Bateson suggested in a letter in 1902 that this new area of biology be called genetics. The word gene followed, coined by Wilhelm Ludvig Johannsen in 1909, and then in 1920 the German botanist Hans Winkler proposed the word genome. The term genomics did not

appear until the mid-1980s and was introduced in 1987 as the name of a new journal (McKusick and Ruddle 1987). The century ended with the genomics revolution, culminating in the announcement of the completion of a draft version of the humane genome in the year 2000.

Realizing the importance of Mendel's papers, William Bateson announced that genetics was to become the most promising research area of the life sciences. One hundred years later one cannot avoid the conclusion that the progress in understanding the role of genes in living systems indeed has been astonishing. The genomics revolution has now expanded beyond genetics, its impact being felt in many other areas of the life sciences, including ecology. In the ecological arena, the interaction between genomics and ecology has led to a new field of research, *evolutionary and ecological functional genomics*. Feder and Mitchell-Olds (2003) indicated that this new multidiscipline 'focuses on the genes that affect evolutionary fitness in natural environments and populations'.

Our definition of ecological genomics given above seems at first sight to include the basic aim of ecology, viewing genomics as a new tool for analysing fundamental ecological questions. However, the merging of genomics with ecology includes more than the incorporation of a toolbox, because with the new technology new scientific questions emerge and existing questions can be answered in a way that was not considered before. We expect therefore that ecological genomics will develop into a truly new discipline, and will forge a mechanistic basis for ecology that is often felt to be missing. This could also strengthen the relationship between ecology and the other life sciences, because to a certain extent ecological genomicists speak the same language and read the same papers as molecular biologists.

Fig. 1.1 illustrates the various fields from which ecological genomics draws and upon which it is still growing. First of all, as indicated by Feder and Mitchell-Olds (2003), ecological genomics is closely linked to evolutionary biology and the associated disciplines of population genetics and evolutionary ecology. Another major area supporting ecological genomics is plant and animal physiology, which have their base in biochemistry and cell biology. A special position is held by microbial ecology, the meeting place of microbiology and ecology, where the use of genomics approaches has proceeded further than in any other subdiscipline of ecology. We consider genomics itself as a mainly technological advance, supporting ecological genomics in the same way as it supports other areas of the life sciences, such as medicine, neurobiology, and agriculture.

The genomics revolution is not only due to advances in molecular biology. Three major technological developments that took place in the 1990s also made it possible: microtechnology, computing, and communication.

Microtechnogy. The possibility of working with molecules on the scale of a few micrometres, given by advances in laser technology, has been very important for one of genomics' most conspicuous achievements, the development of the gene chip.

Computing technology. To assemble a genome from a series of sequences requires tremendous computational power. Extensive calculations are also necessary for the analysis of expression matrices and protein databases. Without the advent of high-speed computers and data-storage systems of vast capacity all this would have been impossible.

Communication technology. Consulting genome databases all over the world has become such normal practice that the scientific progress of any genomics laboratory has become completely dependent on communication with the rest of the World Wide Web. The Internet has become an indispensable part of genomics.

The essence of genomics is that it is the study of the genome and its products *as a unitary whole.* In biology, the suffix -ome signifies the collectivity of units (Lederberg and McCray 2001), as for example in coelome, the system of body cavities, and biome, the entire community of plants and animals in a climatic region. In aiming to investigate many genes at the same time genomics differs from ecology, which although investigating many phenotypes, usually deals with only a few genes at a time (Fig. 1.2). Ecological genomics borrows from these two extremes, investigating phenotypic

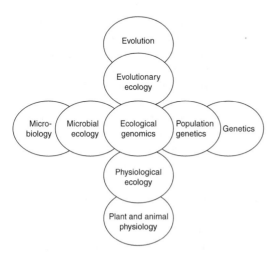

Figure 1.1 The position of ecological genomics in the middle of the other life-science disciplines with which it interacts most intensively.

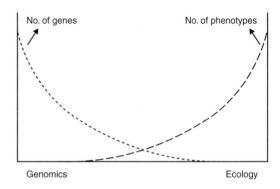

Figure 1.2 The playing field of ecological genomics, in between genomics, with its focus on the single genome of a model organism, studying all the genes that it contains, and ecology, studying a few genes in many species.

biodiversity as well as diversity in the genome. With this new discipline, ecology is enriched by genomics technology and genomics is enriched by ecological questioning and evolutionary views.

Because genomics analyses the genome in its entirety, it transcends classical genetics, which studies genes one by one, relating DNA sequences to proteins and ultimately to heritable traits. Genomics is based on the observation that the impact of one gene on the phenotype can only be understood in the context of the expression of several other genes or, in fact, of all other genes in the genome, plus their products, metabolites, cell structures, and all the interactions between them. This is not to say that every study in genomics deals with everything all the time, but that the mind is set and tools are deployed to maximize awareness of any effects elsewhere in the genome, outside the system under study. Consequently genomics is invariably associated with unexpected findings. The discovery aspect of genomics is expressed aptly in a public-education project of Genome Canada entitled *The GEEE! in Genome* (www.genomecanada.ca).

The work of Spellman and Rubin (2002) and their discovery of *transcriptional territories* in the genome of the fruit fly, *Drosophila melanogaster*, is an example of how the genomics approach can fundamentally alter our way of thinking about the relationship between genes and the environment (see also Weitzman 2002). The authors carried out transcription profiling with DNA microarrays (see Section 2.3) to investigate the expression of almost all of the genes in the fruit fly's genome under 88 different environmental conditions. Their work was in fact a meta-analysis of transcription profiles collected earlier in six separate investigations. Because the complete genome sequence of *Drosophila* is known, it was possible to trace every differentially expressed gene back to its chromosomal position. They concluded that genes physically adjacent in the genome often had similar expression when comparing different environmental challenges. The window of correlated expression appeared to extend to 10 or more adjacent genes and they estimated that 20% of the genome was organized in such 'expression

clusters'. Most astonishingly, genes in one cluster proved to be no more similar in structure or function than could be expected from a random arrangement. Spellman and Rubin (2002) suggested that local changes in chromatin structure trigger the expression of large groups of genes together. Thus a gene may be expressed not because there is a particular need for its product, but because its neighbour is expressed for a reason completely unrelated to the function of the first gene. At the moment it is not known whether such mechanisms lead to unexpected correlations between phenotypic traits, but surely the discovery of transcriptional territories could never have been made on a gene-by-gene basis, and this is due to the genomics approach.

The interactions between the genes within the genome and the dynamic character of the genome on an evolutionary scale have been sketched vividly by Dover (1999) as an *internal tangled bank*. This idea goes back to Darwin (1859) who, after investigating the banks of hollow roads in the English countryside, was intrigued by the great variety of organisms tangled together:

> It is interesting to contemplate an entangled bank, clothed with many plants of many kinds, with birds singing on the bushes, with various insects flitting about, and with worms crawling through the damp earth ...

Darwin considered the way in which all organisms depended on each other as the template for evolution. Inspired by Darwin, Dover (1999) made a distinction between the 'external tangled bank' (the ecology) and the 'internal tangled bank' (the genome), attributing to them complementary roles in the evolutionary process (Fig. 1.3). The concept of the internal tangled bank emphasizes the role of *genetic turbulence* (gene duplication, genetic sweeps, exon shuffling, transposition, etc.) in the genome and it illustrates that there is ample scope for 'innovation from within'. These innovations are then checked against the external tangled bank, and this constitutes the process of evolution. This agrees with François Jacob's famous description of 'evolution through tinkering' (Jacob 1977). It should not surprise us that genetic turbulence leaves many traces in the genome that do not have

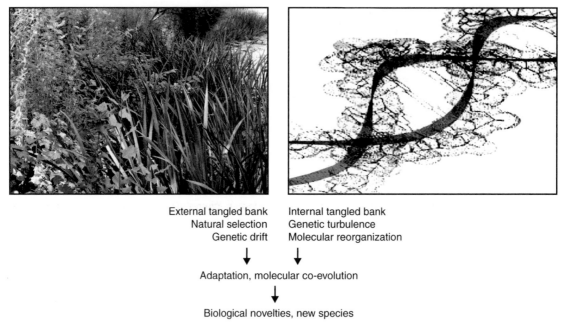

External tangled bank   Internal tangled bank
Natural selection   Genetic turbulence
Genetic drift   Molecular reorganization

↓   ↓

Adaptation, molecular co-evolution

↓

Biological novelties, new species

Figure 1.3 Evolution viewed as an interplay between the two 'tanged banks' of genetic turbulence and natural selection. Modified after Dover (1999), by permission of Oxford University Press.

direct negative phenotypic consequences; these traces from the past provide a valuable historical record for genome investigators to discover.

1.2 Yeast, fly, worm, and weed

A striking feature of genomics is its focus on a limited number of *model species* with fully sequenced genomes and large research networks organized around them. The genomes of these model species have been sequenced completely and the information is shared on the Internet, allowing scientists to take maximal advantage of progress made by others. This explains the extreme speed with which the field is developing. Ecology does not have a strong tradition in standardized experimentation with one species. Thus the genomics approach is all the more striking to an ecologist, who is often more fascinated by the diversity of life than by a single organism, and engaged in a very wide variety of topics, systems, and approaches. In this section we examine the arguments for introducing model species in ecological genomics.

The best-known completely sequenced genomes, in addition to those of mouse and human, are those of the yeast *Saccharomyces cerevisiae*, the 'fly' *Drosophila melanogaster*, the 'worm' *Caenorhabditis elegans* and the 'weed' *Arabidopsis thaliana*. Investigations into the genomes of these model organisms are supported by extensive databases on the Internet that provide a wealth of information about genome maps, genomic sequences, annotated genes, allelic variants, cDNAs, and expressed sequence tags (ESTs), as well as news, upcoming events, and publications. These four model genomes and their relationships with evolutionary related species will be discussed in more detail in Chapter 3. The genomics of the mouse and human are not discussed at length in this book because the model status of these two species has mainly a medical relevance.

The first genome to be sequenced completely was that of *Haemophilus influenzae* (Fleischmann *et al.* 1995). This bacterium is associated with influenza outbreaks, but is not the cause of the disease, which is a virus. Although several years earlier the 'genome' of bacteriophage ΦX174 had been sequenced (Sanger 1977a), 1995 is considered by many as the true beginning of genomics as a science, not in the least because the *H. influenzae* project demonstrated the usefulness of a new strategy of sequencing and assembly (whole-genome shotgun sequencing; see Chapter 2). With 1.8 Mbp the genome of *H. influenzae* was about 10 times larger than that of any virus sequenced before, but still two to four orders of magnitude smaller than the genome of most eukaryotes. Genome sequences of many other prokaryotes soon followed, including that of *Methanococcus jannaschii* an archaeon living at a depth of 2600 m near a hydrothermal vent on the floor of the Pacific Ocean (Bult *et al.* 1996). The genome of this *extremophile* was interesting because of the many genes that were completely unknown before. In 1989, a large network of scientists embarked on a project for sequencing the yeast genome, which was

completed in 1996 and was the first eukaryotic genome to be elucidated (Goffeau *et al.* 1996). Thus, by 1996, the first genomic comparisons were possible between the three domains of life: Bacteria, Archaea, and Eucarya.

The international *Human Genome Project* initiated by the US National Institutes of Health and the US Department of Energy, was launched in 1990 with completion due in 2005. However, in the meantime a private enterprise, Celera Genomics, embarked on a project with the same aim but a different approach and actually overtook the Human Genome Project. The competition was settled with the historic press conference on 26 June 2000, when US President Bill Clinton, J. Craig Venter of Celera Genomics, and Francis Collins of the National Institutes of Health jointly announced that a working draft of the human genome had been completed (Fig. 1.4). Many commentators have

Figure 1.4   From left to right: J. Craig Venter (Celera Genomics), President Clinton, and Francis Collins (National Institutes of Health) on the historic announcement of 26 June 2000 of the completion of a working draft of the human genome. © Win McNamee/Reuters.

Table 1.1 List of complete and published genomes (not including viruses) by June 2005

Taxonomic group	No. of genomes	Remarks on species
Bacteria *total*	211	Many common laboratory models and pathogens
Archaea *total*	21	Several methanogens and extremophiles
Eukarya*		
Myxomycota	1	*Dictyostelium discoideum* (slime mould)
Entamoeba	1	*Entamoeba histolytica* (amoeba causing dysentery)
Apicomplexa	6	Four *Plasmodium* and two *Microsporidium* species
Kinetoplastida	2	*Trypanosoma brucei, Leishmania tropica* (parasites)
Cryptomonadina	1	*Guillardia theta* (flagellated unicellular alga)
Bacillariophyta	1	*Thalassiosira pseudonana* (marine diatom)
Rhodophyta	1	*Cyanidioschyzon merolae* (small unicellular red alga)
Plants	4	*Chlamydomonas reinhardtii* (green alga), *Populus trichocarpa* (black cottonwood), *Arabidopsis thaliana* (thale cress), *Oryza sativa* var. *japonica*, var. *indica* (rice)
Fungi	14	Including *Saccharomyces cerevisiae* (baker's yeast)
Animals		
Nematoda	2	*Caenorhabditis elegans* (free-living roundworm), *Caenorhabditis briggsae*
Insecta	4	*Bombyx mori* (silk worm), *Drosophila melanogaster* (fruit fly), *Anopheles gambiae* (mosquito, malaria vector), *Apis mellifera* (honey bee)
Tunicata	1	*Ciona intestinalis* (sea squirt)
Pisces	3	*Takifugu rubripes* (puffer or fugu fish), *Tetraodon nigroviridis* (puffer fish), *Danio rerio* (zebrafish)
Aves	1	*Gallus gallus* (red jungle fowl)
Mammalia	5	*Rattus norvegicus* (brown rat), *Mus musculus* (house mouse), *Canis familiaris* (domestic dog), *Pan troglodytes* (chimpanzee), *Homo sapiens* (human)
Animals: *total*	16	
Eukarya: *total*	47	
Total	279	

Sources: from www.genomesonline.org, genomenewsnetwork.org, GenBank Nucleotide Sequence Database, and sundry sources.

qualified this announcement as more a matter of public communication than scientific achievement. At that time the accepted criterion for completion of a genome sequence, namely that only a few gaps or gaps of known size remained to be sequenced and that the error rate was below 1 in 10 000 bp, had not been met by far. The euchromatin part of the genome was not completed until mid-2004, although that milestone was again considered by some to be only the end of the beginning (Stein 2004). Nevertheless, the Human Genome Project can be regarded as one of the most successful scientific endeavours in history and the assembly of the 3.12 billion bp of DNA, requiring some 500 million trillion sequence comparisons, was the most extensive computation that had ever been undertaken in biology.

The number of organisms whose genome has been sequenced completely and published is now approaching 300 (Table 1.1). Bacteria dominate the list, as the small size of their genomes makes these organisms well-suited for whole-genome sequencing. By June 2005, no fewer than 730 prokaryotic organisms and 496 eukaryotes were the subject of ongoing genome sequencing projects. The list in Table 1.1 will certainly be out of date by the time this book goes to press, as new genome projects are being launched or completed every month.

The list of species with completed genome sequences does not represent a random choice from the Earth's biodiversity. From an ecologist's point of view, the absence of reptiles, amphibians, molluscs, and annelids is striking, as also is the scarcity of birds and arthropods other than the insects. How did a species come to be a model in genomics? We review the various arguments below, asking whether they would also apply when selecting model species for ecological studies.

Previously established reputation. This holds for yeast, *C. elegans, Drosophila,* mouse, and rat. These species had already proven their usefulness as models before the genomics revolution and were adopted by genomicists because so much was known about their genetics and biochemistry, and, perhaps just as important, because a large research community was interested, could support the work, and use the results.

Genome size. One of the first questions that is asked when a species is considered for whole-genome sequencing is, what is the size of its genome? At least in the beginning, a relatively small genome was a major advantage for a sequencing project. The *genome size* of living organisms ranges across nine orders of magnitude, from 10^3 bp (0.001 Mbp) in RNA viruses to nearly 10^{12} bp (1 000 000 Mbp) in some protists, ferns, and amphibians. The puffer fish, *Takifugu rubripes,* was indeed chosen because of its relatively small genome (one-eighth of the human genome).

Possibility for genetic manipulation. The possibility of genetic manipulation was an important reason why *Arabidopsis, Drosophila,* and mouse became such popular genomic models. The ultimate answer about the function of a gene comes from studies in which the genome segment is knocked out, downregulated, or overexpressed against a genetic background that is the same as that of the wild type. Also, the introduction of constructs in the genome that can report activity of certain genes by means of signal molecules is very important. This can only be done if the species is accessible using recombinant-DNA techniques. Foreign DNA can be introduced using *transposons*; for example, modified P-elements that can 'jump' into the DNA of *Drosophila,* or bacteria such as *Agrobacterium*

tumefaciens that can transfer a piece of DNA to a host plant. DNA can also be introduced by physical means, especially in cell cultures, using electroporation, microinjection, or bombardment with gold particles. Another popular approach is post-transcriptional gene silencing using *RNA interference* (RNAi), also called inhibitory RNA expression. The question can be asked, should the possibility for genetic manipulation be an argument for selecting model species in ecological genomics? We think that it should, knowing that the capacity to generate mutants and transgenes of ecologically relevant species is crucial for confirming the function of genes. Ecologists should also use the natural variation in ecologically relevant traits to guide their explorations of the genome (Koornneef 2004, Tonsor *et al.* 2005). A basic resource for genome investigation can be obtained by using natural varieties of the study species, and developing genetically defined culture stocks.

Medical or agricultural significance. Many bacteria and parasitic protists were chosen because of their pathogenicity to humans (see the many parasites in Table 1.1). Other bacteria and fungi were taken as genomic models because of their potential to cause plant diseases (phytopathogenicity). Obviously, the sequencing of rice was motivated by the huge importance of this species as a staple food for the world population (Adam 2000). Some agriculturally important species have great relevance for ecological questions; for example, the bacterium *Sinorhizobium meliloti,* a symbiont of leguminous plants, is known for its nitrogen-fixing capacities, but it also makes an excellent model system for the analysis of ecological interactions in nutrient cycling, together with its host *Medicago truncatula.*

Biotechnological significance. Many bacteria and fungi are important as producers of valuable products, for example antibiotics, medicines, vitamins, soy sauce, cheese, yoghurt, and other foods made from milk. There is considerable interest in analysing the genomes of these microorganisms because such knowledge is expected to benefit production processes

(Pühler and Selbitschka 2003). Other bacteria are valuable genomic models because of their capacity to degrade environmental pollutants; for example, the marine bacterium *Alcanivorax borkumensis* is a genomic model because it produces surfactants and is associated with the biodegradation of hydrocarbons in oil spills (Röling *et al.* 2004).

Evolutionary position. Whole-genome analysis of organisms at crucial or disputed positions in the tree of life can be expected to contribute significantly to our knowledge of evolution. The sea squirt, *Ci. intestinalis*, was chosen as a model because it belongs to a group, the Urochordata, with properties similar to the ancestors of vertebrates. The study of this species should provide valuable information about the early evolution of the phylum to which we belong ourselves. *Me. jannaschii* was chosen for more or less the same reason, because it was the first sequenced representative from the domain of the Archaea. Many other organisms, although not on the list for a genome project to date, have a strong case for being declared as model species for evolutionary arguments. These include the velvet worm, *Peripatus*, traditionally seen as a missing link between the arthropods and annelids, but now classified as a separate phylum in the Panarthropoda lineage (Nielsen 1995), and the springtail, *Folsomia candida*, formerly regarded as a primitive insect, but now suggested to have developed the hexapod bodyplan before the insects separated from the crustaceans (Nardi *et al.* 2003).

Comparative purposes. Over the last few years, genomicists have realized that assigning functions to genes and recognizing promoter sequences in a model genome can greatly benefit from comparison with a set of carefully chosen reference organisms at defined phylogenetic distances. Comparative genomics is developing an increasing array of bioinformatics techniques, such as *synteny analysis*, *phylogenetic footprinting*, and *phylogenetic shadowing* (see Chapter 3), by which it is possible to understand aspects of a model genome from other genomes. One of the main reasons for sequencing the chimpanzee's genome was to illuminate the human genome, and a variety of fungi were sequenced to illuminate the genome of *S. cerevisae*.

Ecological significance. It will be clear that ecological arguments have only played a minor role in the selection of species for whole-genome sequencing, but we expect them to become more important in the future. Jackson *et al.* (2002) have formulated arguments for the selection of ecological model species, and we present them in slightly adapted form.

Biodiversity. The new range of models should embrace diverse phylogenetic lineages, varying in their physiology and life-history strategy. For example, the model plants *Arabidopsis* and rice both employ the C3 photosynthetic pathway. To complement our genomic knowledge of primary production, new models should be chosen among plants utilizing C4 photosynthesis or crassulacean acid metabolism (CAM). Considering the diversity of life histories, species differing in their mode of reproduction and dispersal capacity should be chosen; for example, hermaphoditism versus gonochorism, parthenogenesis versus bisexual reproduction, etc.

Ecological interactions. Species that take part in critical ecological interactions (mutualisms, antagonisms) are obvious candidates for genomic analysis. One may think of mycorrhizae, nitrogen-fixing symbionts, pollinators, natural enemies of pests, parasites, etc. The most obvious strategy for analysing such interactions would be to sequence the genomes of the players involved and to try and understand interactions between them from mutualisms or antagonisms in gene expression.

Suitability for field studies. The wealth of knowledge from experienced field ecologists should play a role in deciding about new 'ecogenomic' models. Not all species lend themselves to studies of behaviour, foraging strategy, habitat choice, population size, age structure, dispersal, or migration in the field, simply because they are too rare, not easily spotted, difficult to sample quantitatively, impossible to mark and recapture, not easy to distinguish from related species, or inaccessible to invasive techniques. Thus suitability for field research is another important criterion.

Feder and Mitchell-Olds (2003) developed a similar series of criteria for an ideal model species in evolutionary and ecological functional genomics (Fig. 1.5). These authors point out that there is currently a discrepancy between classical model species and many ecologically interesting species. Models such as *Drosophila* and *Arabidopsis* are not very suitable for ecological studies, whereas many popular ecological models have a poorly characterized genome and lack a large community of investigators. In some cases a large ecological community is available, but functional genomic studies are difficult for reasons of quite another nature. For example, many ecologists favour

wild birds as a study object, but there are ethical objections to genetic manipulation of such species and laboratory experiments are restricted by law.

It is not easy to foresee how the list of genomic model species will develop in the future. Obviously, ecologists taking ecological genomics seriously will need to avail themselves of genomic information on their model species, preferably a whole-genome sequence. This is not to say however, that all questions in ecological genomics require the full-length DNA sequence of a species before they can be answered. Some issues may prove to be solvable with the use of less extensive genomic investigations, for example a gene hunt followed by multiplex quantitative PCR, rather than transcription profiling with microarrays of

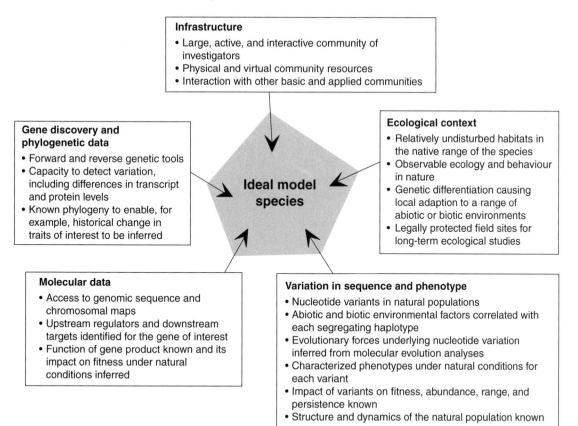

Figure 1.5   Criteria in evolutionary and ecological functional genomics for a model species, according to Feder and Mitchell-Olds (2003). At present few species satisfy all criteria. Reproduced by permission of Nature Publishing Group.

the complete genome (see Section 2.3). In addition, microarray studies with part of the expressed genome are possible even in species lacking a complete DNA sequence. Microarrays can be manufactured at costs that are affordable for small research groups if they are limited to genes associated with a specific function or response pathway (Held *et al.* 2004; see also Section 6.4). Still, the number of species with fully characterized genomes is expected to rise dramatically in the coming years; after a while all the major ecological models will also be genomic models and the saturation point could very well be due to the limited number of molecular ecologists in the worldwide scientific community.

Not all ecological models will enjoy the type of in-depth investigations now dedicated to yeast, fly, worm, and weed. Murray (2000) points out that the development of genome-based tools has a strong element of positive feedback; the rich—that is, widely studied organisms—get richer and the poor get poorer. This development has already been felt in the fields of animal and plant physiology, where many of the species traditionally investigated in comparative physiology and biochemistry have been abandoned in favour of models that can be genetically manipulated to study the function of genes. Murray (2000) predicted that 'the larger its genome and the fewer its students, the more likely work on an organism is to die'. Crawford (2001) has argued, however, that functional genomics should resist this tendency and instead choose species best suited to addressing specific physiological or biochemical processes. For example, the Nobel Prize for Medicine was given to H.A. Krebs for his research on the citric acid cycle, which was conducted on common doves. By modern standards the dove is a non-model species, but it was chosen because its breast muscle is very rich in mitochondria. In animal physiology, *Krogh's principle* assumes that for every physiological problem there is a species uniquely suited for its analysis (Gracey and Cossins 2003). According to this principle, genomic standard species are likely to be suboptimal for at least some problems of physiology, because no model is uniquely suited to answering all questions.

DNA microarrays, with their associated massive generation of data on expression profiles (see Section 2.3), are one of the most tangible features of modern genomics and are often seen as holding the greatest promise for solving problems in ecology. However, not all ecologists are convinced that microarray-based transcription profiling is the best way to advance the genomics revolution into ecology. Thomas and Klaper (2004), for example, argued that commercial microarrays are available only for genomic model species, whereas the interest of ecologists is with species that are important in the environment and amenable to ecological studies; these two interests do not necessarily coincide. This leaves ecologists with two options. One is to develop their own microarrays, starting with spotted cDNAs of unknown sequence, doing a lot of tedious sequencing work, and gradually finding out more about the genome of their study species. Another option is to apply transcriptome samples of non-models to microarrays of model species. In these *cross-species hybridizations* it is assumed that there is sufficient homology between the non-model and the model to allow differential expressions to be assessed reliably. For example, *Arabidopsis* may function as a model for other species of the Brassicaceae, and *Drosophila* as a model for other higher insects. Obviously, how useful such an approach is will depend on how far the sequences of model and non-model diverge. This will not be the same for all parts of the genome and therefore there is some doubt on the validity of cross-species hybridization, although there will certainly be situations where it works well.

Other investigators are less hesitant about the prospects of microarrays in ecology. Gibson (2002) emphasized that today it is feasible to establish a 5000-clone microarray resource within 12 months of a commencing project and that neither the estimated expense nor the availability of technology need to be a major obstacle for progress. We share this optimism. Given the fact that the number of almost completely sequenced organisms is increasing month by month, we can expect that the genome of several species of great interest to ecologists may be completed within a few years.

In addition, we expect that almost all ecologically relevant species will have basic genomics databases—for example, an annotated EST library—sufficient to answer a considerable number of ecological questions.

1.3 -Omics speak

Because of the immediately attractive upswing created by the genomics revolution, and the large financial resources made available in many industrialized countries, adjacent fields of science have adopted similar terms, leading to a great proliferation of designations such as transcriptomics, proteomics, and metabolomics, such that some biologists have complained that what was molecular biology before is now named after one of the '-omics' but in fact is still molecular biology. Zhou *et al.* (2004) proposed a classification of genomics according to three main categories: approach (structural or functional), scientific discipline (evolutionary genomics, ecological genomics, etc.), and object of study (plant genomics, microbial genomics, etc.). An Internet page maintained by Mary Chitty (Cambridge Healthtech Institute) provides a glossary containing no less than 60 single-word entries ending with -omics (www.genomicglossaries.com). The list includes obvious terms such as pharmacogenomics and cardiogenomics, and awkward ones such as saccharomics (the study of all the carbohydrates in the cell) and vaccinomics (the use of bioinformatics and genomics for vaccine development). The three most common extensions of genomics are transcriptomics, proteomics, and metabolomics, and these are introduced briefly here, with reference to Fig. 1.6.

Transcriptomics is the study of all the transcripts that are present at any time in the cell. In principle the transcriptome includes messenger RNAs (mRNAs) in addition to ribosomal RNAs (rRNAs), transfer RNAs (tRNAs), and small nuclear RNAs (snRNAs), but transcriptomics is usually limited to mRNA, the template for translation into protein. The main activity in transcriptomics is to obtain a profile of global gene expression in relation to some condition of interest. Which genes are turned

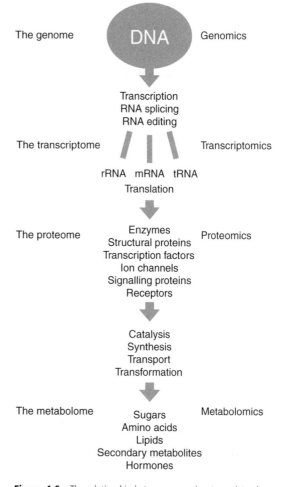

Figure 1.6 The relationship between genomics, transcriptomics, proteomics, and metabolomics.

'on' and 'off' during certain phases of the cell cycle? Which genes are upregulated by certain physiological conditions? Which genes change their expression in response to adaptation to the environment? The study of transcriptomes is part of *functional genomics*, because it does not look at the DNA as such, but at its functions.

In general, it is expected that there are more transcripts than there are protein-encoding genes in the genome, even when considering only those genes that are actually transcribed. This is due to the mechanism of *alternative splicing*: the generation of different mRNAs from the same pre-mRNA during the removal of introns. *RNA editing* (post-transcriptional insertion or deletion of nucleotides, or conversion of one base for another) is another reason for incongruence between the genome and the transcriptome.

There are more reasons why a functional analysis of the genome can provide a different picture than an inventory of genes. Obviously, all cells of an organism have the same genome, but not the same transcriptome. Even when looking at cells of the same type, the transcriptome depends on environmental conditions, physiological state, developmental state, etc. So the transcriptome allows a glimpse of the living cell much more than the genome itself. The argument also holds when making comparisons across species. Classical molecular phylogenetics (see Graur and Li 2000) is based on variation of homologous DNA sequences across species. However, the same structural DNA can be regulated in different ways in different species. We illustrate this argument with an example from Enard *et al.* (2002), who did one of the first studies in what may be called *comparative transcriptomics*.

Enard *et al.* (2002) analysed the expression of 18 000 genes in liver, blood leucocytes, and brain tissue of humans, chimpanzee (*Pan troglodytes*), and rhesus monkey (*Macaca mulatta*). The expression patterns in human blood and liver turned out to be more similar to chimpanzees than to rhesus monkeys, which is in accordance with the phylogenetic distances between the three primate species; however, the expression profiles in the brain were more similar between chimpanzee and rhesus monkey than between either of the two monkey species and human (Fig. 1.7). So, although chimpanzees share 98.7% of their DNA with humans, the human species expresses that DNA in a different manner, especially in the brain. Gene expression in the brain has undergone accelerated evolution compared to gene expression elsewhere in the body, and evolution has resulted in a divergence of humans from chimpanzees, mostly due to regulatory change rather than structural reorganization of the DNA.

Proteomics is the study of all the proteins in the cell. As with genomics, proteomics arose thanks to technological innovation, which in this case is

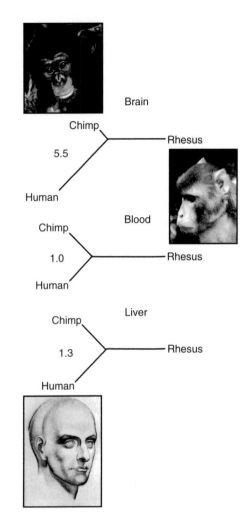

Figure 1.7 Distance trees showing the similarity of gene-expression profiles in brain, blood leucocytes, and liver of human, chimpanzee, and rhesus monkey. Numbers refer to the ratio between the rate of evolution in the human and the chimpanzee lineages, taking the rhesus monkey as an outgroup. Reprinted with permission from Enard *et al.* (2002). Copyright 2002 AAAS.

tandem mass spectrometry (MS/MS) and liquid chromatography coupled to tandem mass spectrometry (LC/MS/MS). The idea is to separate a mixture of soluble proteins by means of chromatography and then to estimate masses, first of the larger peptide and, after a second ionization, of fragments of the same peptide. The fragment patterns provide a fingerprint characteristic of the protein. Interpretation of proteomics data is usually supported by genomic sequence information, in such a way that an observed peptide fragment pattern may be compared to a database of proteins predicted from the genome. Mass spectrometry may also be used to determine the amino acid sequence of a protein. For this application, the protein is cleaved with a protease, for example trypsin, which generates a collection of fragments characteristic of the protein. These fragments may be compared to an *in silico* (computer-simulated) digestion derived from the genome and the known cleavage sites of the protease.

The proteome provides a different picture of a cell's activities to the transcriptome. Several authors have indeed wondered about the lack of correlation between mRNA and protein abundances. One of the reasons for this is the existence of control mechanisms at the ribosomes, where mRNA is translated to peptides. *Translational control* allows the cell to select only certain mRNAs for translation and block others. The selection is often dependent on environmental conditions, so this mechanism allows for physiological adaptation on the level of the proteome, even though the transcriptome remains the same. Another issue is *post-translational modification* or *protein processing*, processes that can greatly affect the function of a protein, for example by acetylation or ubiquitination of the N-terminal residue, hydroxylation of prolines, or cleavage of the molecule into smaller units. The proteome and the genome are linked by many feedback mechanisms, because some proteins are transcription factors necessary for gene activation, others are enzymes involved in transcription or translation, and still others are structural components of chromosomes. So, in a molecular biology context, the living cell can only be understood fully by considering genome, transcriptome, and proteome together.

As an example of a study applying proteomics in an environmental context, consider the work of Shrader *et al.* (2003). These authors studied protein fingerprinting in embryos of zebrafish (*Danio rerio*) exposed to environmental endocrine disrupters. The compound *p*-nonylphenol is a degradation product of certain detergents and is discharged into the aquatic environment through sewage

effluent. Because of its structural similarity to vertebrate steroid hormones, especially oestrogens, nonylphenol has been associated with feminization of male fish. Fig. 1.8 shows a two-dimensional gel of differential protein expression of fish exposed to nonylphenol. This so-called *protein-expression profile* was composed by matching the treatment profile with the control profile and subtracting them from each other. The Venn diagram in Fig. 1.8b provides a pictorial illustration of the number of proteins that are shared between treatments. It is interesting to note that nonylphenol induced several proteins that were also induced by oestradiol (23 in total), but that a

(a)

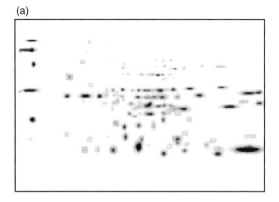

(b)

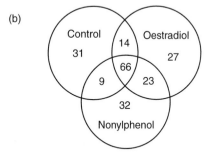

Figure 1.8 (a) Features on images from zebrafish embryos induced by exposure to nonylphenol. Representation of a two-dimensional electrophoresis gel, on which proteins are separated by a combination of isoelectric point and molecular mass, showing only proteins that were differential between the control and nonylphenol-exposed zebrafish. (b) Venn diagram representing the number of proteins shared by two or more treatments. The diagram shows that, from the total of 202 proteins, there were 32 seen only in the nonylphenol treatment and 23 seen in both the nonylphenol treatment and a treatment with the natural hormone oestradiol. After Shrader *et al.* (2003) with permission from Springer.

significant number of proteins (32) were specific to nonylphenol. The study suggests that the two compounds have overlapping but otherwise dissimilar modes of action and that it may be too simple to qualify nonylphenol as only mimicking oestradiol. The functional genomics of endocrine disruption will be discussed in more detail in Chapter 6.

Metabolomics is the study of all low-molecular-weight cellular constituents. Usually only metabolites belonging to a limited category are included, for example all soluble carbohydrates, or all metabolites that can be measured by a certain analytical technique such as pyrolysis gas chromatography or infrared spectrometry. No single method can measure the thousands of different chemical compounds that may be present at any time in a cell, because of the greatly diverging chemical properties (hydrophilic versus hydrophobic compounds, acids versus bases, reactive versus inert compounds, etc.). The metabolome requires a diversity of analytical approaches to obtain a complete picture.

There are still hardly any studies of proteomics and metabolomics that address a truly ecological question and that is why both of these -omics do not play a major role in this book. Their role could grow in the future, when ecology has absorbed the principles of genomics. In Chapter 7 we will address some aspects of metabolomics when discussing metabolic networks. Finally, Table 1.2 describes some other terms used in connection with genomics.

With the further development of ecological genomics, applications will also come within

reach. One can envisage a multitude of issues where a better knowledge of genomes in the environment can support measures to improve ecosystem health, risk assessment of pollution, conservation of endangered species, etc. (Greer *et al.* 2001). Such applications fall outside the scope of this book; however, we mention two examples below, to sketch the range of possibilities.

Purohit *et al.* (2003) suggested that multilocus DNA fingerprints prepared from environmental samples could act as an *indicator DNA signature* (IDS); for example, fingerprints of microbial soil communities could be indicative of soil pollution. Their suggestion can be extended to involve transcription profiles that are characteristic of certain environmental conditions or physiological states. Fig. 1.9 illustrates this principle. When an organism is exposed to polluted soil, this will be accompanied by gene expression that has both a general aspect due to the generality of the stress response and a specific aspect which characterizes the challenge (see also Chapter 6). When the expression profile observed for a suspect soil is compared with a database of reference profiles, the type of pollution and its biological effects may be indicated (Fig. 1.9). This may help to support decisions about the urgency of remediation measures.

As a second example of a possible application consider the case of soil-borne pathogens. Many pathogens attacking economically important crops are difficult to control by conventional strategies such as the use of host resistance and synthetic pesticides. However, some soils have an inherent capacity to suppress diseases and such soils need lower rates of pesticide application to combat

Table 1.2 List of some of the more common -omics designations in addition to those discussed in the text

Term	Object of study
Pathogenomics	Genomes of human pathogens: analysis of genes involved in disease generation
Pharmacogenomics	Genomic responses to drugs, analysis of expression profiles that indicate similarity of action across compounds, analysis of genetic polymorphisms that determine a person's disposition to drug action
Toxicogenomics	Mode of action of toxic compounds, development of expression profiles that indicate similarity of toxic action across compounds
Ecotoxicogenomics	Genomic responses of organisms exposed to environmental pollution
Ionomics	All mineral nutrients and trace elements in an organism, for example using inductively coupled plasma mass spectrometry (ICP-MS)

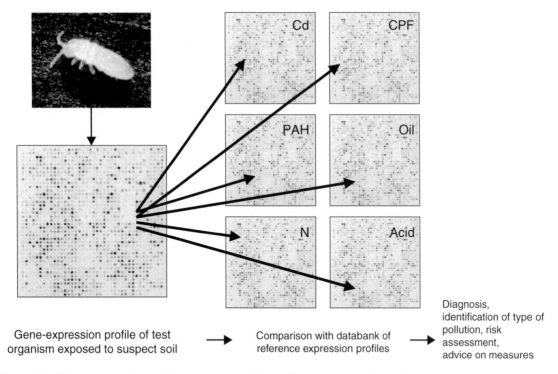

Gene-expression profile of test organism exposed to suspect soil ⟶ Comparison with databank of reference expression profiles ⟶ Diagnosis, identification of type of pollution, risk assessment, advice on measures

Figure 1.9 Risk assessment of soil pollution can be supported by matching the gene-expression profile of an indicator organism, generated after exposure to a suspect soil, with profiles established as a reference and known to be associated with certain types of pollution. Examples are given of soils polluted by specific substances: CPF, chlorpyrifos; PAH, polycyclic aromatic hydrocarbons.

them. *Disease-suppressive capacity* is due to the presence of genes involved with antibiotic production by antagonistic microorganisms (Van Elsas *et al.* 2002; Weller *et al.* 2002; Garbeva *et al.* 2004). In several cases, specific microbial populations have been identified that contribute to disease suppressiveness; however, for most soils, we have little understanding of the consortium of microorganisms and the corresponding genes that are responsible for this critical function. Natural disease-suppressive soils can be regarded as a largely untapped resource for the discovery of new antagonistic microorganisms and antibiotics. We will see several examples of this in Chapter 4. Management strategies can be developed that involve selective stimulation and support of populations of antagonistic microorganisms in the rhizosphere. Genomic methods of soil diagnosis could be used as feedback on agricultural management decisions.

1.4 The structure of this book

We have organized this book to address fundamental questions in three areas of ecology where we believe ecological genomics can make important contributions. Having given a broad introduction to genomics, and ecological genomics in particular, in this chapter, two more specific introductory chapters follow. Chapter 2 explains the most important genomic methodologies, and Chapter 3 gives a survey of what can be learnt from comparing the genomes of model organisms with each other and with those of evolutionarily related species. We also discuss the various properties of both prokaryotic and eukaryotic genomes. Chapters 2 and 3 form the methodological and evolutionary basis for the rest of the book. In the next three chapters, questions relate to different levels of integration, from community ecology down to population ecology, ending with physiological ecology. Each of these chapters ends with an appraisal of how the genomics achievements contribute to answering the basic question of the chapter.

Community structure and function. In Chapter 4 the genomics approach is used to discuss a question fundamental to community ecology, of how biodiversity supports ecosystem function. Most of the examples in this chapter are taken from microbial ecology. We review the ways in which microbiologists use genomics to estimate species diversity in the environment and how functions of uncultured species can be reconstructed from environmental genomes.

Life-history patterns. Chapter 5 discusses the genomic aspects of life-history evolution, an important theme in population ecology. Questions of longevity, reproductive effort, sex, and diapause are discussed, as well as the issue of trade-offs between life-history traits. We show that progress in mechanistic studies of plasticity and optimal timing of reproduction has considerable relevance to ecology.

Stress responses. The many genomic studies of mechanisms that allow plants and animals to survive in harsh environments form the subject of Chapter 6. The way in which plants and animals transduce stress signals into gene expression shows many commonalities across species, as well as stress-specific signatures. We argue that insights in these mechanisms is needed to define the ecological niche of the species.

Integrative ecological genomics. We conclude the book with a short chapter on integrative approaches, discussing some aspects of network analysis and ecological control analysis. These two approaches belong to the realm of systems biology, a new field of research, linking genomics, proteomics, and metabolomics with biochemical modelling. Chapter 7 suggests that integrative approaches are also required in ecological genomics and it discusses some examples to support this claim. Finally, a number of emerging issues are discussed in the outlook section of Chapter 7.

CHAPTER 2

Genome analysis

In this chapter we aim to acquaint the reader with the various molecular techniques that are used in genomics, with an emphasis on those that are of relevance for ecology. We also discuss the nature of the data generated by these techniques and the most common approaches to data analysis.

2.1 Gene discovery

In a fully sequenced genome, genes are found by scanning the sequence using gene-predicting computer programmes and assigning putative functions by searching for similarities in already existing databases (see Section 2.2). For many organisms under investigation in ecological genomics, no genomic database is available and genes must be identified in other ways. This section deals with some of the so-called pre-genomic molecular approaches that may be used to identify ecologically important genes in incompletely characterized genomes.

2.1.1 From gene product to gene

In some cases the primary structure of a gene product (a protein) may be the starting point of gene discovery. This holds especially for proteins that can be isolated relatively easily by some marker or bioassay, or proteins that are highly induced by some experimental treatment. As an example we discuss the isolation of the metallothionein (*Mt*) gene in a species of springtail, *Orchesella cincta* (Hensbergen *et al.* 1999). Attempts to pick up the gene by polymerase chain reaction (PCR) using primers from the then-known *Drosophila Mt* sequence were unsuccessful, which was

explained later by the lack of sufficient homology. Therefore the protein itself was isolated first, using a combination of gel-permeation chromatography and reversed-phase high-performance liquid chromatography (RP-HPLC). Protein isolation was aided greatly by the fact that metallothionein is highly inducible by exposure to cadmium and binds strongly to cadmium at neutral pH. By measuring cadmium concentrations in eluates from a chromatography, the fate of the protein could be monitored. Finally, a purified sample was obtained and a partial amino acid sequence was determined by *N-terminal Edman degradation*. This is a classical technique from biochemistry in which the N-terminal residue of a peptide is labelled and subsequently cleaved off without disturbing the peptide bonds between the other amino acids. The liberated amino acid is then eluted over an HPLC column, detected using the label and identified from the retention time; the cycle is repeated with the next N-terminal residue until the sequence of the peptide is known.

Using a partial amino acid sequence of the purified metallothionein, Hensbergen *et al.* (1999) were then able to develop *degenerate primers* for amplifying the gene by means of the PCR. A degenerate primer is a mixture of DNA sequences all encoding the same amino acid sequence, but allowing for the fact that most amino acids are represented by more than one triplet. These primers were applied to a pool of *complementary* or *copy DNA* (cDNA), which is DNA prepared from mRNAs by *reverse transcription*. The reverse-transcription reaction uses the activity of reverse transcriptase (RNA-dependent DNA polymerase), an enzyme originally isolated from RNA viruses, which can make a complementary strand of DNA using single-stranded mRNA as a template. Reverse transcription is primed by a short oligo(dT) primer, a sequence of 2'-deoxythymine (dT), which anneals to the polyadenosine (poly(A)) tail present at the 3'-end of most eukaryotic mRNAs (Fig. 2.1). To characterize the complete cDNA Hensbergen *et al.* (1999) used a technique known as rapid amplification of cDNA ends (3'- and 5'-*RACE*). A 3'-RACE uses a 3' primer complementary to the poly(A) tail of the cDNA, in combination with a forward gene-specific primer somewhere in the middle of the cDNA (in this case degenerate) to amplify the 3'-end of the cDNA; a 5'-RACE uses a primer complementary to an 'anchor' oligonucleotide RNA which is ligated enzymatically to the 5'-end of the mRNA, in combination with a reverse gene-specific primer. Thus Hensbergen *et al.* (1999) were able to characterize a full-length cDNA starting with

a purified protein. Finally, using the cDNA sequence, Sterenborg and Roelofs (2003) determined the genomic sequence of the *O. cincta* metallothionein, and demonstrated the presence of one intron in the gene. An outline of the complete procedure is given in Fig. 2.2.

Working backwards from protein structure to gene characterization is very laborious and may only be applicable in cases where ecological functions can be associated a priori with a suspected protein (as in the case of metallothionein conferring metal tolerance). The laborious Edman degradation technique has now been replaced mostly by sequencing using mass spectrometry, in which masses are estimated for peptides generated from proteolytic digests of the protein, while the fragment pattern obtained is compared with entries in a database that include sizes predicted from the genomic sequence. The example nevertheless illustrates that it is possible, in principle, to

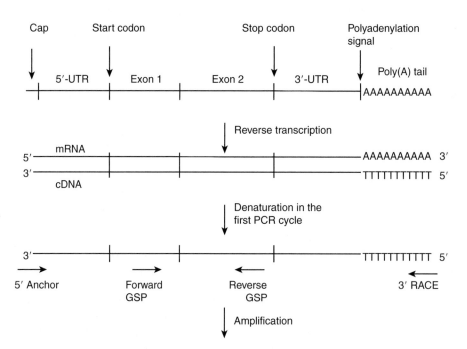

Figure 2.1 The procedure for characterizing a full-length cDNA from partial knowledge of a protein sequence. Top: general structure of a eukaryotic mRNA with two exons, after splicing, capping and polyadenylation. Middle: mRNA–cDNA hybrid synthesized by reverse transcription. Bottom: single-stranded cDNA acting as a template for PCR amplification. Positions are indicated for primers used to amplify cDNA ends in 3'- and 5'-RACE. GSP, gene-specific primer (degenerate), developed from partial knowledge of the amino acid sequence of the protein; UTR, untranslated region.

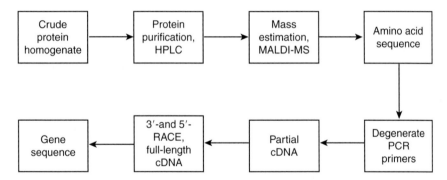

Figure 2.2 Outline of a gene-discovery procedure in a non-model organism, commencing with protein purification and leading to gene characterization. MALDI-MS, matrix-assisted laser desorption ionization mass spectrometry.

work backwards from protein structure to gene characterization, and in some ecological applications this may be the only way to begin genomic explorations.

2.1.2 Differential screening

Liang and Pardee (1992) first proposed that genes whose expression differs between two populations of cells can be discovered by a PCR-based technique called *differential display of mRNA*. When animals or plants are subjected to a certain treatment, for example drought stress, their cells will have a higher or lower abundance of the mRNAs for those genes that respond to the treatment, compared with untreated controls. These *differential genes* can be detected and visualized by a PCR technique that amplifies complementary DNA sequences prepared by reverse transcription from the mRNA pool. The differential display PCR takes advantage of the poly(A) tail to anchor a poly(T) primer at the end of the cDNA. The other primer is a short oligonucleotide, 6 or 7 bp long, with an arbitrary sequence; this primer will anneal somewhere near the end of the cDNA strand, depending on the sequence. Because of the specific (but unknown) annealing site of the forward primer, amplified products from different cDNAs will differ in size and so can be resolved on a high-resolution electrophoresis gel. The presence or absence of a band in one treatment compared to the other is evidence of a differentially expressed

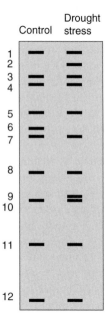

Figure 2.3 Schematic representation of an electrophosis gel displaying differential cDNAs under experimental treatment. The banding pattern indicates that genes 2 and 9 are upregulated and gene 6 is downregulated by drought stress.

gene (Fig. 2.3). Promising bands can be excised and processed for sequencing.

The strength of differential display is illustrated by a study of Liao *et al.* (2002), who used the technique to find genes upregulated by exposure to the toxic metal cadmium in the nematode *Caenorhabditis elegans*. They identified 48 cadmium-inducible mRNAs in the nematode, one of which was a novel protein, not found before in any other organism. The gene, *Cdr-1* (of the cadmium-responsive gene family), is upregulated specifically by cadmium and not, like many other cadmium-induced genes, by general stress factors. The gene encodes a hydrophobic protein, most probably associated with the lysosomal membrane, that pumps Cd^{2+} ions from the cytoplasm into lysosomal vesicles.

A disadvantage of the differential display approach as described above is that most PCR products represent sequences from the 3'-untranslated region (3'-UTR) of the messenger. If the research is conducted with a model organism, as in the case of Liao *et al.* (2002), this presents no problem, because the UTRs located downstream of genes will easily be identified in the genomic database and so will the corresponding genes. For non-model organisms there is a problem, because the UTRs lack homology across species and so the sequence itself does not provide a clue to the function or identity of the gene. To recover the upstream part of the cDNA and characterize the gene one needs to apply 3'-RACE (see above). A technique that overcomes the problem of displaying too many 3'-UTRs is *representational difference analysis* (Pastorian *et al.* 2000). In this method a restriction is applied to the cDNAs before PCR amplification.

Another differential screening strategy, which is often applied in combination with the generation of cDNA libraries and microarray transcription profiling (see below), goes under the name of *suppression-subtractive hybridization* (SSH). The idea is that by two subsequent hybridization templates are made for a PCR that is selective towards those cDNAs that differ in abundance between two samples (Diatchenko *et al.* 1996). The pool of mRNAs to which the procedure is applied should have a homogeneous genetic composition, to avoid false-positive results arising from allelic variation. Differentially expressed cDNAs are enriched by dividing one of the samples (the tester) into two subsamples, and labeling them with different adapters. The control sample (the driver) is then hybridized with each tester sample separately, and the samples are mixed in a second hybridization

with fresh driver, with the result that the mRNAs whose abundance differed between the two samples are over-represented in the population of double-stranded cDNAs with two different adapters (the subtraction effect). The PCR will amplify these cDNAs selectively, whereas the cDNAs with the same adapter will form so-called panhandle structures that are not amplified (the suppression effect). An overview of the procedure is given in Fig. 2.4. The technique is now completely protocolized and available commercially as a kit (Clontech 2002).

The SSH technique enriches for differentially expressed genes, but the subtracted PCR-amplified sample will still contain cDNAs that correspond to mRNAs whose abundance did not differ between tester and driver samples. Therefore, further confirmatory steps are necessary. A recommended procedure is to conduct a differential screening in which dot-blot arrays of clones from the subtracted library are hybridized with labelled probes from either tester or driver populations. To complete the screening, the same procedure should be applied to labelled subtracted and reverse-subtracted probes, and this should confirm the result. As an example, consider the work of Rebrikov *et al.* (2002, 2004), who applied the dot-blot screening method as a means of confirming differential expressions between two closely related strains of the freshwater planarian, *Girardia tigrina*, which reproduce in different ways (Fig. 2.5). One strain of this flatworm is exclusively asexual, whereas the other reproduces both sexually and asexually. Rebrikov *et al.* (2002) were interested in gene-expression patterns specific to asexual reproduction.

In the differential screening, clones that are recognized by the tester probe but not by the driver probe are confirmed to be differentially upregulated (Fig. 2.5, top left panel). In addition, the number of confirmed differentials should increase using the subtracted and reverse-subtracted probes, because they are normalized for low-abundance mRNAs, and their signal will not be detected by the tester and driver probes. Down-regulation can be studied in the same way except that the spotted clones are the result of reverse subtraction (Fig 2.5, top right panel). The whole

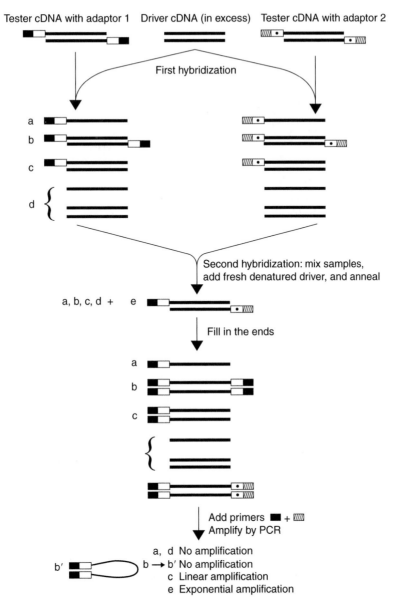

Figure 2.4 Scheme of the suppression-subtractive hybridization (SSH) method for differential screening. Thick, solid lines represent tester or driver sequences, generated by digestion of cDNA with the restriction enzyme *RsaI*. The boxes represent adapter sequences, ligated to the cDNA digests. Letters a–e represent different configurations of DNA molecules. Type e molecules are formed only if the sequence is upregulated in the tester sample compared to the driver sample. Type b molecules are not amplified due to panhandle formation. After Diatchenko *et al.* (1996) by permission of the National Academy of Sciences of the United States of America.

procedure can also be done in an up-scaled manner using microarrays. Rebrikov *et al.* (2002) revealed a novel, extrachromosomal, virus-like element in the asexual flatworm strain.

A critical factor for the success of SSH screening lies with the expression ratio between two treatments. Ji *et al.* (2002) demonstrated that the enrichment effect of SSH is proportional to the

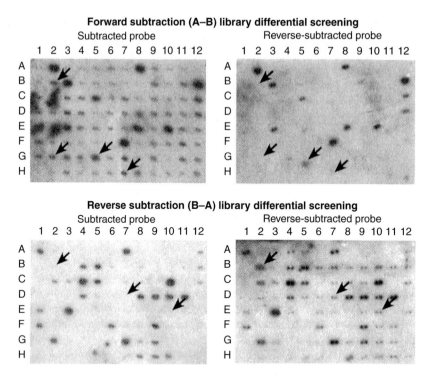

Figure 2.5 Example of results from a differential screening approach applied to clones from a subtracted library (sexual versus asexual strains from the freshwater planarian *G. tigrina*). Clones arrayed on nylon membranes were selected randomly from the subtracted library (top panels) to study upregulation, and the reverse-subtracted library (bottom panels) to study downregulation. The dot-blots were then hybridized to labelled probes prepared from the subtracted (left-hand panels) and reverse-subtracted libraries (right-hand panels). After Rebrikov *et al.* (2004) by permission of Humana Press.

cube of the expression ratio. This non-linear effect implies that genes with highly differential expression (a high expression ratio) will be much more strongly enriched than genes with a lower expression ratio. For example, if genes with an expression ratio of 2 are enriched 10 times, genes with an expression ratio of 5 are enriched by a factor of 156. Theoretical and empirical arguments have demonstrated that only genes differing in expression by a factor of 5 or more will be effectively picked up in current SSH protocols.

Application of SSH in an ecological context is growing rapidly, especially because it is a good preparatory method for generating enriched cDNA clone libraries for spotting microarrays (see below). To illustrate the use of SSH in ecology, a study by Pearson *et al.* (2001) is exemplary. These authors screened for genes differentially expressed in the brown macroalga, *Fucus vesiculosus* (of the

family Phaeophyceae), subjected to drought stress. Many genes were found to have differential expression that could not be identified by homology to known sequences, but those that could included partial sequences for ribulose-1,5-bisphosphate carboxylase/oxygenase, chloroplast-coupling factor ATPase, and a photosystem I P700 chlorophyll *a*-binding protein. This study illustrates the great flexibility of genes encoding components of the photosynthetic apparatus, not only in response to light, but also to the hydration status of the tissues.

2.1.3 From marker to functional gene

Various DNA-fingerprinting methods, which are very popular in molecular ecology to elucidate population structure, geographic variation, or paternity (Beebee and Rowe 2004), can often form the starting point of functional analysis. DNA fingerprinting in general can be done in two ways. One approach is to use identified, often non-coding, polymorphic sequences in the genome, such as *microsatellites* (loci with a variable number of short tandem repeats, e.g. $(GA)_n$, where n varies from, let's say, 5 to 9). Since these are single-locus *codominant* markers (the heterozygotes can be distinguished from either homozygote), they are especially suitable for population analysis (Jarne and Lagoda 1996, Sunnucks 2000). Another approach is *multilocus DNA fingerprinting*, where genotypes are recognized from many markers at the same time, often of unknown sequence. One such multilocus analysis, popular at the beginning of the 1990s when the molecular approach in ecology began its advance, goes under the name of *randomly amplified polymorphic DNA* (RAPD). Williams *et al.* (1990) introduced this technique, which is based on a PCR with 10- to 15-mer primers of arbitrary sequence. Due to variation between individuals in the position or sequence of primer-annealing sites, each individual generates a differ-ent series of bands when PCR products are separated on an agarose gel. A large number of primers (sometimes several hundreds) is used to probe the genome, hence the designation 'random'.

When RAPD markers are found to be associated with certain environmental conditions, important ecological traits, or phenotypes of interest, they are cloned and sequenced, and new primers are developed based on the sequence, allowing a more robust PCR. The DNA segment is then designated with the awkward term *sequence-characterized amplified region*, SCAR, or the more general term *sequence-tagged site*, STS. The use of SCARs has become very popular in plant breeding and crop science, because it allows rapid screening of many individuals for certain traits of interest and it aids marker-assisted selection programmes. In such breeding programmes, SCARs are designed to link with resistance/susceptibility genes or other genes that determine the commercial value of the plant or animal. For example, Haymes *et al.* (1997) developed a SCAR for resistance of strawberry (*Fragaria* × *ananassa*) to the fungus *Phytophthora fragariae* (Oomycota), which causes a form of

root rot. With a combination of primers, a reliable identification could be made for a resistance allele of the *Rpf* (regulation of pathogenicity factors) gene, which encodes a small excreted protein with a signalling function.

Because the RAPD procedure produces only a limited number of bands from each primer and the banding pattern is sensitive to the amount of template DNA and PCR conditions (e.g. Mg^{2+} concentration), more reliable fingerprinting tech-niques have been developed, one of the most popular being *amplified fragment length polymorph-ism* (AFLP; Vos *et al.* 1995). In this technique spe-cific adapter sequences are ligated to DNA digests obtained with two restriction enzymes before a PCR is done. One of the restriction enzymes is a frequent cutter—it binds to a short, common sequence of nucleotides—that will ensure that sufficient fragments are obtained with a size range that allows easy separation by electrophoresis; the other is a rare cutter—it binds to a longer, less-common sequence—used to limit the number of fragments amplified by the PCR (only the frag-ments with a frequent cut on one side and a rare cut on the other side are amplified). The PCR uses primers targeted to the adapter sequences, with one, two, or three bases extending in the amplicon, to select a subpopulation of the fragments. The reaction products are resolved by electrophoresis on a polyacrylamide gel (Fig. 2.6). AFLP can also be applied to cDNAs, in which case it can be used as a differential screening method (see above) when the fingerprints from two pools of cDNA are compared for differential bands (cDNA-AFLP).

In complex genomes the number of bands obtained can be very large, sometimes leading to difficulty in interpretation when AFLPs are used for resolving population structure. The number may be decreased by extending the selective bases. For example, using an overhang of four rather than three selective bases, as in Fig. 2.6, reduces the expected number of bands by a factor of 16. Another strategy is the use of three rather than two endonucleases, while retaining only two adapters. Depending on the recognition sequence of the third enzyme, this leads to a reduction of the number of amplified fragments by a factor of

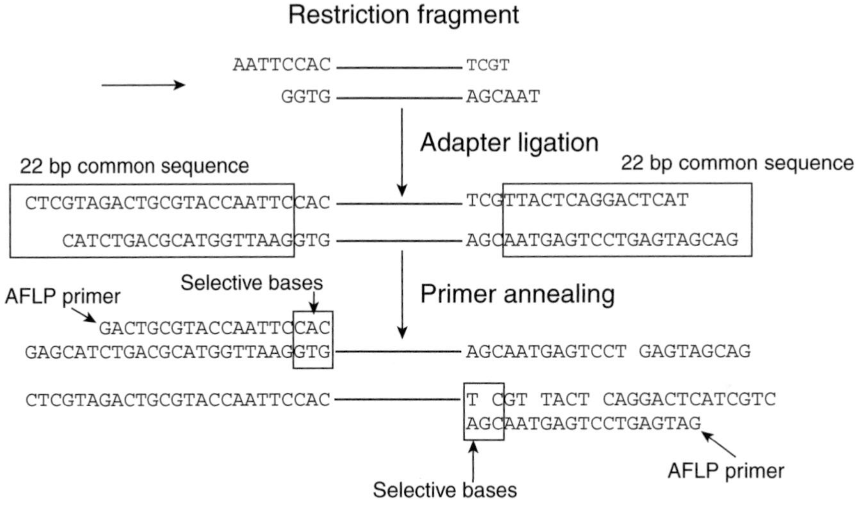

Restriction fragment

Figure 2.6 Outline of the AFLP DNA-fingerprinting method. Top: double-stranded *Eco*RI–*Mse*I restriction fragment with 'sticky ends'—that is, overhanging single-strand sequences. Centre: the same fragment after ligation of the *Eco*RI and *Mse*I adapters (boxed). Bottom: both strands with AFLP primers annealed. Each primer has three parts: the 5′ part corresponding to the adapter, the restriction site sequence, and three selective bases at the 3′ end, which make sure that on average $(1/4)^3 = 1.6\%$ of the fragments generated are amplified. The number of selective bases can be adjusted to genome size. One of the primers is labelled, with either γ-^{32}P or a fluorophore, to allow detection of the bands. After Vos *et al.* (1995), by permission of Oxford University Press.

around 10. This variant of the technique is called three-enzyme AFLP (Van der Wurff *et al.* 2000).

Because AFLP generates a large number of bands (50–150) for each genotype, it became the preferred method for mapping ecologically relevant traits in the 1990s. Many traits that determine the fitness of an organism in the environment are measurable in the phenotype as a quantitative score (body size, clutch size, flowering time, longevity, disease resistance, etc.). The genomic segment underlying such quantitative traits is called a *quantitative trait locus* (QTL). The identification of QTLs in the genome is a major area of research in ecological genetics and plant and animal breeding (Tanksley 1993, Jansen and Stam 1994). QTL mapping uses controlled crosses, preferrably starting with two inbred parents that differ in the trait of interest, to correlate the segregation of bands in AFLP (or other) fingerprints with the segregation of the trait. When the offspring from two inbred parents are sib-mated for several generations, recombination breaks up the linkage between traits in the parental chromosomes and *recombinant inbred lines* develop, each of which

contains a nearly homozygous segment from one of the parental chromosomes. The degree of precision that may be obtained is obviously dependent on the recombination frequency around the locus. In general it proves to be very difficult to pinpoint individual genes in this way; often a region of several thousand to some millions of base pairs remains for molecular analysis, so a QTL is not a genetic locus in the strict sense.

The genomic revolution has opened up new prospects for QTL mapping by using *single nucleotide polymorphisms* (SNPs; Borevitz and Nordberg 2003). SNPs are positions in the genome at which at least some individuals of a species have a base pair different from the most common form (see Section 3.1). SNPs are contrasted with other types of genetic variation, such as insertions/deletions and duplications. SNPs and insertions/deletions constitute the predominant source of variation in a population. Depending on the species, there is an SNP every 50 (in *Drosophila*) to every 1000 (in humans) base pairs in the genome. High-throughput genomics technology for SNP genotyping has provided a very powerful instrument

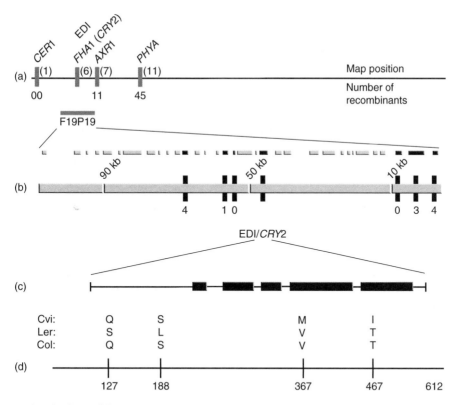

Figure 2.7 Map-based isolation of the EDI QTL in *A. thaliana*. (a) Linkage map of chromosome 1 showing the position of various molecular markers and the EDI QTL. F19P19 is the designation of the clone containing the locus. (b) Physical map of the F19P19 clone. The shaded boxes represent open reading frames according to the *Arabidopsis* genome sequence. The black markers represent seven newly developed molecular markers, used to localize the QTL further. (c) Genomic structure of the *CRY2* gene, including the 5′- and 3′-UTRs and five exons (black boxes). (d) Part of the CRY2 protein sequence, showing four variable amino acids. Q, glutamine; S, serine; L, leucine; M, methionine; V, valine; I, isoleucine; T, threonine; Cvi, Cape Verde mutant (with EDI); Ler, laboratory strain (not EDI); Col, Colombia strain (not EDI). After El-Assal *et al.* (2001), reproduced by permission of Nature Publishing Group.

for QTL mapping. The genotyping is usually applied to a large population of unrelated individuals, taking advantage of natural genetic variation, and an analysis is made of the *linkage disequilibrium parameter* (*D*) between pairs of sites in the genome. For a full treatment of the (conceptually complicated) linkage disequilibrium analysis we refer the reader to population genetics textbooks such as that by Hartl and Clark (1997).

An example of successful identification of the molecular mechanism underlying a quantitative trait is provided by the case of flowering time in *Arabidopsis* (El-Assal *et al.* 2001). *A. thaliana* from the Cape Verde islands flower much earlier than laboratory strains from temperate regions, and are

hardly sensitive to day length. Mapping with a variety of molecular markers had located an 'early day length insensitivity' (EDI) QTL to a 50 kbp region at one end of chromosome 1. The genomic sequence showed that this region contained 15 open reading frames (ORFs); however, the gene *cry2* (cryptochrome-2; encoding a photoreceptor protein) was considered a good candidate as the causal agent of the EDI syndrome. Sequence analysis showed that the Cape Verde mutant of this gene differed from the laboratory strain at 12 nucleotide positions: four in the promoter, one in the 3′-UTR, and seven in the coding region (Fig. 2.7). One of these mutations, leading to the substitution of a valine residue for a methionine in the protein, was proven to be the cause of photoperiod insensitivity. The CRY-2 protein appears to control a signalling pathway involving genes that promote flowering (see Section 5.3.4); under short-day conditions, the amount of CRY-2 protein is greatly downregulated during the photoperiod and this suppresses early flowering in plants from temperate regions under short day length. The mutated protein is less sensitive to the light-induced downregulation and that is why the Cape Verde plants flower earlier. It is obvious that in the tropical Cape Verde islands there is less need for suppression of flowering in response to short photoperiods, so the plants with mutated CRY-2 would have increased in frequency on the Cape Verde islands and natural selection finally drove the mutation to fixation.

This study is remarkable for three reasons. First, it shows how a QTL, with the aid of molecular markers and genomic sequence information, can allow a designated gene to be traced. Second, it is interesting that a very simple genetic change, such as mutation of a single base pair, can have such dramatic effects on the life cycle of a plant. Third, it is one of few examples of a molecular mechanism underlying selection in the wild being unravelled in such detail.

2.2 Sequencing genomes

All genome-sequencing projects employ the sequencing principle developed by Sanger (1977b), which makes use of the fact that in a PCR the extension by DNA polymerase is terminated if a dideoxynucleotide (ddNTP) is incorporated in the sequence, rather than a normal deoxynucleotide (dNTP). The trick is to make a reaction mix including normal nucleotides and chain-terminator nucleotides in such a way that amplicons are generated that are terminated randomly at all positions of the sequence. The result of the reaction is a collection of DNAs, each with the same sequence starting at the 5′-end of the template, but each differing in length so that the every nucleotide in the entire sequence is represented by a fragment that terminates at that nucleotide. Four similar reactions are conducted, each having one

of four radioactively labelled ddNTPs. Separation of the amplified fragments by electrophoresis and reading the labelled bases in four different lanes allows reconstruction of the sequence of the template. A further breakthrough came from Leroy E. Hood in 1986 with the use of fluorescent labels, allowing a single sequencing reaction containing all four ddNTPs, and detection of DNA fragments using a laser. Machines were developed that could perform DNA sequence analysis completely automatically and send the sequence to a computer (Smith *et al.* 1986). The principle of sequencing by *dideoxy chain termination* is treated in all molecular biology textbooks and so is not discussed here in any more detail. Other sequencing principles are emerging, such as sequencing by microarray hybridization, mass spectrometry, atomic force microscopy, and nanopore electrophoresis, but these are not yet applied on a large scale (Gibson and Muse 2002). In this section we will discuss the two main strategies for organizing sequencing projects and the analysis that follows from the data.

Laboratories of molecular ecology usually have their own sequencing facilities, allowing sequencing operations on a limited scale; however, it is not likely that they will embark on a whole-genome sequencing project in-house. Instead, whole-genome sequencing is usually contracted out to a commercial sequencing centre or is organized in collaborative networks of many different laboratories. Because of these conditions, we restrict our discussion of genome sequencing to the main principles and avoid too much technical detail.

2.2.1 Library construction

The first, and often most crucial, step in a sequencing project is the construction of a recombinant DNA library. A *genomic library* is a collection of clones that together encompasses all DNA sequences in the genome of the species. The library is prepared by fragmentation of the genome and insertion of all fragments into a suitable vector. The vector is kept in a host bacterium (usually *Escherichia coli*) and DNA from the target species can be cloned by growing the bacteria. The initial fragmentation can be done in two ways: restriction with endonucleases or mechanical shearing. In the case of enzymatic restriction, the cleavage sites of the endonuclease are the same as the cloning sites in the vector and this allows direct insertion in the vector. In the case of mechanical shearing additional enzymatic manipulation is necessary to ligate adapters to the fragments, which will allow insertion into the cloning site of the vector. An advantage of mechanical fragmentation is that it is essentially random and avoids the possible bias arising from the fact that some regions of the genome may be poor in the pertinent restriction sites. In general, it is difficult to make sure that all fragments from the genome have an equal representation in the library; regions of non-coding, repetitive DNA tend to be under-represented, especially if the library is prepared with enzymatic restriction.

The number of clones (n) that needs to be prepared to cover the complete genome is given by the following formula (Russell 2002):

$$n = \frac{\ln(1-p)}{\ln(1-s/G)}$$

where p is the probability that at least one copy of a DNA fragment from the target organism is in the library, s is the average insert size (bp) of fragments in each clone, and G is the size of the genome (bp). For example, to sequence the genome of a bacterium with a genome size of 2 Mbp, using a vector that accepts inserts of 1.8 kbp and aiming for a 99% chance that a genomic fragment is in the library, 5115 clones need to be made. Obviously, n increases with decreasing insert size, increasing genome size, and increasing values of p. Note that this formula specifies the required number of clones in terms of probability. To minimize the chance of missing genomic fragments, most of the fragments will be present more than once in the library. In fact, the average fragment frequency may be around four or five.

Many different types of cloning vector are available commercially. Traditionally, the most common vectors used in recombinant DNA technology are *plasmid cloning vectors* such as pUC19.

These vectors are derived from naturally occurring plasmids (extrachromosomal elements that replicate autonomously in bacteria; see Section 3.2), which are further manipulated to suit the purpose of cloning. The manipulated plasmid includes an antibiotic-resistance gene to allow selection of only those host cells that actually have the plasmid. It also has multiple cloning sites inserted in a copy of the *lacZ* gene to allow plasmids with succesful inserts to be selected on the basis of β-galactosidase activity (white/blue screening of bacterial colonies). Plasmid cloning vectors will accept foreign DNA fragments of a few kilobase pairs in size; from the formula above it can be seen that for most eukaryotic genomes ($G = 10$–$100\,000$ Mbp) the use of these vectors for genome sequencing would need an insuperable number of clones in many cases.

Another group of vectors, called *cosmids*, can accept DNA fragments of around 40 kbp. Cosmids are completely artificially constructed molecules that contain a number of features (Fig. 2.8a). Like in a pUC vector, there is an origin of replication (ori) sequence, which is recognized by the DNA polymerase of *E. coli* and ensures efficient replication in the host. Likewise, there is an *amp*^{R} gene, which confers resistance to ampicillin and allows selection of plasmid-containing hosts. Then there are several cloning sites, where foreign DNA can be inserted, as in normal plasmid cloning vectors. The most characteristic feature of a cosmid is a so-called *cos* site; this is a recognition site for a phage λ endonuclease, which will cleave the plasmid to prepare it for assembly into a phage head. The phage can, however, only be packaged when the plasmid has a length of about 45 kbp. The cosmid itself has been made small, around 5 kbp, so is not packaged by itself; when it carries an insert of around 40 kbp it is the right size. Upon adding the appropriate proteins, phage λ is assembled *in vivo*. Then the phages are used to infect *E. coli*, which transfers the foreign DNA to the final host.

Cloning vectors that accept still larger DNA fragments are *bacterial artificial chromosomes* (BACs) and *yeast artificial chromosomes* (YACs). BACs are derived from an extrachromosomal plasmid

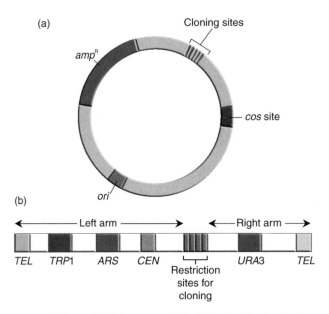

Figure 2.8 Schematic representations of (a) a cosmid cloning vector and (b) a YAC. *ori*, origin of replication; *amp*^{R}, ampicillin-resistance gene; *cos*, recognition site for phage λ endonuclease; *TEL*, yeast telomere; *TRP1*, gene conferring tryptophan independence; *ARS*, autonomous replicating sequence; *CEN*, centromeric sequence; *URA3*, gene conferring uracil independence. From Russell (2002), with permission from Pearson Education Inc..

normally involved in conjugation, the F factor, or sex factor. BACs utilize the origin of replication from the F factor and have several other features such as cloning sites and selectable markers. They can accept DNA fragments of up to 200 kbp. One version of the BAC vector is termed *fosmid* (for F$_1$-origin-based cosmid-sized vector), and is introduced in the host via phage particles, as for cosmids. Very large DNA fragments, up to 500 kbp, are allowed when using YACs. A YAC is a linear molecule, engineered to resemble a yeast chromosome, with a centromere in the middle and a telomere at either end. The construct includes a sequence that is recognized by the yeast DNA polymerase machinery and allows autonomous replication in yeast. There are selectable markers on each arm (to test for intact chromosomes after insertion of the foreign DNA and restriction sites for cloning; Fig. 2.8b). Obviously, the YAC is grown using budding yeast as a host, rather than *E. coli*.

A genomic library is a valuable resource for any laboratory, because it can be used not only for sequencing but also for gene identification by *library screening*. For example, when a cDNA of interest has been picked up by differential display or another gene-discovery method (see Section 2.1), a labelled probe can be made from that sequence and the library screened for any complementary sequences, which can then be picked up and characterized. To do this, the cDNA probe is labelled radioactively and hybridized to a membrane upon which a replica of the library is printed. After washing, remaining radioactivity is detected by autoradiography and one or more spots will indicate the clones in the library that have the sequence of the probe. The same technique can be used for identifying microsatellites, in which case the probe has repeat sequences typical for these loci. Finally, it is also possible to use probes from other species and detect homologous genes by cross-species hybridization. Any clones in which the probe demonstrates a gene of interest can be subcloned and sequenced. For genomic models, library screening in this way has now mostly been replaced by microarray-based techniques (see Section 2.3); however, for ecological laboratories working on non-model organisms the importance of a good genomic library cannot be overestimated.

In addition to whole-genome libraries there are also *cDNA libraries* and *chromosome libraries*. A chromosome library is a genomic library of only one chromosome, which is a valuable resource in genome-sequencing projects that use the hierarchical method (see below). A cDNA library is a collection of clones containing reverse-transcribed mRNAs, usually representing a specific physiological condition (e.g. messengers collected after exposure to drought) or a certain tissue (e.g. messengers expressed specifically in the gonads). Fragments of sequenced cDNAs are called *expressed sequence tags* (ESTs). ESTs are usually produced in a high-throughput, single-pass pipeline, leading to a large collection of sequence reads of 400–700 bp. Development of an *EST library* is often the first thing to do when commencing a genomics project on a novel organism. Even though most organisms investigated in ecological genomics lack a full genomic sequence, they usually do have an EST library.

2.2.2 Hierarchical sequencing of a genome

The principle of hierarchical sequencing is that the clones in a clone library are ordered with respect to their position in the genome before commencing sequencing. An ordered series of clones will produce a *physical map* of the genome. In this type of map the distance between markers is measured in physical units: nucleotides. The ultimate physical map is a complete genome sequence. A physical map contrasts with a *genetic map* that is developed from linkage disequilibrium data, in which distances between markers are derived from inheritance and recombination, and are measured in centiMorgans (cM). A physical map of the genome can be constructed in three ways (Gibson and Muse 2002):

Restriction-fragment fingerprinting. Clones are aligned based on their restriction digest patterns. Restriction enzymes cleave DNA at well-defined recognition sites and so each digested clone will produce a characteristic fingerprint of fragment lengths upon electrophoresis; these fingerprints are compared with one another. When there is a correspondence between two bands, one in fingerprint A, the other in fingerprint B, it is likely that they have the same sequence and that clones A and B overlap partly. If another band in fingerprint B is similar to a band in fingerprint C, while A has no such band, a relationship between B and C is established. When many restriction digests are generated and compared using computer programs, a physical map of clone markers results.

Terminal sequencing. A further step to identify interconnections between the clones in a library, often used to span the gaps remaining after fingerprinting, is to sequence both ends of the clones. The idea is that at least one end of a clone matches an already assembled part of the sequence. The other end will then extend into the gap or will match an adjacent clone. Terminal sequencing, or end sequencing, is also done as part of the verification procedure after assembly of a genome (see below).

Chromosomal walking. Starting with the sequence of one of the clones, a short labelled probe is made using the terminal sequence of that clone. The library is then screened for other clones with that sequence; from the ones that show hybridization, one clone is chosen for further sequencing. Then the terminal sequence of the new clone is used to develop a terminal probe that will pick up the next overlap.

After the physical relationship between clones is established, a so-called *minimal tiling path* is defined. This is an alignment of minimally overlapping clones (e.g. BACs) in such a way that the complete sequence of, for instance, a chromosome is covered (Fig. 2.9). Then each BAC is fragmented by shearing and subcloned for automatic sequencing. The sequence reads obtained are ordered in a series of contiguous sequences, which results in a so-called *contig*, which is essentially the reconstructed sequence of a BAC. Closing the overlap between contigs allows the construction of larger pieces of the genome, so-called scaffolds.

Usually the complete genome sequence cannot yet be assembled from the scaffolds because of

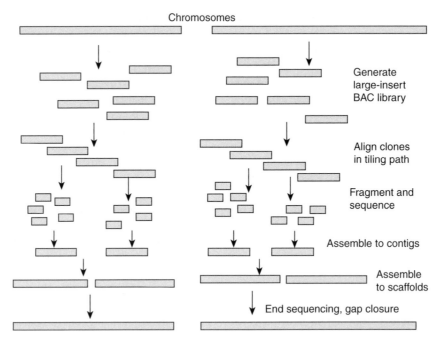

Chromosomes

Generate large-insert BAC library

Align clones in tiling path

Fragment and sequence

Assemble to contigs

Assemble to scaffolds

End sequencing, gap closure

Figure 2.9 Scheme of the hierarchical approach to whole-genome sequencing.

several gaps, which have to be filled in by further sequencing. This is often done by sequencing both ends of a collection of clones (end sequencing) and looking for identity with parts already sequenced. If an end sequence happens to fall into an already sequenced contig that is adjacent to a gap, there is a good chance that the other end of the clone extends into the gap. In this way an attempt is made to fill in all gaps. Gap closure can also be supported by sequence information from other sources, such as existing cDNAs or ESTs. Finally, comparison of the sequence-based physical maps with the corresponding genetic map (if available) can also be of high value. The gene order in the genetic map should correspond to the gene order in the physical map, although the genetic map may locally expand or contract the physical map due to unequal rates of recombination across the chromosome. The final result is a sequence assembly of the entire genome. This usually still requires further editing to remove errors. For example, after publication of the draft sequence of the human genome in early 2001, it took another 3 years before the sequence (and only the euchromatin part of it)

was considered to be 99% complete, in October 2004.

An interesting aspect of the hierarchical approach to genome sequencing is that the work can be distributed among laboratories, each focusing on designated parts of a chromosome or on a certain collection of BACs. Sequencing the yeast genome is the prime example of a project that was mostly completed using the hierarchical approach. Started in 1989, a group of 35 laboratories embarked on the task of sequencing chromosome III, which was completed in 1992. Then in the meantime new projects were formulated, which led to collaboration between 92 laboratories over the years, involving 600 committed scientists, until the completion of the sequence was announced in 1996 (Goffeau *et al*. 1996). Looking back, Dujon (1996) mentioned that two aspects are critical in a genome programme: construction of clone libraries 'upstream' of the sequencing and quality control of the sequence 'downstream' of the sequencing. The average accuracy of the yeast genome at the time was estimated as 99.9%, which seems a high figure, but Dujon (1996) noted that even this figure allows only one-third of all protein-coding genes in the yeast genome to be completely error-free. With a sequence accuracy of 99.99% the proportion of completely error-free proteins rises to 85%.

2.2.3 Whole-genome shotgun (WGS) sequencing

The principle of WGS sequencing was introduced in 1995, when the genome of the bacterium *H. influenzae* was published (Fleischmann *et al.* 1995). The term shotgun evokes the image of a cloud of shot fired at short range to hit the genome more or less at random. The strategy is to skip the ordering of clones and the construction of physical maps and to just sequence clones in random order until it may be assumed that all genomic fragments have been covered at least once (Fig. 2.10). The average number of times that a fragment is sequenced is called the *depth of coverage*. The idea is that the likelihood that a segment is not represented at all should be as small as possible by increasing the mean coverage. It may be assumed

that the probability of a base position being sequenced *r* times, P(*r*), follows a Poisson distribution, which is given by:

$$P(r) = \frac{\mu^r}{r!e^\mu}$$

where μ is the mean depth of coverage. When the genome size is *G* and the sequencing has delivered *N* bases, $\mu = N/G$. The probability that a base is then still not sequenced is

$$P(0) = e^{-N/G}$$

With 6-fold coverage, the expected fraction of bases not yet sequenced is 0.00248, or 0.25% of the genome. So, even with a high degree of redundancy, there will always remain gaps in the genome sequence; increasing the sequencing effort helps very little after 5-fold coverage because of the principle of diminishing returns inherent in the exponential function.

The theory of WGS sequencing goes back to Lander and Waterman (1988). The principle is that the preparation of genome fragments is essentially random, which is approximated by applying shearing, rather than enzymatic digestion of DNA,

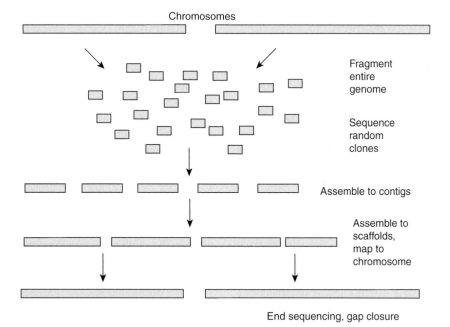

Figure 2.10 Scheme of the WGS approach to sequencing genomes.

to construct the library. It is also assumed that the sequencing itself is random; that is, random clones are picked from the library. Non-random elements in the procedure may be due to the fact that not all segments from the genome are cloned with the same efficiency and that some parts of the genome are unstable in the cloning vector. This was recognized in the shotgun sequencing of the *Drosophila* and human genomes, which contain long stretches of repetitive, non-coding DNA. Around one-third (60 Mbp) of the *Drosophila* genome consists of simple repetitive sequences occasionally interrupted by inserted transposable elements and tandem arrays of RNA genes, with some islands of coding sequences. The long stretches of repetitive DNA are not transcribed, do not uncoil in the interphase of the cell cycle, and are microscopically recognized as *heterochromatin*. In *Drosophila* they are located mainly in the centromeric regions of the autosomes and in the entire Y chromosome. *Euchromatin* is defined as DNA that is found in an uncoiled stage during the interphase and contains actively transcribed genes. The sequencing of the *Drosophila* genome was restricted to euchromatin, whereas heterochromatin was operationally defined as that portion of the genome that could not be cloned stably in BACs (Adams *et al.* 2000). Another issue of non-randomness is that some clones in the library may contain deleterious genes, which are expressed when growing the host, with the result that the clone is lost. To minimize this effect, Fleischmann *et al.* (1995) selected a narrow range of DNA fragments, between 1.6 and 2.0 kbp in size, for inclusion in the library. By chosing a maximum of 2 kbp, the probability that a complete gene is present in a single cloned fragment is small.

The WGS approach delivers an enormous number of reads, from which the sequence of the whole genome has to be assembled (Myers 1999). The *assembly phase* makes use of computer programmes that in principle compare every sequence read with every other read to identify overlaps. From these overlaps contigs and scaffolds can be constructed, as for hierarchical sequencing (see above). Large scaffolds of more than 100 kbp can be mapped onto a chromosomal position. This is

done by matching the scaffold sequence with previously cloned genes whose position is known from linkage maps, or by developing probes that can indicate a scaffold's position by fluorescent *in situ* hybridization (FISH) in chromosome preparations. Still considerable effort has to be devoted to verification and accuracy checks, including additional end sequencing of large insert libraries to close the gaps.

The WGS strategy was introduced by J.C. Venter and co-workers, then at the Institute for Genomic Research (TIGR). After the *H. influenzae* project had proven that the strategy was highly effective and could lead to complete genomic sequences in record time, Venter founded Celera Genomics in 1998 with the primary mission to sequence and assemble the human genome by 2001 (Weber and Myers 1997, Venter *et al.* 1998). As a test system and to explore the applicability of WGS sequencing to large eukaryotic genomes, Celera Genomics first sequenced the *Drosophila* genome, which was published in March 2000, while the completion of the human genome was announced later in the same year (see Section 1.1). Nowadays, the WGS strategy is applied by all large genomic sequencing centres. The organization of a WGS project is different from a hierarchical sequencing project; WGS sequencing is typically done in a single centre, since it relies on massive parallel sequencing and enormous computational power to assemble the genome. Goffeau *et al.* (1996) contrasted the two approaches by naming them the 'factory approach' and the 'network approach'. Nothwithstanding this contrast, the two approaches are actually mutually supportive. Hierarchical sequencing employs a 'shotgun' element at the stage where large-insert libraries are fragmented and subcloned (Fig. 2.9), whereas of course the assembly of WGS-generated scaffolds can hardly be done without the information on genetic and physical maps. The finishing of the human genome sequence in 2004 showed that a hybrid strategy is necessary to prevent WGS sequences from skipping duplicated regions and so 'simplifying' the genome (She *et al.* 2004). Nowadays large genomes (>30 Mbp) are not sequenced by the WGS method alone, but by combining this with shotgun sequencing of separate BACs (the BAC-by-BAC approach).

2.2.4 Gene finding and annotation

Finding genes in a fully sequenced genome would seem to be a trivial task at first sight, but is more complicated in practice. In fact, the concept of gene as a 'complete chromosomal segment responsible for making a functional product' emphasizes that a gene is not just a piece of DNA but a unit that must be judged by its deeds. So the concept of gene includes structural as well as regulatory elements (promoter, transcription factor-binding sites, etc.). The following five criteria may support the recognition of genes in a genomic database (modified from Snyder and Gerstein 2003).

Identifying ORFs. A DNA sequence that can be read as a series of amino acid-encoding triplets bounded by a start and a stop codon is called an ORF. Computer programs have been developed to locate ORFs in a genomic sequence; however, these programs sometimes have difficulty finding genes if these are small or consist of several *exons* (the protein coding parts of the gene) interrupted by large *introns* (non-coding sequences that are spliced out of the RNA after transcription). There are exceptional cases, such as genes embedded in the intron of another gene, that can make the life of a gene hunter particularly difficult. Gene-finding programs have to be trained for every new organism. Gene identification just by scanning the sequence and establishing homologies with known sequences in a database is called *ab initio* gene discovery.

Codon bias. Not all ORFs in the genome are actually transcribed. A *pseudogene* is a gene that bears resemblance to a functional gene but contains defects such as frameshift or nonsense mutations. Pseudogenes are common in large gene families, groups of genes that underwent repeated duplication leading to a collection of genes with similar but often slightly differentiated function (see Section 3.1). Most pseudogenes are not transcribed; some of them are, but

do not lead to functional peptides. Whether a gene is transcribed actively can sometimes be judged from *codon-usage bias*. This term refers to the fact that although for most amino acids two or more codons are available in the genetic code, some codons occur much more frequently than others. For example, for the amino acid leucine six different triplets are available (CUU, CUC, CUA, CUG, UUA, and UUG), but the majority of leucines are encoded by UUG (in the yeast genome it is 80%, rather than the expected 16.7%). This phenomenon is usually attributed to the fact that the UUG tRNA of leucine is more abundant than other leucine tRNAs and so translation of a messenger in which that codon is used proceeds more efficiently. Codon bias has shed a new light on the neutrality of *synonymous mutations*. These are mutations that change a codon into another codon for the same amino acid. Such mutations were long thought to be neutral to selection, but the phenomenon of codon bias indicates that there might be subtle preferences even in synonymous mutations. If a gene is transcribed actively there is a continuous selection pressure for the use of preferred codons. Conversely, if many uncommon codons are seen in an ORF it may indicate that the gene is not actively transcribed.

Homology search. Many genes in a genome may be 'identified' from their similarity to known genes from other species. Using the Basic Local Alignment Search Tool (or BLAST; see Section 2.4.1) or other alignment procedures putative protein-encoding sequences may be compared with databases of sequences from other species, and if there is evidence that in the other species the same ORF encodes an active gene, the case for the new ORF being a real gene is strengthened. Of course, this will only work with genes whose sequence is relatively conserved over species, and this is not always the case. An interesting aspect of genome sequencing is that every new genome comes with 20–60% previously unknown genes. When the yeast genome was published in 1996 about one-half of the ORFs were unknown in other species. In line with molecular biology's predilection for wordplay in technical terms, these ORFs are called *orphans*. The most striking aspect was that the same orphans were not only unknown in yeast, but also in other species (Dujon 1996).

Association with promoter elements. If there are characteristic sequences upstream of an ORF that match with known transcription factor-binding sites, this will increase the likelihood that the gene is functional. In addition, the nature of the transcription factor-binding sites may shed light on the physiological context in which the gene is expressed. Genome-wide discovery of transcription factor-binding sites using conserved sequences and probes from other species is known as *phylogenetic footprinting* (see Chapter 3).

Match with transcript or protein sequences. For many model organisms there is not only a genomic database but also a large collection of cDNA sequences, ESTs, or serial analysis of gene expression (SAGE) tags (see Section 2.3.3). If the sequence of an ORF matches the sequence of a transcript of the same species, the status of the ORF as a true gene is obvious. The same holds for evidence from microarray hybridization (see Section 2.3.1) or matches with protein databases of the same species.

The conclusion from these criteria is that the genome sequence itself does not allow solid enumeration of genes. Only functional studies (transcript analysis, proteins, knockout studies) can firmly qualify an ORF as a gene. There is a tendency for the number of genes to be over-estimated when the genome sequence of a species

first becomes known. For instance, in the *Saccharomyces cerevisiae* genome, the initial estimate was 6274 possible genes, but this number has gone down over the years. Three large-scale studies (Kellis *et al.* 2003, Cliften *et al.* 2003, Brachat *et al.* 2003) compared the yeast genome with other fungal species for which a genomic sequence became available by the beginning of the twenty-first century. This allowed several previously unknown ORFs in yeast to be characterized and several spurious ORFs to be removed from the gene catalogue. The Saccharomyces Database (SGD) now classifies genes as verified, uncharacterized, or dubious, according to the degree of certainty that an ORF actually encodes a functional protein. Snyder and Gerstein (2003) used a more extensive classification and estimated the number of genes in the yeast genome to be 6128 (Table 2.1).

After an enumeration of the genes has been made, the next most common activity in genome analysis is to name the genes and classify them according to function. This process is called *annotation*. Without annotation, a genomic database is useless. Annotation is based on a system of *gene ontology* (GO): a controlled vocabulary used to describe the molecular function and cellular location of gene products and the biological processes in which they are involved. The Gene Ontology Consortium (2000), a collaborative project between the major genomic databases, has proposed to characterize genes according to three major categories: biological process, molecular function, and cellular component. Each of these main categories has numerous denominations to qualify the gene.

Table 2.1 Classification of ORFs in the yeast genome according to Snyder and Gerstein (2003)

Type of ORF	Status	Number
Essential ORFs	Well-characterized and essential metabolic function	1106
Known ORFs, but no essential ORFs	Well-characterized function, sequence longer than 100 codons	2289
Homology ORFs	Only validated by homology to genes in other organisms	2060
Other ORFs	Includes short ORFs and ORFs identified by transposon tagging	536
New short ORFs	101 Transposon-identified ORFs and 36 homology-identified ORFs, added since 1996	137
Total	Protein-encoding genes (2003 estimate)	6128
Questionable ORFs	No evidence for transcription, function, or sequence conservation	283
Disabled ORFs	Pseudogenes	221

The system ensures that homologous genes in different databases are assigned the same functional category. The GO system has now developed into an extensive hierarchical structure by which a multitude of aspects of a gene and its product can be described in a way that is valid for all biological species. Conversely, the user of the database finds significant support in the GO classification of the gene. Table 2.2 provides some examples of GO designations of a random set of genes in the database for *D. melanogaster* (FlyBase).

When genes conduct more than one function, application of the GO system may become problematic, especially for the higher categories. This will be the case for genes that have strong pleiotropic effects on the phenotype. To meet these difficulties, Fraser and Marcotte (2004) proposed that the hierarchical GO system should be supplemented with a bottom-up network analysis, in which the function of a gene is not assigned by a human curator, but is determined by statistical analysis of expression profiles. In the system proposed by Fraser and Marcotte (2004) gene functions are defined in terms of the topological properties of the network in which the gene sits;

that is, the way it is linked and interacts with other genes.

The naming of new genes comes with extensive guidelines that researchers have to follow before a name is accepted by the authority that supervises the genomic database. Each database has conventions for naming and annotating genes and often a nomenclature committee exists to develop internationally accepted standard genetic nomenclature for genes, as well as designations for markers, QTLs, mutations, and allelic variants of genes. There is no universal system for gene names and their abbreviations. Different traditions prevail in different databases. In the human genome gene symbols should be written in upper-case italics (e.g. *PGM* is the gene symbol for a gene named phosphogluconate mutase), whereas in the *Drosophila* genome gene symbols are written in italics, in lower case except for the first letter (e.g. *Pgm*); in bacterial genetics, gene symbols are written in lower case throughout (e.g. *pgm*). There is bewildering diversity of conventions for denoting alleles, splice variants, and mutants. Different genes of the same family are usually written with the same core symbol followed by a number or

Table 2.2 Examples of application of the GO classification, illustrated by gene annotations in the genome of *D. melanogaster* (www.flybase.org)

Gene name (annotation)	Gene identifier	Molecular function	Biological process	Cellular component
Acetylcholinesterase	*Ace*	Acetylcholinesterase activity	Neuromuscular synaptic transmission	Synapse
Aconitase	*Acon*	Aconitate hydratase activity	Amino acid biosynthesis	Mitochondrion
ATP citrate lyase	*ATPCL*	ATP citrate synthase activity	Acetyl-CoA biosynthesis, tricarboxylic acid cycle	Cytoplasm
Croquemort	*Crq*	Scavenger receptor ativity	Apoptosis, macrophage activation	Integral to plasma membrane
Cytochrome *c* oxidase subunit Va	*CoVa*	Cytochrome *c* oxidase activity	Electron transport	Mitochondrial inner membrane
Glucose dehydrogenase	*Gld*	Glucose dehydrogenase (acceptor) activity	Cuticle biosynthesis, glucose metabolism	Extracellular
α-Mannosidase II	*α-Man-II*	Mannosyl-oligosaccharide 1,3–1,6-α-mannosidase activity	N-linked glycosylation	Golgi membrane
Myosin heavy chain	*Mhc*	Cytoskeletal protein binding, structural constituent of muscle	Cytokinesis, striated muscle contraction	Striated muscle thick filament
Rhodopsin 2	*Rh2*	G-protein-coupled photoreceptor activity	Phototransduction, visual perception	Integral to membrane

a letter (e.g. *Pgm1* and *Pgm2*). The protein is often written in upper-case, regular font style (e.g. PGM); however in bacterial genetics, only the first letter of a protein name is in upper case; for example, Nif is dinitrogenase, an enzyme consisting of two subunits, Nifα and Nifβ, which are encoded by genes *nifD* and *nifK*, respectively, of the *nif* operon.

2.3 Transcription profiling

Transcription profiling has proven to be one of the genomics approaches most attractive to ecologists. The aim of transcription profiling is to develop a complete overview of all the genes in a genome that are upregulated or downregulated in response to some factor of interest, in comparison with a designated reference expression. It is important to realize that transcription profiling is essentially relative; that is, comparisons in gene expression are made between a challenged object and a reference object, for the same object at different points in time, or among a series of objects given different treatments.

2.3.1 Microarrays

To develop a genome-wide image of gene expression, all the RNAs present in the cell at a certain time point need to be assessed, as well as their relative abundance. As noted in Section 1.3, the collective pool of RNA (rRNAs, tRNAs, mRNAs, and non-coding iRNAs) in a cell is called the transcriptome. Gene expression is usually focused on mRNA, also called *poly(A)-RNA*, because mRNA of eukaryotes is often isolated by taking advantage of the characteristic poly(A) tail on each messenger (Fig. 2.1). The genomics way to get an overview of all transcripts is to generate two labelled samples, one from the challenged organism or cell and one from the reference, and hybridize these in competition with each other to a large number of DNA sequences, immobilized on a coated glass plate in an ordered array. Such a device, usually denoted as *microarray*, first described by Schena *et al.* (1995), is the cornerstone of transcription profiling in genomics. The term

microarray is contrasted with *macroarray,* which uses essentially the same technology but is conducted on a nylon membrane with fewer spots and using radioactive labels rather than fluorescence. The spots on a microarray, which are packed very close together, are called probes or sometimes features. The sample of transcripts that is interrogated by the array is called the target.

Microarray hybridization is sometimes called *reverse hybridization,* because the probe represents the immobile phase. This contrasts with traditional Southern blotting, in which the target is immobilized (usually on a membrane) and the (labelled) probe is mobile.

The hybridizations on a microarray are visualized by labelling the transcripts—the reference with a green fluorescent label and the test sample with a red label—and the array is scanned using a laser. To check on dye-specific artefacts it is necessary to include an experiment in which the two dyes are interchanged (*dye-swapping*). Each spot on the array that holds a DNA sequence of sufficient homology to one of the sequences in the collection of mRNAs will be labelled red or green if the abundance of that messenger was greater or less in the the tested sample compared with the reference. In this way, genes that are upregulated relative to the reference will be labelled with one colour and genes that are downregulated will be labelled with the other. For all spots that correspond to messengers whose abundance in the sample is similar to the reference, equal amounts of label will bind and the spot will be perceived as yellow. Spots for which there is no corresponding messenger will not be labelled at all.

The use of a microarray for transcription profiling can be illustrated with a classical paper by DeRisi *et al.* (1997). These investigators were interested in the changes in physiology that occur in yeast (*S. cerevisiae*) when cells switch from anaerobic growth (fermentation), using glucose as a carbon source and producing ethanol, to aerobic growth, which occurs when glucose is depleted after fast growth and the cells turn to respiring ethanol. This *diauxic shift* is accompanied by a fundamental reorganization of the cell's physiology, in which the expression of many genes changes. Since at the time the microarray had just been introduced (Schena *et al.* 1995) and the genome sequence of yeast completed (Goffeau *et al.* 1996), the authors could apply one of the first genomics approaches to transcription profiling. DeRisi *et al.* (1997) amplified ORFs from the yeast genome using PCRs with primers specific for each gene. These DNA fragments, approximately 6400, corresponding to nearly all the genes in the yeast genome, were printed onto glass slides using a robotic device. In the experiment, cells were harvested at different stages in the growth phase and mRNA was isolated. These mRNAs were reverse-transcribed to cDNAs and labelled with a red carbocyanin (Cy5) label, and this was mixed with a green (Cy3) labelled sample prepared from cells harvested directly after the start of the experiment. After hybridization and washing the array was scanned using a fluorescent confocal microscope and images generated like the one shown in Fig. 2.11.

Microarray scans are often represented in colour, where a red spot indicates upregulation, a green spot indicates downregulation, and a yellow spot constant expression; however, the fluorescence intensities of the two carbocyanine labels are not actually 'seen' in colour by the detector, which just records digital values for intensities at different wavelengths. Therefore a colour representation of a microarray image is also referred to as having *false colours*. In Fig. 2.11 and elsewhere in this book we represent microarray scans in grey tones, although the difference between upregulation and downregulation is not always discernible in this way. The reader should consult the original publications to obtain complete views of microarray images.

DeRisi *et al.* (1997) discussed their data in terms of the biochemical pathways for carbon and energy metabolism (pentose phosphate pathway, glycolysis, Krebs cycle, glycoxylate cycle). They were able to show that the changes in mRNA abundance during diauxic shift could be mapped onto this biochemical framework and indicated that a significant redirection of metabolites was taking place. It was also apparent that groups of genes responded in a coordinated fashion and seemed to be regulated by a common factor. This

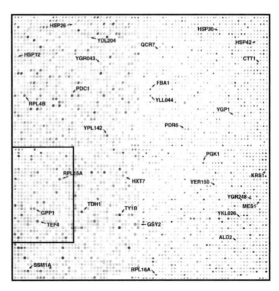

Figure 2.11 Scan of a microarray as used by DeRisi *et al.* (1997). Each spot represents one of 6500 DNA sequences from the yeast genome, printed on a glass plate of 18 × 18 mm. The array was used to detect gene-expression changes occurring when the cells had depleted glucose from the medium after 9.5 h and shifted to aerobic metabolism. In this negative greyscale picture, hybridization intensities are indicated by different shades of grey. In full-colour images upregulation and downregulation is indicated in red and green, *resp.* Some of the spots are indicated by codes designating the genes. The rectangle on the left indicates a section of the microarray singled out for further analysis (not shown). Reprinted with permission from DeRisi *et al.* (1997). Copyright 1997 AAAS.

observation triggered further studies into the mechanisms by which transcription factors activate sets of genes collectively.

In the course of the 1990s new techniques for the fabrication of microarrays were introduced, which led to so-called high-density oligonucleotide arrays or *gene chips* (Lockhart *et al.* 1996, Lipschutz *et al.* 1999, Lockhart and Winzeler 2000). In the these systems, short sequences of 20–25 nucleotides are synthesized directly on the substrate using photolithography. In such arrays, all the probes are synthetic and genomic sequence information alone is sufficient to construct the array. The probes can be made in such a way that the most unique part of a transcript is represented; this does not need to be the coding part of the gene; in fact, often the 3′- or 5′-UTR of the mRNA is used, since this is more specific for a transcript

Table 2.3 Overview of features of the two most common microarray approaches in genomics

	Spotted cDNA microarrays	High-density oligonucleotide microarrays (GeneChips)
Probes	Known and unknown cDNAs	Oligonucleotides designed on the basis of genome knowledge
Probe manufacture	PCR-amplified from EST libraries or complete cDNA libraries	Synthesized by photolithography
Probe deposition	Spotted (printed) by means of a robot	Synthesized directly on the substrate
Labelling of target sample	Fluorescent labels attached to the cDNAs during reverse transcription	Biotinylated cRNAs, produced by T7 RNA polymerase from cDNA; biotin recognized by fluorescently labelled streptavidin
Manufacture of arrays	Can be home-made in an academic research center	Manufactured by specialized companies (Affymetrix, Agilent)

than the coding region. It is also common to use multiple probes designed to hybridize to different regions of the same transcript. In addition, each probe is supplemented by a control sequence that has one mismatched base in the middle of the sequence. Because the target cDNA should only bind to the perfect probe and not to the mismatch probe, an accurate measure of transcript abundance can be obtained by subtracting the match signal from the mismatch signal. Hybridization with these arrays is non-competitive; only a single sample of target is applied to each array and comparison with a reference is made across chips. In Table 2.3 a contrast is made of the characteristics of spotted cDNA microarrays and gene chips.

Oligonucleotide microarrays allow a greater coverage of the genome and may be more repeatable across laboratories. The extremely large number of synthetic oligonucleotides that may be packed on an array (some technologies allow for 195 000 probes on an area of hardly more than 2 cm^2) allow each gene to be represented by multiple probes (Nuwaysir *et al.* 2002). This is useful when aiming to detect polymorphisms and different splice variants of the same gene, and it also allows for within-array replication of gene expression. However, such arrays are obviously only possible for true model organisms, whose genome is completely known. The technology to manufacture these oligonucleotide arrays is advanced and applied only by a few specialized companies. The costs of commercially available microarrays is sometimes prohibitive to their use

in academic research laboratories. The fact that many species of interest to ecologists are not genomic models at the moment, added to the modest amount of financial resources available to ecologists in comparison with medical and agricultural research, implies that spotted cDNA microarrays will probably remain the main platform of choice for ecological genomics. Fig. 2.12 provides an overview of the two approaches to transcription profiling (Schulze and Downward 2001).

2.3.2 Microarray-based transcription profiling in ecological genomics

At which stage in an ecological research project can microarray-based transcription profiling be worthwhile? There are several strategic considerations. Microarrays can be considered as an instrument of diagnosis; that is, transcription profiling can provide insights into the functional performance of an already well-known genome. For example, one might use a microarray to characterize the physiological state of an organism in the environment and use this information to draw inferences about its fitness or about certain characteristics of the environment. We call this the *diagnostic use* of microarrays. This type of use is expected to follow from an earlier developmental phase in which various probes are tested for their indicative value to the properties of interest, until the optimal design of the microarray is decided. Another use of microarrays in transcription

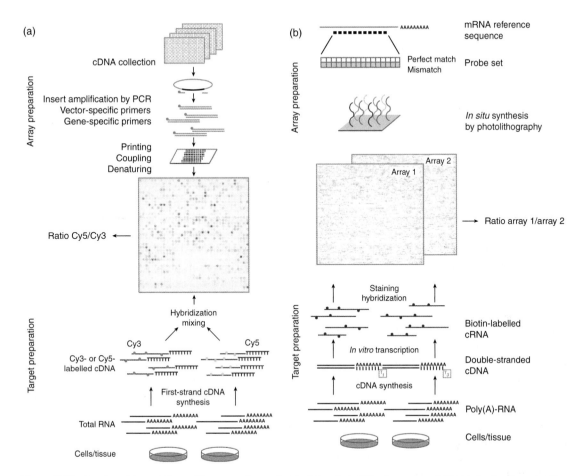

Figure 2.12 Overview of transcription-profiling approaches applied for (a) spotted cDNA microarrays and (b) high-density oligonucleotide microarrays. In the cDNA approach (a), the microarray is printed from a cDNA library (often developed from ESTs), and the probes are isolated by PCR amplification using primers specific to the gene or the vector. PCR products are printed using a high-precision robotic device. The target sample is obtained from RNA isolated from two groups of cells or tissues. This is used for reverse transcription in the presence of nucleotides with fluorescent labels (e.g. Cy3 and Cy5). The two samples are mixed in a hybridization buffer and brought into contact with the array under competitive conditions, such that for each probe the most abundant transcript (with either a Cy3 or a Cy5 label) binds most to the array. Scanning of the array with wavelengths corresponding to the excitation spectra of the two dyes will provide a picture of the transcript abundance profile in the sample, relative to the reference. In the high-density oligonucleotide microarray approach (b), sequence information of the transcriptome of a genomic model species is used to develop probes of 25 nucleotides which provide a perfect match with a unique part of a each transcript. In addition, control probes are developed with a single base mismatch. Each transcript sequence is represented by 16–20 different probes. The probes are synthesized *in situ* and fixed directly to the array. The target sample is prepared from poly(A)-RNA isolated from the cells or tissues of interest, which is reverse-transcribed to generate double-stranded cDNA, using a poly(dT) with a transcriptional start site for T7 RNA polymerase; this polymerase is then used to synthesize cRNA using biotinylated nucleotides. The two pools of amplified RNAs are hybridized with two different arrays and target binding is detected by staining with a fluorescent dye coupled to streptavidin, which recognizes the biotin label. Signal intensities of probe sets of the two different arrays are used to calculate relative transcript abundance. After Schulze and Downward (2001), reproduced by permission of Nature Publishing Group.

profiling is the *explorative use* (see Feder and Mitchell-Olds 2003). With this term we indicate that arrays may be manufactured from genomes that are not completely sequenced; the array is

then used to discover transcripts that respond to some factor of interest, and these transcripts are only sequenced after the arraying experiment has identified them as promising or interesting.

One might also envisage a *shuttle approach,* for example when explorative transcription profiling in a non-model species of great ecological relevance is alternated with more detailed studies using microarrays of genomic model species, aimed to identify functions of genes in the non-model by analogy with the model.

Another issue of strategic nature is the distinction between *local* and *global* transcription profiling (Schulze and Downward 2001). In the local approach, the interest is with a limited set of genes, for example in mutants with a specific over-expression or knockout, and the microarray is used to identify the collective set of genes upregulated or downregulated in association with this particular condition. The problem with this analysis is that microarray experiments can easily deliver lists of hundreds of genes whose expression is changed but who may not be all equally relevant to the phenotype of interest. In the global approach, expression profiling is used to generate a genome-wide picture of co-regulated genes, signatures of certain pathways, etc. This is often followed by statistical analysis (see Section 2.4) aimed at identifying clusters of genes with similar function.

Finally, we emphasize that the interpretation of data obtained from microarray transcription profiling is greatly improved when a biochemical framework or other a-priori knowledge is available for the phenomenon of interest. This was exemplified in the study discussed above where the upregulation and downregulation of genes under diauxic shift in yeast was related to the known network of carbon metabolism in the cell. Various biochemical systems for interpretation of transcription profiling are presented in Chapters 5 and 6; network analysis is discussed in Chapter 7.

2.3.3 Serial analysis of gene expression (SAGE)

Microarrays are not the only approach to transcription profiling. No less than 27 different methods for transcription profiling are listed by Shimkets (2004). A technique that deserves mention here goes under the name of SAGE (Velculescu *et al.* 1995, 2000, Matsumura *et al.*

2003). The principle is that short sequences of 9–15 bp contain sufficient information to uniquely identify a transcript, provided that the sequence is isolated from a defined position in the transcript. This is obvious from the fact that the frequency of a 9 bp sequence in a genome is 4^{-9}, which equals $1/262\,144$, a number which exceeds the number of transcripts in even large genomes. In SAGE, such short sequences (called tags) are excised from cDNAs and linked one after the other (concatenated) in a 1 kbp clone for sequencing. Then simply counting the number of tags in the sequences provides an absolute measure of the abundance of the corresponding mRNA.

The experimental procedure of SAGE is a hallmark of ingenious design in molecular biology. In short, mRNA is isolated and reverse-transcribed to cDNA including a biotin label at the 3′-end. The cDNAs are then cleaved with a so-called anchoring enzyme and the 3′-ends are isolated by means of streptavidin-coated beads (biotin attaches specifically and strongly to streptavidin). Then the sample is divided into two subsamples and two different PCR primers are ligated to the sticky ends of the cDNAs. The primer sequence includes a recognition site for a type IIS restriction enzyme. The crucial element of the technique is that this type of restriction enzyme cleaves DNA 20 bp downstream from the recognition site. This enzyme is called the tagging enzyme. Consequently, when the enzyme digests the bead-attached cDNAs, it cleaves off a short sequence, which consists of the PCR primer site at the 5′-end, the recognition site of the tagging enzyme, the recognition site of the anchoring enzyme, and a transcript-specific short tag at the 3′-end (Fig. 2.13). The two samples are then mixed, the blunt ends left by the tagging enzyme are ligated to each other, and a PCR is applied using the two primer sites in the linkers. This leads to a population of so-called ditags, which are purified and concatenated (serially linked to each other) into larger clones for sequencing. The arrangement of two tags bordered by anchor enzyme restriction sites can be recognized unambiguously by software and counted. Comparison to a genomic database will reveal the identity of the genes whose tags are in the sequenced pool.

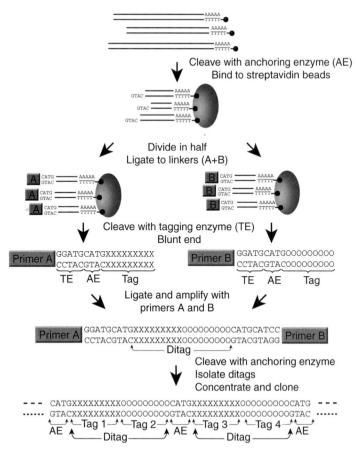

Figure 2.13 Schematic outline of the SAGE procedure. Double-stranded cDNA is synthesized from mRNA with a biotinylated oligo(dT) primer. The cDNAs are cleaved with an anchoring enzyme (AE) and then captured on streptavidin-coated beads. The anchoring enzyme recognizes four bases (it is a four-cutter) and is expected to cleave each cDNA at least once. The sample is divided into two halves and linker sequences are ligated to each end, containing primer sites for a PCR and a recognition site for a type IIS restriction enzyme (tagging enzyme, TE). Digestion with this enzyme liberates a sequence consisting of the primer site, the two restriction sites, and a short transcript-specific tag. The two samples are then mixed again, and the tag ends ligated so as to produce a PCR template with two tandomly arranged tags (ditag). Finally the ditag sequences are purified, contatenated, cloned, and sequenced. Reprinted with permission from Velculescu et al. (1995). Copyright 1995 AAAS.

SAGE has developed into a well-established methodology with its own community that organizes workshops, exchanges experiences, and maintains a website (www.sagenet.org). The great advantage of SAGE is that it is very sensitive and performs better than microarray analysis when identifying genes with low expression levels (Gibson and Muse 2002). Applications at the moment are mostly in the medical sector (cancer research) and yeast transcriptomics; ecological applications are still rare (but see Chapter 5). A disadvantage of SAGE is that it relies on massive sequencing, which may be prohibitive to some laboratories, and the recognition of tags requires knowledge about the genome.

2.3.4 Quantitative PCR

In a normal PCR, the amplification product is only a qualitative indicator of the template; the amount of DNA produced after a certain number of cycles, which is usually assessed on a gel, is hardly dependent on the initial amount of template. To make the PCR quantitative, systems have been

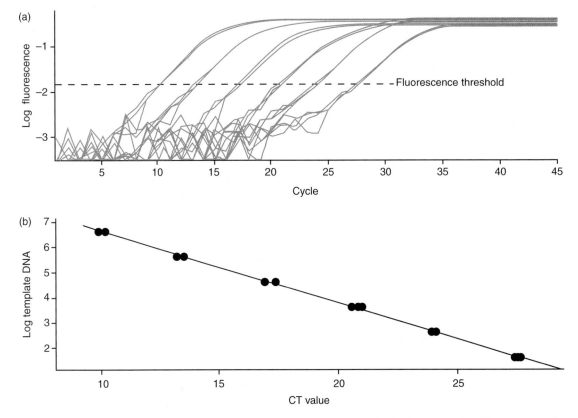

Figure 2.14 (a) Time course of fluorescence during a quantitative PCR reaction monitored with Syber Green, using different amounts of template. Note that fluorescence is plotted on a logarithmic scale, so the linear increase of the lines indicates exponential amplification. The cycle number at which a pre-set fluorescence threshold is reached is designated as the cycle threshold (CT). This value is plotted in (b) as a function of the known amount of template DNA estimated by spectroscopic methods. The linear relationship indicates that CT can be used as a measure for the initial amount of DNA in the reaction vessel. After Rutledge and Côté (2003) with permission from Oxford University Press.

developed by which the build-up of product in a PCR can be monitored during the reaction (Rasmussen *et al.* 1998). This requires a thermocycler coupled to a device that can read optical signals coming from the reaction vessels or capillaries. Two principles are applied. One is the use of a dye, Syber Green, which becomes fluorescent when intercalated in double-stranded DNA. Since after a few cycles the bulk of DNA in a PCR mix consists of the amplified product, the increase of fluorescence is a measure of product build-up. The other principle is to use a template-specific primer that carries two fluorophores, one an emitter and the other a quencher, extinguishing the fluorescence signal by means of *fluorescence resonance energy transfer* (FRET). As long as the fluorophores are

close together there is no fluorescence, but when the PCR proceeds, polymerase activity destroys the primer and the loss of physical proximity between the two dyes allows the fluorescence of the emitter to become detectable. As more and more FRET primer is digested, free emitter fluorophore accumulates in the solution and the fluorescence signal increases with time.

Monitoring of PCR amplification with fluorescent dyes in real time produces a sigmoidal curve (Fig. 2.14a). Usually it takes more than 10 cycles for the level of fluorescence to be detectable. Then an exponential phase is entered, in which the signal amplifies by a fixed factor with every cycle, until a plateau is reached, when the signal hardly increases with the increasing number of cycles.

Two parameters describe the essence of such a curve, *cycle threshold value* and *PCR efficiency*. The cycle threshold value is defined as the cycle number at which the curve increases above a baseline or error variance. This is a measure of the amount of template in the original mixture. Calibration curves can be made to correlate the cycle threshold value with the number of template copies and usually a linear relationship is observed (Rasmussen *et al.* 1998, Rutledge and Côté 2003). In mathematical terms, the relationship between the initial number of DNA molecules, N_0, and the cycle threshold is described by:

$$N_0 = \frac{N_C}{(e+1)^C}$$

where N_C is the number of DNA molecules at the threshold fluorescence, e is the PCR efficiency, and C is the cycle threshold (Rutledge and Côté 2003). PCR efficiency is given here as a fraction, so if efficiency is 100%, $e = 1$, and the products build up by a factor of two with every cycle. From this formula it can be seen that

$$\log N_0 = \log N_C - C \log(e+1)$$

and because N_C and e are constants, a plot of $\log N_0$ versus C should show a straight line with slope $-\log(e+1)$. This is indeed the case, as shown in Fig. 2.14b. The amplification efficiency (e) can be estimated from this slope and $e+1$ should be around 2.

When real-time PCR is applied to cDNA generated by reverse transcription, the cycle threshold value becomes a measure of the amount of mRNA upon which the reverse transcription reaction was done. In this version it is called quantitative reverse-transcription PCR. An important consideration, however, is the calibration of the reaction, especially since the amount of total RNA will not be exactly the same from one sample to another. Therefore the reactions always include expression measurements of so-called *housekeeping genes,* which are assumed not to vary from one sample to another and so represent an unbiased measure of the amount RNA isolated. Stürzenbaum and Kille (2001) discuss the suitability of control genes for quantitative reverse-transcription PCR.

The most commonly used genes are 28S rRNA, β-actin, elongation factor 1α, albumin, tubulin, and hypoxanthin phosphoribosyl transferase. Obviously, the sequences of these genes need to be known in the species of interest and reliable PCR primers developed. Stürzenbaum and Kille (2001) emphasize that control genes must be checked in every new set-up and preferably more than one control gene should be analysed. Genes that appear to have invariable expression during one stage of life (e.g. β-actin in the adult) may prove to vary in another stage (e.g. β-actin during juvenile morphogenesis). When a reliable housekeeping gene is obtained, division of the cycle threshold value of the target gene by the cycle threshold value of the housekeeping gene leads to a *fold regulation* (FR) value. In a multiplex set-up, FR values of many genes together constitute a transcription profile.

2.4 Data analysis in ecological genomics

Data generated by genomics programmes, be it sequences or gene-expression profiles, are invariably voluminous in nature. Management and analysis of genomics, proteomics, and metabolomics data has developed into a science of its own, *bioinformatics.* Attwood and Parry-Smith (1999) and Lesk (2002) give a succinct overview of this field, with practical exercises focused on consulting genome databanks and conducting alignments. Since this type of bioinformatics is already covered in several textbooks, only a very short treatment will be given here, while most emphasis will be paid to data analysis associated with transcription profiling.

2.4.1 Alignment and homology search

One of the first questions that an investigator asks when a DNA sequence becomes available is whether similar sequences have been published by others. Similarity of the new sequence with already existing sequences is usually ascribed to *homology.* In evolutionary biology, homology is defined very strictly and reserved for those cases where the similarity between two positions in a sequence is due to their common descent from the same ancestral sequence. In genomic analysis the term homology is often used more loosely to indicate any similarity exceeding a certain threshold. In principle, similarity may also be due to convergence: the development of the same pattern through positive selection in different lineages. Base-pair similarities that have arisen in different taxa independently through convergence or reversal are called *homoplasies*. Of course, computer programs cannot discriminate between the various causes of similarity; often supplemental information (not forgetting visual inspection) is needed to distinguish homology from similarity.

The most important tool for answering the question of similarity is the Basic Local Alignment Search Tool, abbreviated to *BLAST* (Altschul *et al.* 1990). This computer program has become so influential that *to BLAST* is now a common verb among molecular biologists. One of the most important functions of the BLAST algorithm is to search a large database, such as the GenBank Nucleotide Sequence Database, using a *query sequence* provided by the investigator and to produce a table of sequences deposited by others, with an indicator of the similarity of the query sequence

to such sequences. There is a whole suite of different BLAST programs; BLASTP aligns a query amino acid sequence with database protein sequences, BLASTX aligns a query translated nucleotide sequence with database protein sequences, and TBLASTX aligns a query translated nucleotide sequence with database translated nucleotide sequences (Lesk 2002; see www.ncbi.nlm.nih.gov/BLAST).

Table 2.4 and Fig. 2.15 give an example of a BLAST search. D. Roelofs (unpublished data), using suppression subtractive hybridization (see Section 2.1), found a partial cDNA in the springtail *O. cincta* (Collembola) that had a differential expression between two populations, one metal-tolerant and the other not. The sequence was BLASTed to GenBank, which produced high similarities to genes annotated as ATP-binding cassette (ABC) transporters in *D. melanogaster*, *Anopheles gambiae* (mosquito, Diptera), and *Apis mellifera* (honey bee, Hymenoptera). The fact that the most significant hits were all insects and all hits indicated the same type of gene put great confidence in the conclusion that the unknown cDNA was a collembolan version of an ABC transporter. The ABC transporters are a large family of proteins involved in the transport of a

Table 2.4 Example of a table in a BLAST report, showing the eight most significant alignments of a query sequence of an SSH clone from *O. cincta* (Collembola) that was upregulated in a metal-tolerant population, compared to a reference population (D. Roelofs, unpublished data)

Code	Species and gene identification	Score (bits)	*E* value	*N*
gi | 24650854 | ref | NM_170376.1 |	*Drosophila melanogaster* CG9990-PB	251	3e-78	3
gi | 28626479 | gb | BT004906.1 |	*Drosophila melanogaster* LD15982	251	3e-67	3
gi | 24650852 | ref | NM_143371.1 |	*Drosophila melanogaster* CG9990-PA	251	3e-78	3
gi | 31208990 | ref | XM_313462.1 |	*Anopheles gambiae* ENSANGP00000019635	245	5e-78	3
gi | 48113775 | ref | XM_393164.1 |	*Apis mellifera*, similar to ENSANGP00000019635	237	5e-75	3
gi | 28317030 | gb | BT003531.1 |	*Drosophila melanogaster* RE14039	230	2e-70	2
gi | 31206646 | ref | XM_312290.1 |	*Anopheles gambiae* ENSANGP00000020975	222	4e-69	3
gi | 31241126 | ref | XM_320987.1 |	*Anopheles gambiae* ENSANGP00000007803	217	9e-67	3

Notes: All these hits are annotated as ATP-binding cassette (ABC) transporter complexes in three different species of insects. See text for details. Score is a measure of similarity, defined by the BLAST algorithm, expressed in bits; the higher the score, the better the alignment. *E* is the *Expect value*, a parameter describing the number of hits one expects by chance when searching a database of a particular size. It depends on the size of the database as well as on the sequence length. Expect values are given as powers of ten, so 3e-78 means 3×10^{-78}. *E* decreases exponentially with the bit score. The lower *E*, the better the alignment. *N* is the number of *high-scoring segments pairs* (separated by non-matching segments and gaps).

Key for alignment scores

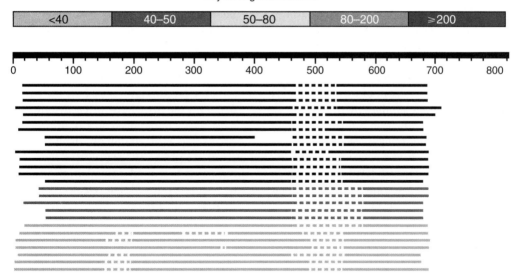

Figure 2.15 Figure in a BLAST report indicating which parts of sequences in the database have a similar score to a query sequence (top black bar, showing sequence positions 0–820). The grey tones of the bars below indicate the degree of similarity of sequences in the database. The figure shows that in this case the scores of the first 14 alignments are good, and that the optimal alignments assume a gap of around 80 bp in the region between positions 460 and 540 of the query sequence. See Table 2.4 for an explanation of the example.

wide variety of cellular compounds, such as sugars, ions, peptides, and more complex organic molecules. In this case the upregulation of an ABC transporter might indicate that the capacity for removing ions out of the cell or into lysosomal vesicles is increased constitutively in the metal-tolerant population, to avoid cellular damage from free metal ions in the cytoplasm when exposed to polluted soil.

One of the most important parameters in the BLAST report is the *E value* or Expect value, a statistical measure of the likelihood that the query sequence is found in the database by chance if there were no true match. The codes in the table link to web pages on which the gene is described, with information on the genomic position, gene classification, cellular function, and original publications.

As the name of BLAST indicates, similarities between sequences are established after *alignment*. This extremely important procedure in bioinformatics, which underlies all genomic analysis, allows for a 'sliding' of one sequence relative to another until the number of corresponding bases is maximized. To maximize the correspondence it is

often necessary to assume that the two sequences differ not only by simple substitutions, but in places also by deletions and insertions, leading to gaps in the alignment. The alignment is optimized by allowing positive scores for each match and negative scores for gaps, while maximizing the final score using a computer algorithm. Gap penalties may have a large influence on the optimal alignment produced by BLAST. Aligned sequences of different species can be used further to develop a *phylogeny*, a hypothetical tree of evolutionary descent, for the sequence of interest. Such phylogenies can use the nucleotide sequence itself or, in the case of an ORF, the amino acid sequence implied by it. The reader is referred to Graur and Li (2000), Gibson and Muse (2002), and Lesk (2002) for a more detailed treatment of sequence-alignment procedures and phylogenetic analysis.

2.4.2 Processing of microarray data

As indicated in Section 2.3, one of the most prominent types of genomics approaches, with great potential for ecological applications, is transcription profiling using microarrays. The raw data from such experiments come in the form of 16-bit TIFF images in which each pixel is assigned an intensity score between 0 and 2^{16} ($=65\,536$). These files need to be transformed into a gene-expression matrix before the true analysis can begin (Fig. 2.16). The data-processing phase is conducted by software supplied with microarray scanners, but is important to understand how the images are actually transformed into a gene-expression matrix. The reader is referred to Brazma and Voli (2000), Quackenbush (2001), Butte (2002), Causton *et al.* (2003), and Kim and Tidor (2003) for a detailed discussion of this matter.

Image analysis. The raw data generated by a scanner consist of a picture of spatially explicit fluorescence intensities (see Fig. 2.11). The image-processing software must recognize the spots, determine their boundaries, measure the signal coming from each spot, compare it with a local background, and assign the result to the correct probe. Usually the software makes use of the fact that the elements are arranged in subgroups; in the case of spotted microarrays these are so-called pen groups, which derive from the arrangements of pens in the spotter.

The user needs to indicate where approximately these subgrids lie, and the software adjusts a grid to overlay the spots. So-called landing lights can help the correct placement of the grid over the spots. These landing lights may represent spiked controls at known positions with a strong signal. Once the grid is placed, the pixels that make up the spot have to be determined. For most scanners the pixel size is $10\,\mu m$, so that a circular spot of $200\,\mu m$ in diameter contains about 314 pixels. The signal from these pixels is compared to a similar number of pixels within the same grid cell but outside the spot. A frequency distribution of these signals is made, which ideally shows two separate intensity peaks: a peak at high intensity for the spot pixels and one at low intensity for the surrounding pixels. The intensity of the lowest peak is then subtracted from that of the highest peak to arrive at a *background-corrected spot intensity*. In this way, each spot is corrected for its own local background; this avoids variation in the background across the array and is better than global background correction, where only a single value for the background is chosen to correct all spots on the array. If there are several probes

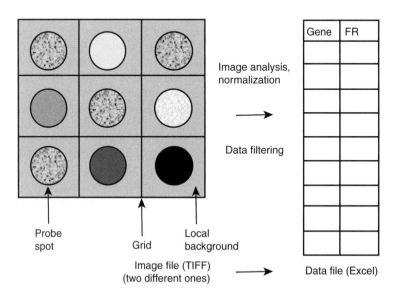

Probe spot   Grid   Local background

Image file (TIFF) (two different ones) $\longrightarrow$ Data file (Excel)

Image analysis, normalization

Data filtering

Figure 2.16 Illustrating the steps that have to be taken to transform a microarray scan into a gene-expression data matrix. FR, fold regulation.

on the array that represent the same gene (this is always the case with oligonucleotide gene chips), the corrected spot intensity is averaged over all spots for the same gene.

Estimating expression ratios. As explained in Section 2.3, transcription profiling with the use of microarrays is essentially relative. In spotted cDNA arrays a query sample is compared with a reference sample by competitive hybridization and two signals, corresponding to two different fluorescent dyes, are read from each spot. When using gene chips, the reference and query samples are hybridized to different arrays, but also in this case spots corresponding to the same probe are compared. The quotient of the two signals is defined as the expression ratio T_i for each gene i:

$$T_i = \frac{Q_i}{R_i}$$

where Q_i is the signal for gene i in the query sample and R_i is the signal for the same gene in the reference sample. In spotted cDNA microarrays, Q derives from the Cy5 signal (red) and R from the Cy3 signal (green), or vice versa. The expression ratios are always logarithmically transformed and generally the logarithm to the base 2 is applied. This transformation results in a quantity known as *fold change* or *fold regulation* (FR). As a result of the transformation, *FR* takes a value of 0 if there is no change, a value of 1 if there is a 2-fold increase, and a value of -1 if there is a 2-fold decrease in expression. A 4-fold increase results in FR $= 2$ and a 4-fold decrease in FR $= -2$. So:

$$FR_i = {}^2\log T_i = \frac{{}^{10}\log Q_i - {}^{10}\log R_i}{{}^{10}\log 2}$$

It should be noted that the use of expression ratios has a disadvantage, namely that information about the actual signal intensity is lost. So genes that are expressed weakly in both the Q and R samples are treated similarly to genes with an overall strong expression, if the relative up- or downregulation is the same. Whereas taking relative measures is common in transcription profiling, in other microarray applications, for example the detection of microbial diversity in the environment (see Chapter 4), absolute values are taken. Kerr and Churchill (2001) took a stand against the argument

that transcription profiling is necessarily relative and they proposed that the two readings could be considered as two correlated measurements, as in a traditional incomplete block design, common in agricultural experimentation.

Normalization. For various reasons, the FR values obtained cannot be directly compared across replicate measurements or different experiments. The most common source of variation is the use of different amounts of RNA as the starting material from which the target sample was prepared. Another important source is unequal incorporation of dyes in the cDNAs. There are various strategies for normalization (Causton *et al.* 2003). One approach is to calculate the mean FR value of all probes on the array and to subtract this mean value from all other values (*total intensity normalization*). This will make sure that the mean expression ratio over the whole array is unity. Another approach is to consider a regression of $\log Q_i$ against $\log R_i$. If the initial amount of RNA is exactly the same for the Q and R samples and if the labelling and detection efficiencies are also identical, such a plot would show a cluster of points around a straight line through the origin with slope 1 (of course, individual genes will lie apart from the line due to up- or downregulation). However, often the data do not fall exactly on such a straight line and show a curving trend (Fig. 2.17, left-hand panel). Since the interest lies in deviations from the diagonal, insight may be increased by rotating the plot by 45° and re-scaling the axes. This can be done by plotting $M = {}^2\log Q - {}^2\log R$ over $A = {}^2\log Q + {}^2\log R$, and in such a plot M should be independent of A (Fig. 2.17, right-hand panel). If this is not the case, one can correct the data by subtracting a quantity c, which depends on A and is defined as the difference between the local deviation of the data from a horizontal line. This correction term is estimated for each value of A by means of *local weighted regression* (loess; Smyth *et al.* 2003).

The third approach in data normalization is to use expression ratios of housekeeping genes as a basis for normalization. One can also use spiked controls that do not cross-hybridize with the

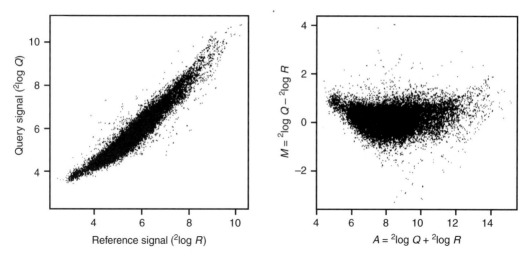

Figure 2.17 (a) Scatterplot of two signal intensities, log Q_i (from the query sample, e.g. fluorescence from Cy5) and log R_i (from the reference sample, e.g. Cy3) over all genes i in a microarray expression analysis. Ideally one expects that the average of the data falls along a straight line with slope 1 through the origin. The slightly curved shape indicates intensity-dependent bias. (b) When plotted as $M = {}^2\text{log } Q - {}^2\text{log } R$ over $A = {}^2\text{log } Q + {}^2\text{log } R$, the bias is visualized more clearly. The data can be corrected by subtracting a term which depends on A and is estimated by local weighted regression (loess). From Smyth *et al.* (2003) by permission of Humana Press.

target; for example photosynthesis genes from *Arabidopsis* on an insect microarray. These probes are then queried with a well-known amount of added RNA and so their signals provide a stable reference. Unlike in quantitative PCR, the use of housekeeping genes is not the preferred approach in the case of microarrays, because it ignores the multitude of information on the array (expression of only a few genes is used to correct for expression of thousands of others) and microarrays do not allow very precise measurement.

Data filtering. In addition to normalization it is also recommended to filter the data to remove dubious expression measurements. The most important data-filtering operation is to screen for low-intensity measurements that have a large inaccuracy. As a criterion, each fluorescence signal should be at least twice the standard deviation of the local background. Another issue in data filtering concerns the case where probes targeting the same transcript produce inconsistent results.

2.4.3 Statistical analysis of microarray experiments

The result of a single microarray experiment is a data file that can be viewed as a matrix with one very long column in which the FR values of all genes are noted. Usually one experiment involves several samples and these are taken together in one gene-expression matrix with a number of columns; for example different points in time, different physiological states of the organism, or different environments from which the RNA was isolated (Fig. 2.18).

Because gene-expression matrices are valuable resources for statistical analysis and a single investigator is often not able to exploit all possible data-analysis techniques, the data are often published on the Internet. This allows other researchers to compare expression profiles across studies, in much the same way as a genomic database is consulted by different people. Brazma *et al.* (2001) considered the requirements that such data matrices should have in order to be valuable for the research community. They developed a standard known as minimum information about a microarray experiment (*MIAME*). This standard stipulates that publication of gene-expression matrices should be accompanied by details about:

• Experimental design (conditions, doses, replication, quality-control measures, etc.)

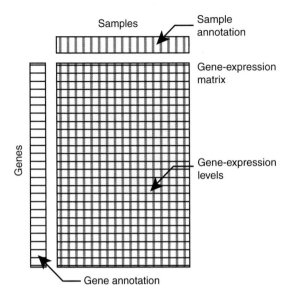

Samples — Sample annotation

Gene-expression matrix

Genes

Gene-expression levels

Gene annotation

Figure 2.18 Conceptual view of a data matrix which forms the starting point of statistical analysis of microarray-based transcription profiling. Gene annotation is done using standard codes and links to gene sequence databases; there is no universally accepted guideline for annotation of the horizontal axis (samples). Each cell contains a (log-transformed) fold regulation value. After Brazma *et al.* (2001), by permission of Nature Publishing Group.

- Array design (type of array, source of probes, clone identifications, etc.)
- Hybridization conditions (buffer, washing procedure, hybridization time, temperature, etc.)
- Measurements (quantification, normalization, data filtering, etc.)

As part of MIAME it is required that the raw image file from the scanner, with relevant scanning parameters, is supplied with the publication.

An interesting aspect of gene expression is that it often accompanied by a significant degree of noise. Gene expression is of a fundamentally quantitative nature and it depends on factors such as physiological state of the organism, developmental phase, tissue or cell population, and genetic background, factors that cannot always be controlled in an experiment. For the ecologist who is accustomed to the sometimes extremely large fluctuations in patterns and processes in nature, quantitative variation does not come as a surprise, but for molecular biologists this is a relatively new

phenomenon, which contrasts greatly with the deterministic nature of DNA sequences and biochemical pathways. As a consequence, we see in transcription profiling the use of statistical techniques, such as analysis of variance and cluster analysis, that are also used in ecology for analysing population fluctuations and community composition.

One of the first considerations when dealing with noise is *replication*. There are various elements of pseudoreplication in microarray analysis; for example, several signal readings from one spot, or several probes representing one gene. Sometimes investigators replicate the cDNA synthesis but draw from the same pool of RNA, which is also to be considered pseudoreplication. True replication should come from the use of different animals and plants for RNA extraction, or, even better, from repeating the experiment in its entirety. Since labour and costs associated with microarray experimentation are factors not to be neglected, a careful consideration of the type and degree of replication is of great importance.

Another issue of concern is the microarray platform itself. Tan *et al.* (2003) compared the performance of three different commercial microarrays (Amersham, Agilent, and Affymetrix) to profile the same cell line and noted that the majority of genes identified as differentially expressed by each technology were identified uniquely by that technology. Of 185 differentials, only four genes were identified by all three platforms. So gene-expression results may depend to a large degree upon the type of microarray used in the experiment. This could be due to differences in probe sequence, variation in labelling and hybridization conditions, and the lack of standards across multiple technologies. Obviously, significant improvements in reproducibility are still needed to remove these sources of noise (Marshall 2004).

One approach to analysing microarray data is to apply an *analysis of variance* (ANOVA) or an equivalent linear model to each gene separately (Jin *et al.* 2001). For example, if two factors are considered (such as sex and age in a study of gene expression in *Drosophila*), the analysis takes the form of a two-way analysis of variance applied to each gene, for which the model is (Sokal and Rohlf 1995):

$$FR_{ijk} = \mu + \alpha_i + \beta_j + (\alpha\beta)_{ij} + \varepsilon_{ijk}$$

where FR_{ijk} is the FR value of this gene expected for replicate k of sex i at age j, μ is the overall mean expression, α_i is the effect of factor A (e.g. sex) at level i, β_j is the effect of factor B (e.g. age) at level j, $(\alpha\beta)_{ij}$ is the interactive effect of A and B together, and ε_{ijk} is the error term. There must be at least two replicate measurements for each combination of factors for such an analysis to be useful. The ANOVAs will lead to a few thousand F tests, one for each gene, and these tests will indicate the genes that are differential between sex, that change with age, and that change with age in a sex-dependent manner.

One should be careful, though, in attaching an absolute value to the outcome of a significance test. Because tests are applied to the same larger database, the P values for the F tests may not represent the true type I errors. To avoid taking too many false-positive results on board, a significance level of 10^{-4} or even smaller is chosen. New statistical methods are being developed for controlling the false-positive discovery rate (Cheng *et al.* 2004). In addition, the significance of the effect must be balanced against the magnitude of the effect. In judging magnitudes of effects in transcription profiling, up- or downregulation by a factor of 2 is usually chosen as a threshold (FR value of greater than 1 or smaller than −1). Not all effects exceeding this threshold will be significant and not all significant effects will exceed the criterion of 2-fold change. To illustrate this, Jin *et al.* (2001) developed a presentation known as a *volcano plot*, in which the apparent P values are plotted as a function of the FR value, both on a logarithmic scale (Fig. 2.19). Such a plot shows that some genes with a highly significant effect do not fulfil the criterion of 2-fold change, while on the other hand there are also genes that do fulfil the criterion of 2-fold change but are not significant. The plot may help to identify these different groups of genes, and focus further research on the most promising among them.

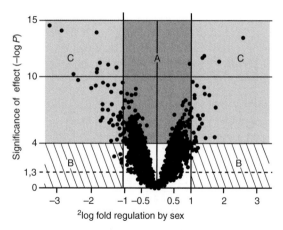

Figure 2.19   Volcano plot showing the relationship between the P values of F tests applied to gene expressions of *D. melanogaster* as a function of sex, and the magnitude of the effect. Each point is a separate gene. The horizontal axis gives the fold regulation value of the gene with respect to sex (genes to the left are downregulated in males compared to females, genes to the right are upregulated). The ^{2}log value of the gene-expression change is plotted, so a value of 1 implies a 2-fold upregulation and a value of −1 a 2-fold downregulation. The vertical axis gives the ^{10}log-transformed reciprocal P value, so the line at 1.3 corresponds to $P = 0.05$ and the line at 4 to $P = 10^{-4}$. The latter value was taken by Jin *et al.* (2001) as the 'preset experiment-wise false positive acceptance level'. There are several genes that are significant but have a fold regulation lower than a factor of 2 (region A) and there are also genes that are more than 2-fold up- or downregulated, but are not significant (regions B). Only genes in regions C are beyond doubt. After Jin *et al.* (2001), by permission of Nature Publishing Group.

In addition to analysis of variance, various non-parametric methods have been proposed, some of the most popular being *SAM*, Statistical Analysis of Microarrays (see www-stat.stanford.edu/~tibs/ SAM; Tusher *et al.* 2001) and *RDAM*, Rank Difference Analysis of Microarrays (Martin *et al.* 2004). In non-parametric methods the raw signals from a microarray scan are replaced by ranks and variation among replicates by rank differences.

Gene-by-gene analysis of microarray data, be it parametric (ANOVA) or non-parametric (rank-based), has the disadvantage that it does not consider the correlation structure among the expressions of different genes, otherwise than by controlling the false-discovery rate. Actually, the gene-expression matrix is of a multivariate nature (see Fig. 2.18). Each gene may considered as a *case* (also called an object) and the samples represent a number of measurements made on that case. Various multivariate statistical techniques can be applied to data organized in this way and one of the most common is some form of *hierarchical clustering*. The logic of clustering is evident from the fact that groups of genes will have similar expression patterns over samples, because they are induced by the same environmental conditions or regulated by the same transcription factors. The most common clustering algorithm to apply to microarray data comes from Eisen *et al.* (1998).

Clustering starts by developing a matrix of pairwise distances between the genes. There are different ways to calculate distances, one of the most straightforward being *Euclidean distance*. Suppose we are considering genes A and B, and we have observations on gene expression of a_i for gene A and b_i for gene B in sample i, then the Euclidean distance D_{Eucl} between the genes is:

$$D_{Eucl}(A,B) = \sqrt{\sum_{i=1}^{n}(a_i - b_i)^2}$$

where n is the number of samples. This distance measure is calculated for each pair of genes, resulting in a distance matrix, which is input to the clustering algorithm. To illustrate the calculation, Table 2.5 provides a hypothetical example of a very simple gene-expression matrix and the calculation of Euclidean distance between the three genes. This example suggests that the distance between genes B and C is smaller than between either A and B or A and C.

The Euclidean distance is not the only way of defining distances between genes. Other measures are Minkowski distance, Manhattan distance, and Hamming distance. In addition, the clustering may be based on a similarity measure, such as Pearson correlation, rather than distance. The reader is referred to Causton *et al.* (2003) and textbooks of multivariate statistical analysis for more information.

The object of clustering analysis is to develop a dendrogram that groups together genes with similar expression patterns. There are several principles that can be applied to achieve clustering. In an influential paper on gene-expression data analysis, Eisen *et al.* (1998) applied the so-called average linkage method. In this method a computer algorithm screens the matrix of pairwise distances for the smallest value (in the case of the genes sampled in Table 2.5, this would be 2.53, between genes B and C; see Table 2.6). Then a node is defined between these genes and gene-expression values are calculated for the node by averaging over the two genes involved. The distance matrix is then updated and a new smallest distance is identified. The procedure is repeated until $g-1$ nodes have been made, where g is the number of genes. Software packages such as developed by Eisen *et al.* (1998) not only provide a computational procedure but also a pictorial presentation of the clustered gene-expression pattern; each gene is qualified by a colour code, where red is used for upregulated expression and green for downregulated expression.

Cluster analysis is usually done in conjunction with other multivariate statistical techniques, such as *principal component analysis* (PCA; also known as singular-value decomposition). The aim of PCA is to find combinations of genes that jointly contribute most to the variability in the data. Technically speaking, one aims to find axes in the

Table 2.5 Hypothetical gene-expression matrix, illustrating the calculation of Euclidean distances (D_{Eucl}) between genes (see also Tables 2.6 and 2.7)

Gene	Sample		
	1	2	3
A	0.2	−2.3	−1.9
B	3.6	2.1	1.2
C	1.2	1.3	1.1

Table 2.6 Euclidean distance (D_{Eucl}) between the three genes in Table 2.5 over the three samples

Gene	D_{Eucl}		
	A	B	C
A	1	6.37	4.79
B	–	1	2.53
C	–	–	1

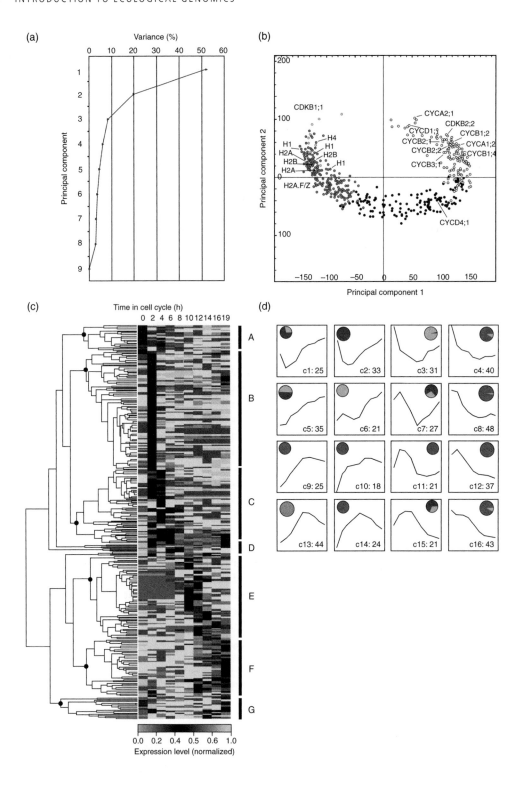

multidimensional space of gene expressions in such a way that if the data points are considered with respect to these axes, the dispersion of points is greatest. These axes or principal components are then said to explain a particular fraction of the variation. The operation is also seen as a *dimensionality reduction*.

A convenient way to represent the results of a PCA is a *biplot* (Chapman *et al*. 2001), a graph in which the data points are projected onto the two axes that explain the greatest amount and the second greatest amount of the variance in the data. In such a graph each gene is represented by a dot, and genes positioned close together have similar expressions over treatments. Usually an interpretation can be given to the axis, based on the treatment variable with which it is most strongly correlated, and from this interpretation a biologically meaningful summary of the effects can be made up.

It turns out that for gene-expression matrices only a few principal components can actually explain the data. This is due to the fact that many gene expressions are correlated with each other, but also that the number of samples is usually much smaller than the number of genes. So the amount of information in a gene-expression matrix is not so large as would seem from the long list of genes. By plotting the data in a two-dimensional graph a lot of the information is captured already. Holter *et al*. (2000) express the situation by analogy with spectral analysis of music: 'the complex "music of the genes" is orchestrated through a few underlying patterns', and 'the genes in a microarray comprise a set of identically tuned strings'. Kim *et al*. (2001) and Kim and Tidor (2003) showed that dimensionality reduction leads to a limited number of building blocks that can be scaled and added together in various combinations to best reconstruct the data. The authors developed a system in which the genome of *C. elegans* is dissected into 43 *expression mountains*, where each mountain represents a functional group of 5–1818 genes with a high internal correlation of expression.

Another popular approach to analysing gene-expression data is the technique of *self-organizing maps*. A self-organizing map starts with a set of nodes with a simple topology, for example in a two-dimensional grid, and continues by applying an algorithm in which the nodes are mapped onto the highly dimensional space of the data points in such a way that the distance from data points to nodes is as small as possible (Tamayo *et al*. 1999). The number of nodes needs to be specified beforehand, and so the procedure is equivalent to placing a fixed number of flags in a landscape in such a way that the scatter of data points is organized around the flags in an optimal way. The algorithm can dissect a complex data set in a limited number of genes clusters, where each cluster has a characteristic response over treatments.

As an example of a full statistical analysis of a microarray study we discuss the analysis of genome-wide gene-expression changes during the cell cycle of *A. thaliana* (Menges *et al*. 2002, 2003). An oligonucleotide microarray of the genome of *Arabidopsis* (Affymetrix ATH1) was used to look at gene-expression changes during a synchronized cell culture (Fig. 2.20). Plant cells were grown in suspension under continuous agitation and were synchronized by release from a chemical inhibitor. Gene expressions were first analysed by PCA and this demonstrated that the first principal axis already explained more than 50% of the variation, while the second added another 20%. In dimensionality-reduced representation two main clusters appeared (Fig. 2.20b). Hierarchical cluster analysis (Fig. 2.20c)

Figure 2.20 Results of statistical analysis of gene expression changes during the cell cycle of a synchronized cell suspension culture of *Arabidopsis*. (a) PCA showing the percentage of explained variance by successive axes. (b) PCA biplot of the data in a state space reduced to two dimensions. (c) Result of hierarchical clustering, showing gene expressions as a 'heat diagram', where different shades of grey indicate upregulation and downregulation. Note the oblique pattern of dark fields indicating peak expressions at different times in the cell cycle. (d) Sixteen patterns of gene expression change over the cell cycle, identified by self-organizing map analysis. The number of genes involved in each pattern (numbered c1–c16) is indicated. The small pie charts show the composition of each cluster with respect to four gene classes characteristic for certain cell-cycle phases (S, M, G_1, G_2). After Menges *et al*. (2002), with permission from Springer.

confirmed the existence of two main clusters, each of which could be subdivided further. In total, 16 different clusters were recognized, each representing a particular pattern of expression during the cell cycle (Fig. 2.20d). For example, cluster 5 (35 genes) represents genes that increase continuously in expression during the experiment whereas cluster 16 (43 genes) represents genes with decreasing expression over time. Using different ways of synchronizing the cell culture, Menges *et al.* (2003) finally developed a refined list of 1082 genes (out of some 14 000 genes that could be detected under the conditions chosen) that were cell-cycle regulated, of which 371 have no known function at the moment.

2.4.4 Towards an analytical framework for ecological genomics

The statistical data analysis discussed above applies equally to genomics in general as to ecological genomics in particular. One important issue arises, however, which is that ecologists may be interested in the variation across samples (the horizontal axis of a gene-expression data matrix) more than the variation across genes (the vertical axis of a gene-expression data matrix). This seems especially relevant if the samples come from different habitats, different treatments, or different genetic lines of the same species. The question in ecology could be which samples have similar expression profiles over genes?, rather than which genes have similar expression profiles over samples?. Of course, it is perfectly possible to apply the same statistical techniques to sample-based profiles. The only thing that needs to be done is that distances between samples rather than between genes are calculated from the primary data. For example, the matrix of Euclidean distances between samples, calculated from Table 2.5, is given here as Table 2.7.

The distances indicate that samples 2 and 3 are most similar and 1 and 3 are most different. Because a gene-expression matrix has many more genes than samples, such an analysis would enforce a much greater information reduction than is achieved by the gene-clustering approaches

outlined above. The situation in ecology may be similar to that in pharmacology and toxicology, where the aim often is to classify drugs on the basis of similarities of expression profiles across genes (Hamadeh *et al.* 2002, Waters and Fostel 2004). In an ecological application, the data points in a biplot of PCA would represent conditions (environments, treatments) rather than genes.

The classification of ecological categories using transcription profiling may eventually lead to a genomics-based characterization of an ecological system (Fig. 2.21). Since the state of any ecological system (organism, population, community) will fluctuate in time and space without obvious adverse consequences, there is a range of operation which could be considered as 'normal'. The 95% confidence space of such undisturbed states may

Table 2.7 Euclidean distance (D_{Eucl}) between the three samples in Table 2.5 over the three genes

Sample	D_{Eucl}		
	1	2	3
1	1	2.92	3.19
2	–	1	1.00
3	–	–	1

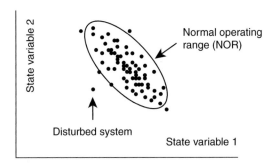

Figure 2.21 Hypothetical picture of the states of an ecological system in two dimensions (e.g. PCA axes) resulting from dimensionality reduction of gene-expression profiles. Each point on the diagram represents a composite characterization of clustered gene expression (state) under undisturbed conditions. The state fluctuates in time and space due to the inherent noise in gene expression. The 95% confidence ellipsoid of undisturbed states is defined as the normal operating range (NOR). If a system finds itself out of this range, this is indicative of disturbance or stress.

be defined as the *normal operating range* (NOR) of the system. This concept, already mentioned by Odum (1979), was used by Kersting (1984) to characterize the fluctuations of *Daphnia* and *Chlorella* populations in a closed microcosm, monitored for a long time. In an ecological genomics context, it can be used to characterize the state of an organism on the basis of its gene-expression profile (Fig. 2.21). When observations are made for transcription profiles under several different undisturbed conditions, these can be used to define a reference profile and its variability defines the NOR. Then, when a new situation is encountered, the distance between the new state and the border of the NOR is a measure of disturbance or stress (Fig. 2.21).

A consistent data-analysis framework for ecological genomics still has to be developed. The classification of samples using transcription profiling, as suggested above, is only one possible approach. The point we want to make is that ecological genomics does not necessarily have to follow the pathway that has been laid out by genomics in medical or agricultural sciences. Ecological genomics will have to find ways of data analysis to suit its own, ecological, questions.

CHAPTER 3

Comparing genomes

In this chapter we will deal with the first step that usually follows completion of a full genome sequence, which is to inspect the genome for its general properties, such as G/C content, number of genes, gene distribution, etc., and compare it with other species, related or unrelated. Genetic model species will be the starting point for our comparisons, but we will explore, wherever possible, links with species from the same clade and highlight the ecological significance of models and their wild relatives.

3.1 Properties of genomes

Once the genome of a species is elucidated the researcher is able to analyse the properties that characterize the genome as a whole. This is usually not done in isolation, but in comparison with other species, and therefore this part of genome science is called *comparative genomics*. The term *structural genomics*—the study of genome sequences, genetic and physical maps, etc.—is also appropriate here, as a contrast with functional genomics; however, some scientists use this term for the analysis of structural, three-dimensional, properties of proteins, using X-ray diffraction and nuclear magnetic resonance. Comparative genomics draws heavily on bioinformatics and uses computer programs to find patterns in genome sequences across species, to estimate similarities that can support the assignment of gene functions, and to develop phylogenetic trees by which the evolutionary relationships among genomes are visualized (Hedges 2002). Arguments from comparative genomics have become crucial in selecting new species for whole-genome sequencing. Sequencing efforts can be

optimized to discover what information the new species can provide about functionalities in already-sequenced genomes and to determine at what evolutionary distances such species should be placed to maximize that information (Eddy 2005). Comparative genomics is now a recognized sub-discipline of genome science and the first hand-books are appearing (Saccone and Pesole 2003). The field is closely related to molecular phylogenetics, for which we refer the reader to Hughes (1999) and Graur and Li (2000). Before discussing the genomes of species in more detail, in this section we provide an overview of the properties of genomes.

3.1.1 Genome size

Usually the size of a genome is known to some accuracy before its complete sequence is elucidated, because genome size is essential knowledge for optimal construction of a genomic library and the design of a genome-sequencing project. Genome size may be estimated using biochemical methods or flow cytometry, and is usually expressed in picograms of the haploid genome per cell. This is easily converted to nucleotides by the general formula $G = 0.987 \times 10^9\,C$, where G is genome size in basepairs and C is genome size in pg (roughly, 1 pg is equivalent to 1000 Mbp). Genome size in pg is also known as the C *value*. Estimates of genome size are now available for more than 3800 species of animals (www.genomesize.com; Gregory 2005) and 3900 species of land plants (www.rbgkew.org.uk/cval). The size of a genome varies dramatically across species. Table 3.1 provides a few examples of fully sequenced genomes that illustrate this diversity.

Table 3.1 Genome size and number of genes for organisms with completely sequenced genomes

Species	Total size of the genome (kbp)	Estimated no. of protein-encoding genes
Bacteriophage φX174	5.4	10
Mycoplasma genitalium	580	468
Methanococcus jannaschii	1665	1738
Haemophilus influenzae	1830	1743
Escherichia coli	4639	4288
Agrobacterium tumefaciens	5670	5419
Pseudomonas aeruginosa	6264	5570
Saccharomyces cerevisiae	12 610	6128
Caenorhabditis elegans	95 500	18 424
Drosophila melanogaster	123 000	13 601
Arabidopsis thaliana	125 000	25 498
Oryza sativa	466 000	50 820

There is an obvious increase of genome size going from viruses to prokaryotes, and further to unicellular eukaryotes and multicellular eukaryotes. This increase can be related to the increasing complexity of the cells and tissues involved. Viruses do not need their genomes to encode all the proteins that they require for their maintenance and propagation, because they exploit the molecular machinery of the host. The minimal number of genes required for an autonomous self-replicating entity was estimated by Graur and Li (2000) as 256 (based on a comparison of prokaryotic genomes) or 254 (based on a review of knockout studies). This is about one-half of the number of genes in *Mycoplasma genitalium*, which has 468 genes, the lowest number of any independently living organism. Unicellular eukaryotes have a genome size one order of magnitude greater than the average prokaryote; however, the extremes meet each other; the yeast genome is only twice as large as the largest genome of a prokaryote sequenced so far (*Pseudomonas aeruginosa*) and even smaller than that of some Cyanobacteria. Another order of magnitude lies between unicellular and multicellular eukaryotes, but the variability among the latter group is enormous.

A second trend in genome sizes across species is the reduction of the genome (*genome miniaturization*) in endosymbiotic organisms and parasites. In the ultimate endosymbiotic entity, the mitochondrion, many genes were lost compared to its alphaproteobacterial ancestor, partly due to deleting functions that were not necessary in the symbiotic life style, and partly by migration of genes to the nuclear genome of the host. The same holds for chloroplasts; however, chloroplasts (stemming from Cyanobacteria) have retained a significantly larger genome than mitochondria. Genome size reduction is also seen in parasites, and this may explain the very small genome of *M. genitalium*. However, it must be pointed out that there is also an opposite tendency: parasites need to have specialized proteins for adhesion to tissues of their host and to thwart the host's immune response, so they have larger genomes.

The two trends noted above are about the only ones that can be observed when comparing genome sizes across species. In fact, there is a remarkable lack of correspondence between genome size and organism complexity, especially among eukaryotes. For example, the marbled lungfish, *Protopterus aethiopicus*, has more than 40 times the amount of DNA per cell than humans! Fig. 3.1 provides an overview of the ranges for the various taxonomic groups.

The lack of correspondence between genome size and organism complexity has become known as the *C value paradox*. It is obvious that the enormous increase of genome size in some lineages is not accompanied by a proportional increase of the number of genes. This is already evident from Table 3.1, which shows that the number of genes tends to reach a plateau of some 20 000–50 000 with increasing genome size. In fact, the non-genic fraction of the DNA is the main responsible factor for the *C* value paradox, such that in eukaryotes anything between 30 and 99% of the genome can consist of non-coding DNA (repetitive sequences, mobile elements, introns, intergenic spacers, etc.). Ohno (1972) introduced the term *junk DNA* to stress the fact that, according to him, non-coding DNA is a useless but mostly harmless part of the genome and persists because of the replicative nature of DNA and linkage with functional genes.

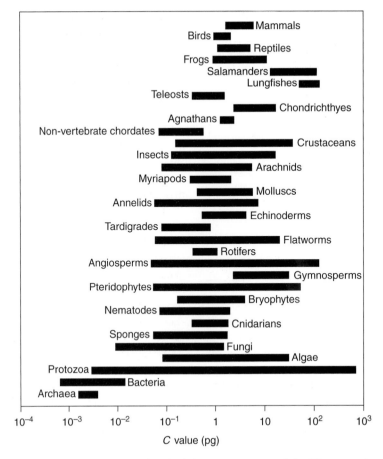

Figure 3.1 Ranges of reported genome sizes (*C* value of the haploid genome, in pg per cell) for different organism groups. From Gregory (2005), with permission from Elsevier.

How does a genome enlarge itself? *Polyploidization* is one of the most important mechanisms. This can concern a duplication of the entire genome (global polyploidization), or parts of the genome (regional polyploidization). Many global polyploidy mutations are highly deleterious because they interfere with cell division and meiosis. In mammals, polyploidy destroys the mechanism of dosage compensation, which normally inactivates one X chromosome in the female to compensate for the lack of complementary genes on the Y chromosome in the male. Still, polyploidies that concern an even number of chromosomes, such as *tetraploidy* (doubling of the diploid genome), sometimes do not have adverse effects on the phenotype and may be a mechanism for

evolutionary innovation. A famous example is the tobacco species, *Nicotiana digluta* ($2n = 72$), which is assumed to originate from genome doubling in a sterile hybrid ($2n = 36$) between *Nicotiana tabacum* ($2n = 48$) and *Nicotiana glutinosa* ($2n = 24$).

Polyploidy is especially common in angiosperms and pteridophytes, but rare in gymnosperms and bryophytes. There are also large differences in ploid levels between plant families. Polyploidy is relatively rare in animals; however, some families of fish, most notably in the Cypriniformes (carp, minnows) and several species of the amphibian order Anura (frogs and toads) are known for their relatively high occurrence of polyploid species. The only known cases of polyploidy in mammals are two species of octodontid rodent,

Figure 3.2 The red viscacha rat, *Tympanoctomys barrerae* (Octodontidae, Rodentia), representing a rare case of tetraploidy in mammals. The animal has 100 autosomes plus two sex chromosomes (X and Y). Courtesy of M.T. Gallardo, Universidad Austral de Chile.

Tympanoctomys barrerae and *Pipanacoctomys aureus*, which inhabit the salt deserts of Central Argentina (Gallardo *et al*. 1999, 2004; Fig. 3.2). *T. barrerae* has a C value of 16.8 pg, which is twice the amount of its closest relatives (8.2–7.6 pg) and out of the normal range of mammals (see Fig. 3.1).

Triploidy (arising from hybridization of a haploid and a diploid gamete) always leads to sterility, although it does not prevent growth and asexual reproduction in many plants, as shown by the triploid banana plant (*Musa acuminata*). This condition may actually be exploited by plant breeders to produce commercially attractive seedless fruits, for example of watermelon (*Citrullus lanatus*). The polyploid nature of many commercially interesting crops—for example, corn (*Zea mays*), with a genome size of 2500 Mbp, and wheat (*Triticum aestivum*), with a genome of 16 000 Mbp—is a serious obstacle for the commencement of genome-sequencing projects on these species.

In polyploid organisms the term genome size may become ambiguous. Greilhuber *et al*. (2005) have proposed that in addition to the C value, which denotes the haploid genome with chromosome number n, another term, *Cx value*, should be used to denote the chromosome base number (x) in a polyploid organism. For a tetraploid, $2n = 4x$, and for a triploid, $2n = 3x$. We should also note that when a tetraploid organism comes into existence, the C value doubles but the Cx value is not changed, because the number of different chromosomes remains the same. However, when

the organism evolves further, paralogous genes may differentiate in function, the two chromosomal complements undergo rearrangements and eventually the tetraploid situation will no longer be distinguishable from a normal diploid. This situation is termed *cryptopolyploidy*.

According to several authors (e.g. Spring 1997) the vertebrates as a whole may be considered a cryptopolyploid lineage, originating from two successive rounds of duplication, one prior to the divergence of jawless fish (Cyclostomata) and one thereafter (Gnathostomata and higher vertebrates). This hypothesis, also known as the *2R hypothesis*, seems to be supported by a variety of evidence, including the occurrence of several genes in 4-fold copies in vertebrate compared to invertebrate genomes. However, Hughes (1999) pointed out that under the 2R hypothesis one expects the phylogeny of duplicated genes in vertebrates to show four clusters arising from three splits, one dividing the tree in two branches, the other two splitting each branch. Hughes (1999) tested this expectation using several gene families, one of which, the *Notch* genes, is reproduced in Fig. 3.3. *Notch* genes encode transmembrane signalling proteins, and were first discovered in *Drosophila* where mutations in this gene are recessive and embryonic lethal due to hypertrophy of neural tissue at the expense of epidermal tissue. In the heterozygous condition these mutations cause notches at the edges of the wings, hence the name of the gene. *Drosophila* has only one *Notch* gene, but vertebrates have four. Despite the fact that the one/four comparison would support the 2R hypothesis, the phylogeny clearly does not (Fig. 3.3). In the tree, the *Notch*4 variant even clusters outside the insect *Notch* and outside vertebrate *Notch*1, 2, and 3. One gene duplication seems to have taken place prior to the split between vertebrates and insects, whereas two successive duplications followed in the vertebrate lineage. A similar situation holds for seven other gene families. So, there is no phylogenetic evidence for two rounds of whole-genome duplication as assumed by the 2R hypothesis.

The second mechanism of genome enlargement, regional genome duplication, will lead to localized

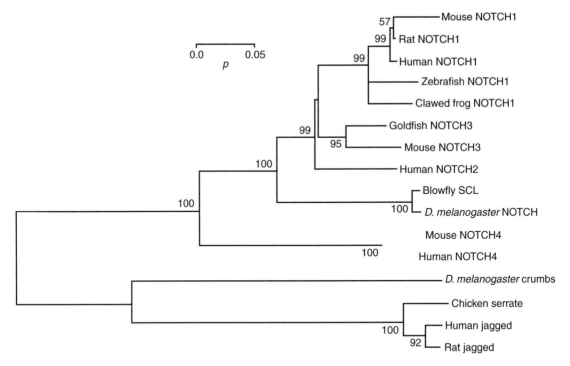

Figure 3.3 Phylogeny of the *Notch* gene family in some vertebrates and insects, constructed on the basis of proportional amino acid differences (*p*). *Serrate*, *Jagged*, and *Crumbs* were used as an outgroup to root the tree. The values along the branches are percentages of support in 1000 bootstrap samples. After Hughes (1999) by permission from Oxford University Press.

repetitive sequences. This may occur through *unequal crossing-over*, in which case whole genes or even substantial segments of a chromosome are duplicated in one of the two gametes, while the other gamete is left without the corresponding segment, or *replication slippage*, which is the most common mechanism in the case of microsatellite sequences. The reader is referred to Goldstein and Schlötterer (1999) for an extensive treatment of mutation mechanisms leading to the dispersion of repetitive sequences in the genome.

A third mechanism by which genomes may be enlarged is *duplicative transposition*. Transposable elements are sequences that have an intrinsic capacity to change their genomic location, either by physically moving from one position to another, or by making a copy of themselves that 'jumps' to another place in the genome. The mobile element carries its own genes, necessary for catalyzing the transposition (including an enzyme

called transposase), but in addition to these it may also carry genes or non-coding DNA that have nothing to do with the transposition itself and are thus relocated in the genome. There is an enormous variety of such mobile elements in many animal genomes. In *Drosophila*, one of the transposable elements is the well-known *P element*. It is assumed that a P element was introduced into *D. melanogaster* from another *Drosophila* species by a parasitic mite feeding on the eggs (Hoy 1994). Subsequently, P elements have been exploited by fruit fly geneticists to develop engineered P-element vectors that can introduce exogenous DNA into the germline of the fruit fly.

How could the increase of genome size in eukaryotes, and especially the dramatic proliferation of non-coding DNA, be maintained during evolution? Lynch and Conery (2003) have argued that genome complexity evolved mainly by neutral mechanisms, which apply to small populations

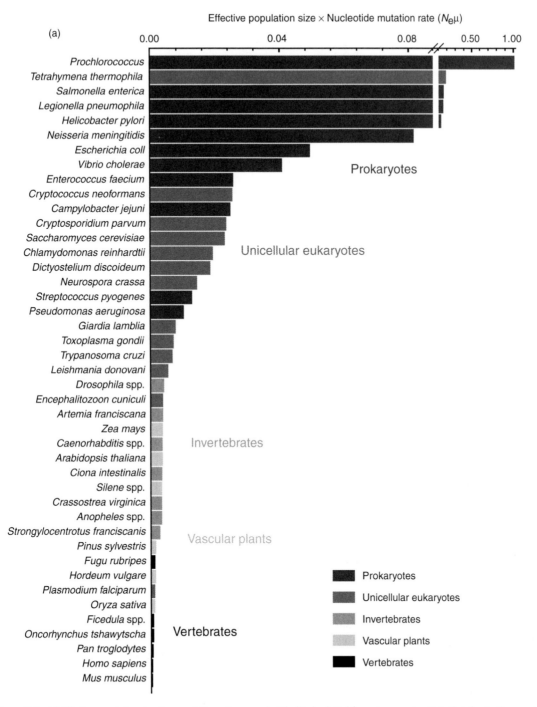

Effective population size × Nucleotide mutation rate ($N_e\mu$)

(a)

	0.00	0.04	0.08	0.50	1.00

Prokaryotes

Unicellular eukaryotes

Invertebrates

Vascular plants

Vertebrates

Prochlorococcus
Tetrahymena thermophila
Salmonella enterica
Legionella pneumophila
Helicobacter pylori
Neisseria meningitidis
Escherichia coli
Vibrio cholerae
Enterococcus faecium
Cryptococcus neoformans
Campylobacter jejuni
Cryptosporidium parvum
Saccharomyces cerevisiae
Chlamydomonas reinhardtii
Dictyostelium discoideum
Neurospora crassa
Streptococcus pyogenes
Pseudomonas aeruginosa
Giardia lamblia
Toxoplasma gondii
Trypanosoma cruzi
Leishmania donovani
Drosophila spp.
Encephalitozoon cuniculi
Artemia franciscana
Zea mays
Caenorhabditis spp.
Arabidopsis thaliana
Ciona intestinalis
Silene spp.
Crassostrea virginica
Anopheles spp.
Strongylocentrotus franciscanis
Pinus sylvestris
Fugu rubripes
Hordeum vulgare
Plasmodium falciparum
Oryza sativa
Ficedula spp.
Oncorhynchus tshawytscha
Pan troglodytes
Homo sapiens
Mus musculus

- Prokaryotes
- Unicellular eukaryotes
- Invertebrates
- Vascular plants
- Vertebrates

Figure 3.4 (a) Effective population size times mutation rate per nucleotide ($N_e\mu$), plotted for various species. Note that despite the general decreasing trend from prokaryotes to vertebrates, there is some mixing of species across taxonomic categories (e.g. *Tetrahymena thermophila* is a ciliate but has a value of $N_e\mu$ comparable to bacteria). (b) $N_e\mu$ and gene number as a function of genome size. Note the saturation curve for gene number, which is due to the preponderance of non-genic DNA in large genomes and the very good correlation between $N_e\mu$ and genome size. Reprinted with permission from Lynch and Conery (2003). Copyright 2003 AAAS.

(b)

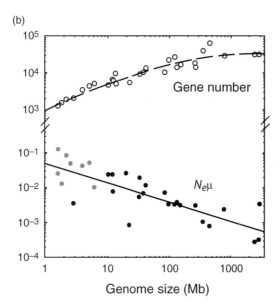

Figure 3.4 *(Continued.)*

composition of the parental generation. In small populations, random genetic drift is the most important factor for changes in gene frequencies. Population genetic theory shows that the effect of drift depends not only on N_e, but also on the mutation rate per nucleotide (μ); this is why Lynch and Conery (2003) calculated $N_e\mu$ for every species and correlated this with genome size (Fig. 3.4).

According to the theory of Lynch and Conery (2003) there is nothing adaptive in the increase of genome complexity; it can be considered an effect of drift governed by the laws of neutral population genetics. Although this conclusion has been challenged and examples to the contrary have been pointed out (Vinogradov 2004), convincing adaptive explanations for the increase in genome complexity have not been given; the neutralist hypothesis remains the most likely.

more than large populations. The theory is supported by an obvious negative correlation between genome size and effective population size (Fig. 3.4). The *effective population size*, N_e, is defined in population genetics as the number of individuals of a species in a theoretical ideal population that would have the same magnitude of random genetic drift as an actual population has (Hartl and Clark 1997). N_e is smaller than the real population size for two reasons. First, the actual population fluctuates over time; the effective size is the harmonic mean of abundance over time and so low numbers have a disproportionally large effect on N_e. Second, not all members of the population may actually participate in reproduction and transfer their genes to the next generation. If the sex ratio of the breeding pool departs from unity, for example due to the fact that a few males sire all females and many males have no offspring, this will decrease N_e far below the census population. Population size has a pervasive influence on the genetic structure of a population because small populations are conducive to sampling errors arising from the fact that the genetic material of the offspring generation is a sample from the genetic

3.1.2 Gene families

Limiting the comparison to genes rather than the whole genome, we note that a large number of genes are similar to each other due to their common descent from a duplication event (like the *Notch* genes in Fig. 3.3). Such genes are said to be *paralogous* to each other. Genes in different species sharing a common ancestor are called *orthologous*. These two different types of homology are illustrated in Fig. 3.5. All the paralogous genes in a genome constitute a *gene family*. Examples of gene families are the globulins, homeobox proteins, esterases, trypsin-like peptidases, G-protein-coupled receptors, cytochrome P450s, and proteins involved in the immune response.

After a duplication event, there are four possible scenarios for the fate of duplicates:

1. Both copies remain active with the same function and together produce twice the amount of mRNA that the ancestral gene did before; this situation may be maintained because of the selective advantage of a large amount of gene product. We call this *superfunctionalization*.
2. One of the copies is silenced by degenerative mutation and the other continues its function. This is called *nonfunctionalization*.

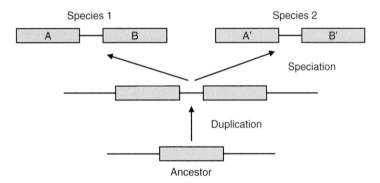

Figure 3.5 Illustrating the difference between paralogous and orthologous homologies. Genes A and B in species 1 are paralogous to each other, as are A′ and B′ in species 2; however, gene A in species 1 and gene A′ in species 2 are orthologues, as are genes B and B′.

3. One copy acquires a new advantageous function and is preserved by natural selection while the other retains the original function (*neofunctionalization*).

4. Both copies continue to be active but are compromised by deleterious mutations in their regulatory regions to the effect that each gene has a lower expression than the ancestral gene and and both duplicates are required to produce the full complement of functions (*subfunctionalization*). This situation is also described as *duplication–degeneration–complementation* (the DDC model).

Analysis of the genomic databases for several eukaryotic species has suggested that the vast majority of gene duplications eventually leads to nonfunctionalization; neutral evolution seems to dominate the fate of duplicated genes (Lynch and Conery 2000). Still, many duplicated genes are preserved and the most important mechanism for their survival seems to be subfunctionalization; the *Hoxb1* genes in zebrafish closely approach this model (Prince and Pickett 2002). Super-functionalization is well-known in the case of pesticide resistance in insects where tolerant strains often have duplicated or quadruplicated copies of genes encoding detoxification enzymes (Devonshire and Field 1991). It is assumed that such adaptive gene amplifications contribute significantly to the maintenance of paralogues in the early phase after a duplication (Kondrashev *et al.* 2002). Neofunctionalization is often considered to be a rare evolutionary pathway, but recent work

by Kellis *et al.* (2004) has shown that it may be more common than thought previously. Kellis *et al.* (2004) analyzed the genome of baker's yeast, *Saccharomyces cerevisiae*, by comparison with a related fungus, *Kluyveromyces waltii*, and obtained proof for a whole-genome duplication in the *Saccharomyces* lineage. Moreover, they could track the various gene pairs and were able to show that in the majority of cases one of the paralogues retained the ancestral function, while the other underwent accelerated evolution. Interestingly, deletion mutations of the ancestral paralogue were often lethal, whereas mutations in the derived copy never were. These data support the neofunctionalization model and suggest that duplication may indeed be a mechanism for creating evolutionary novelty.

With many genes present as members of a family, phylogenetic analysis becomes more complicated. Should we include all the genes of a large gene family when comparing them with a set of homologous genes in another species? One possible solution is to limit comparisons to the *core proteome*: the number of distinct families in each organism, counting a cluster of paralogues as one. Rubin *et al.* (2000) made a genome-wide comparison of gene families among the three model eukaryotic species, yeast, nematode, and fruit fly, and compared them with the bacterium, *H. influenzae* (Table 3.2). It is remarkable that *Drosophila* has a core proteome only twice that of yeast. One would not expect that just doubling the number of genes could make up for

Table 3.2 Estimated fraction of genes arising through duplication and estimates of the core proteome in three eukaryotes and one prokaryotic organism

Species	Total no. of predicted genes	No. of genes duplicated	Percentage of paralogues	Total no. of distinct families
Haemophilus influenzae	1709	284	17	1425
Saccharomyces cerevisiae	6241	1858	30	4383
Caenorhabditis elegans	18 424	8971	49	9453
Drosophila melanogaster	13 601	5536	41	8065

Reprinted with permission from Rubin *et al.* (2000). Copyright 2000 AAAS.

the increase in complexity associated with multi-cellularity, the development of a spatially differentiated body plan, the need for communication networks between organs, and the processing of sensory information and locomotion! It is also interesting to note that the core proteome of the nematode is the same size as that of the fruit fly, which again is astonishing given the large differences in development and body plan that exist between the two species. The lesson from this comparison must be that complexity in metazoans is not achieved by sheer number of genes. Instead, it is the regulation of these genes, in place and time and in response to the environment, added to the way they are organized in networks, that determines the differences between the species.

3.1.3 Skew, GC content, and codon usage

Turning our attention from genes to nucleotides, we will now explore patterns of the genomic occurrence of the four bases, adenine (A), guanine (G), thymidine (T), and cytosine (C). Obviously, A, G, T, and C are not distributed randomly in a DNA molecule. For the moment, disregarding the sequence itself, we will discuss two quantities in which deviations from random occurrence are expressed, GC skew and GC content.

If there is no mutation bias between the two strands of a DNA molecule, the orientation of any base pair with respect to the leading and lagging strand is arbitrary and one expects that the number of As on a strand will be equal to the number of Ts; the same is expected for G and C. However, in practice deviations are seen, especially in bacterial

genomes. The bias is expressed in a quantity called *GC skew* (S_{GC}), which is defined as:

$$S_{GC} = \frac{f_G - f_C}{f_G + f_C}$$

where f_G is the frequency of G and f_C the frequency of C in a certain segment (window) of DNA. S_{AT} is defined in an analogous way. The expected value of S_{GC} when the frequencies of G and C are the same is zero. Lobry (1996) found that GC skew showed a consistent switch across the origin of replication in three bacterial genomes, from a negative value (bias towards C) leftward of the origin of replication, to a positive value (bias towards G) rightward of the origin. This was confirmed by Blattner *et al.* (1997) when the genome of *E. coli* was completely sequenced (Fig. 3.6). The most likely explanation for skew is mutational bias associated with DNA replication; that is, in some way or another the way in which the leading and lagging strands are replicated causes the leading strand to become more rich in G. This interpretation is supported by the fact that GC skew is most prominent in intergenic positions and third codon positions (Fig. 3.6), where selective pressure is relaxed compared to the first and second codon positions. That the leading and lagging strands of prokaryotic DNA can differ in mutation rate was confirmed by Tillier and Collins (2000), who showed that sequences of the same gene can vary depending on their orientation in the genome; that is, whether they are encoded in the leading or the lagging strand.

Another phenomenon that characterizes many genomes is bias in GC base pairs over AT base pairs. This bias is valid for different segments of a

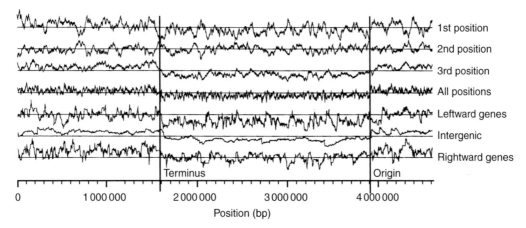

Figure 3.6 GC skew over the entire genome of *E. coli*, expressed as $(f_G - f_C) / (f_G + f_C)$ in 10 kbp windows of the genome, drawn in linear representation (in fact, the molecule is circular). GC skew is plotted separately for the three codon positions, for leftward genes, rightward genes, and non-protein coding regions. A negative value (below the line) indicates bias towards C, and a positive value means bias towards G. The two vertical lines indicate the origin of replication (right) and terminus of replication (left). Reprinted with permission from Blattner *et al.* (1997). Copyright 1997 AAAS.

chromosome, between coding and non-coding DNA, and across species. In the majority of cases, the GC content in coding regions of the DNA is higher than in flanking regions of a gene. The 5′ flanking region tends to be more GC-rich than the 3′ flanking region, which is due to the fact that promoter sequences tend to be relatively rich in GC. The GC content is also biased over longer stretches of DNA (around 300 kbp). Pieces of DNA with similar GC content are called *isochores*. This term refers to fractions of equal density obtained when DNA is subjected to CsCl gradient ultra-centrifugation. When mammalian DNA is sheared mechanically it tends to fall apart into fragments belonging to five discrete families, called L1, L2, H1, H2, and H3, each with a different GC contents (P_{GC}; Bernardi 2000). The GC-poor families L1 and L2 ($P_{GC} < 44\%$) are poor in genes, whereas the GC-rich families H1 ($44 \leq P_{GC} < 47$), H2 ($47 \leq P_{GC} < 52$), and H3 ($P_{GC} \geq 52\%$) are increasingly rich in genes. The human genome consists of a mosaic of such isochores (Fig. 3.7) and the same holds for other mammals. The distribution of isochores is correlated with the banding pattern of metaphase chromosomes seen in microscopic preparations using fluorescent dyes. The short isochores may indicate so-called *CpG islands*, GC-rich stretches of DNA of at least 200 bp with

an overrepresentation of GC dinucleotide repeats (the p in CpG indicates that it is a dinucleotide, not a base pair). Such CpG islands are indicative of genes that are switched on frequently.

In invertebrates and plants there is less hetero-geneity of GC content in the genome; very long isochores are observed in the centre of chromo-some I in *Arabidopsis* and in the centre of chro-mosome III of *C. elegans* (Fig. 3.7). There is no correlation between gene density and GC content in the *C. elegans* genome; however, in *Drosophila* there is such a correlation, which is consistent with the greater variability of GC content over the genome (Oliver *et al.* 2001).

In addition to variation along the chromosome, there are also conspicious differences between species in GC content. The GC content of Bacteria and Archaea varies between 25 and 75%, depend-ing on the species, whereas the GC content of eukaryotes varies within a much smaller range, between 40 and 55%. GC contents in coding regions of the DNA are now known for more than 20 000 species; a database, fed by GenBank sequences for full-length proteins, is maintained by Nakamura *et al.* (2000). Table 3.3 lists some examples of GC content taken from this database.

There is no good explanation why the GC con-tent of a genome should vary between species.

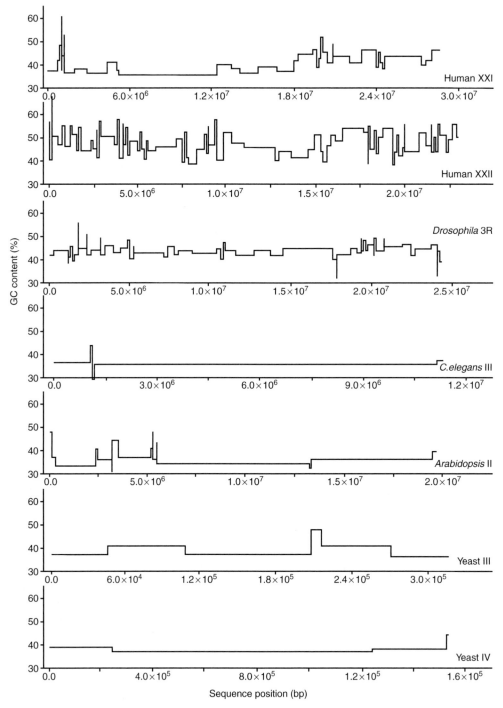

Figure 3.7 Isochore maps of some eukaryotic chromosomes, calculated from the genomic databases with an improved 'compositional segmentation' algorithm. For human chromosomes XXI and XXII the largest available contigs were taken. Note that there is much more structure in the human and fruit fly genomes, indicating variable gene density along the chromosome, than in the *C. elegans, Arabidopsis*, and yeast genomes, indicating a more even spread of genes. After Oliver *et al.* (2001), with permission from Elsevier.

Table 3.3 GC contents of coding regions of genomes of some selected species from the three domains of life

Species	GC (%)			
	Overall	First codon position	Second codon position	Third codon position
Archaea				
Methanococcus maripaludis	34.08	44.70	32.29	25.26
Halobacterium salinarum	64.85	68.07	43.04	83.45
Thermoplasma acidophilum	47.37	49.55	37.55	55.02
Sulfolobus solfataricus	36.49	43.22	33.65	32.59
Pyrococcus furiosus	31.09	49.46	34.56	39.24
Bacteria				
Sinorhizobium meliloti	63.14	64.88	45.23	79.30
Escherichia coli	51.39	58.27	40.84	55.06
Leptospira interrogans	36.42	44.70	34.62	29.94
Actinomyces naeslundi	67.51	63.74	49.50	89.27
Thermotoga maritima	46.40	52.24	34.40	52.55
Eukarya				
Saccharomyces cerevisiae (budding yeast)	39.76	44.60	36.61	38.09
Caenorhabditis elegans (roundworm)	42.93	49.95	38.93	39.90
Arabidopsis thaliana (thale cress)	44.60	50.90	40.52	42.37
Drosophila melanogaster (fruit fly)	53.97	55.87	41.51	64.52
Fundulus heteroclitus (mummichog fish)	53.93	55.10	40.89	65.81
Danio rerio (zebrafish)	50.94	54.50	40.82	57.52
Xenopus laevis (clawed frog)	46.97	52.75	39.94	48.22
Vipera aspis (aspis viper)	52.20	44.05	43.17	69.38
Anas platyrhynchos (mallard duck)	51.27	53.89	41.16	58.78
Sus scrofa (wild pig)	54.44	56.26	41.81	65.25

Source: Data from www.kazusa.or.jp/codon (Nakamura *et al.* 2000).

One of the forces acting upon the GC content is asymmetry in substitution rates. If the rate of substitution from G/C to A/T is denoted as u and the rate of substitution from A/T to G/C as v, then the equilibrium content of GC is expected to take the value P_{GC}, where (Graur and Li 2000)

$$P_{GC} = \frac{v}{u + v}$$

If the two rates are equal, P_{GC} is expected to be 1/2, or 50%. A high GC ratio is evidence for v being larger than u. The asymmetry of substitutions is called *GC mutational pressure*. However, GC content is not only shaped by mutation, but to some extent also by selection. One of the selective constraints arises from codon usage. We know that several amino acids are encoded in the DNA by more than one triplet, but this does not imply that all possible triplets are used in proportion. In most amino acids there is a bias towards the use of certain codons because the tRNAs of these codons are more abundant (*codon usage bias*, see Section 2.2). Depending on the number of Gs in the preferred codon, selection can constrain the GC content. Another source of selection is due to the superior stability of the G–C bond, which uses three hydrogen bridges, whereas the A–T bond uses two. It has been argued by several researchers that high ambient temperatures would favour high GC contents and that endothermic ('warm blooded') animals (birds and mammals) would have higher GC contents than ectothermic ('cold-blooded') animals (most reptiles, amphibians, fish, and invertebrates).

If selection is a major factor influencing the GC content and if GC mutational pressure drives the GC content upwards, one would expect that P_{GC} is higher in the third codon positions of a protein-encoding gene, compared with the first and second positions. There is some evidence that this is indeed the case both for prokaryotes and eukaryotes (Table 3.3); however, whether this is indeed indicative of selection is questionable; bias introduced by mobile elements with a high GC content is usually ignored (Duret and Hurst 2001). The possible role of temperature stability in the evolution of GC content was critically examined by Hughes (1999) and he concluded that the hypothesis is not supported strongly by the data. Among thermophilic prokaryotes there are species with low and high GC contents and within the vertebrates the picture is also not consistent (Table 3.3). The strongest effect in the GC content seems to stem from phylogenetic constraints: taxonomically related species tend to have similar GC contents. Although purifying selection may play a modest role, the dominant factors acting upon GC content seem to be neutral processes, GC mutational pressure, and random drift.

3.1.4 Gene order

In genetics, two loci are called *syntenic* if they are located on the same chromosome (Russell 2002). In genomics, however, the term *synteny* is used to indicate a situation where a series of genes is arranged in the same order on different genomes (Gibson and Muse 2002). Passarge *et al.* (1999) have rightly pointed out that this new usage is incorrect and etymologically awkward. Another term that may be more appropriate is *colinearity*; however, in this book we will comply with the most common usage and use synteny in the genomics understanding. The presence of synteny between two genomes is somewhat dependent on the scale of analysis. On the level of the chromosomes, synteny may be demonstrated by techniques such as *chromosome painting* (using inter-species fluorescent *in situ* hybridization); however, this does not exclude the presence of extensive rearrangements on the level of individual genes. The term *microsynteny*

is used to indicate detailed sequence comparisons between individual genes within a chromosomal segment.

In any genome, genes are found to be organized in clusters and these clusters sometimes maintain the same order across species, even across groups as far apart as mammals and fish. Well-known examples of synteny are histone genes, *Hox* gene clusters, and the genes of the major histocompatibility complex (MHC). There are also large synteny blocks, covering hundreds of kilobase pairs, between the genomes of rice and *Arabidopsis* (see Section 3.2).

The *Hox* genes are a famous example of long-range synteny. Indeed, the same order of genes can be found in the *Hox* clusters of nematodes, insects, and mammals. *Hox* genes encode transcription factors that regulate developmental patterns across the anterior-posterior axis of bilaterian animals (Carroll *et al.* 2005). Macro-evolutionary relationships in the animal kingdom can be partly understood as duplications followed by neo- and nonfunctionalization of essentially the same pattern of *Hox* genes (Amores *et al.* 1998; Carroll 2000).

Synteny analysis is an important tool in comparative genomics. The relative order of genes in one species can provide clues about the presence or even the function of genes in another species. Similarly, by looking at the order of genes in a cluster one can discover genes by homology to another species that were missed by automatic gene-finding algorithms. Synteny analysis can also be a tool to reveal duplications across species, for example by searching two regions in one genome that have the same gene order as one region in another genome (*doubly conserved synteny*; Kellis *et al.* 2004).

How could such blocks of gene order be maintained while other regions of the genome are reshuffled extensively by recombination? How can it be that some genes are free to move through the genome while others are tied, for millions of years, to the same neighbours? One of the reasons could be selective pressure acting upon the cluster as an integrated whole. This is certainly the case if genes are organized in operons, as in all prokaryotic and some eukaryotic genomes (see Sections 3.2 and 3.3).

Another functional constraint is illustrated by the *Hox* genes. In most animals, the order of these genes along the chromosome is the same as the order of their expression domains along the anterior-posterior body axis. In addition, the genes at the front end of the complex are expressed earlier in development than the ones at the back end. These observations suggest that it is the requirement for coherent temporal expression that is maintaining colinearity of the *Hox* cluster (Patel 2004). The genetic developmental system that governs the basic positional information of tissues in all animals was called the *zootype* (Slack *et al.* 1993).

Another reason for conservation of gene order, proposed more recently, is *interdigitization* of regulatory elements. We know that the expression of genes is controlled by regulatory elements, usually in the 5′ region of the gene. It turns out that some regulatory elements may be physically linked to genes close by, or even be located in the introns of other genes. The fixation of regulatory elements inside the territory of neighbouring genes thus forges a physical bond resulting in close linkage between the genes. Another way in which regulatory elements may link genes together occurs when the expression of a group of genes is controlled by a single *locus-control region*. The principle of the locus-control region was first discovered in the β-globulin cluster of the human genome, but similar regions, presumably participating in dynamic chromatin alteration, have now been found in other gene clusters. In all these cases genes cannot move independently from each other without gaining a severe selective disadvantage. The shuffling and reorganization of the genome during evolution, as highlighted in Chapter 1 and indicated by the term turbulence, is inevitably limited to some extent by such processes.

An example of synteny analysis is provided by a comparison between two distantly related species of nematode, sharing a common ancestor 300–500 million years ago (Guiliano *et al.* 2003). A nematode parasite of vertebrates, *Brugia malayi* (order Ascarida), was compared with the well-known model species *C. elegans* (order Rhabditida). Whereas the genome of the latter species is completely known and many genes are annotated, the genome of *Brugia* is only known incompletely; the comparison was undertaken partly to reveal more of the function of genes in *Brugia* from knowledge of *C. elegans*. Figure 3.8

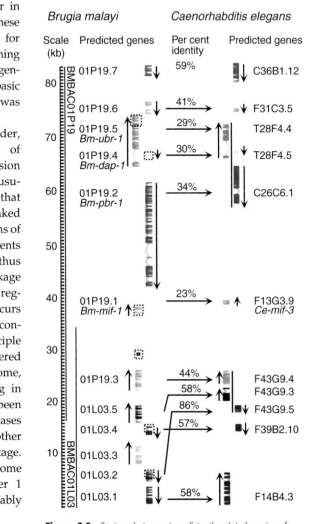

Figure 3.8 Synteny between two distantly related species of nematode, *Brugia malayi* and *Caenorhabditis elegans*. An alignment is shown of two overlapping contigs from a genomic library of *Brugia* (BMBAC01P19 and BMBAC01L03) with the *C. elegans* genome. Exons of genes are indicated by boxes, with brackets between them to indicate introns. Genes are designated by code names from top to bottom (01P19.7, etc.), but have different codes in the different species. The direction of transcription is indicated for each gene by a vertical arrow. The horizontal arrows indicate matches of putative *B. malayi* genes in the *C. elegans* genome. From Guiliano *et al.* (2003), by permission of BioMed Central.

shows an alignment of two overlapping contigs of *B. malayi* with a homologous portion of the genome of *C. elegans*. Eleven putative *Brugia* genes in this region can be homologized to genes in the *C. elegans* genome and, except for one case, the order of the genes is the same in the two species; in two cases the direction of transcription was reversed. Although earlier comparisons between two related species, *Caenorhabditis briggsae* and *C. elegans*, had shown that nematodes have a high rate of intrachromosomal rearrangement, the present observations demonstrate that there may still be local clusters of synteny across relatively large phylogenetic distances. To explain this, Guilliano *et al.* (2003) could not identify a common function for the gene cluster but suggested that promoters or *cis*-acting regulatory elements could be embedded within other cluster members, so synteny could be maintained by interdigitization of regulatory sequences.

3.1.5 Patterns of synonymous and nonsynonymous substitutions

Spontaneous mutations in the genome are the primary source of all genetic variation. This section discusses the different types of mutation that can occur in the genome and what effect they may have on adaptive evolution of the organism. We show that genomics tools have facilitated new and accurate estimations of mutation rates and their effects on phenotypic change.

Basically, DNA can mutate in four ways: *substitution*, *deletion*, *insertion*, and *inversion*. Substitution is the change of a nucleotide for another and is generally caused by errors in DNA replication. Substitutions can be divided into two classes: *transitions* and *transversions*. A transition is the change of a purine into another purine (A/G) or of a pyrimidine into another pyrimidine (C/T). Change from a purine to a pyrimidine or vice versa is called a transversion. If a substitution causes a nucleotide site to become polymorphic we call this site a *single nucleotide polymorphism* (SNP, pronounced *snip*). If substitutions take place in coding regions they can either be *synonymous* or *nonsynonymous*. A synonymous substitution does

not change the amino acid sequence of the encoded protein and is called silent. Nonsynonymous substitutions do change the amino acid composition of the encoded protein. Finally, substitution can result in a stop codon, which is called a nonsense mutation. If we look closely at the properties of the genetic code we can conclude that synonymous substitutions occur at the third codon position. Most (but not all) substitutions at the first codon position cause an amino acid replacement and all nucleotide substitutions at the second codon position cause either an amino acid replacement or a nonsense mutation. Substitutions and the resulting SNPs are scattered randomly throughout the genome with a higher abundance in non-coding regions. The distribution of SNPs is a welcome source for high-density mapping studies in genomic model organisms (see the example discussed at the end of Section 2.1).

Insertions and deletions (*indels*) involve a variable number of nucleotides ranging from one base to large blocks of DNA. They occur at quite high frequency in non-coding regions of the genome. The main source of small indels is again errors in DNA replication, whereas large indels seem to be caused by unequal crossing-over and transposition. Transposons or transposable elements cause DNA segments to change chromosomal position thereby generating indels. Unequal crossing-over is believed to be very important in the generation of multigene families such as rRNA genes.

Measuring mutation rates is extremely difficult, because the frequency of mutation within each generation is very low. The traditional way to study genomic mutation rate is to perform *mutation accumulation* (MA) experiments. The standard way to design such an experiment is described very well by Keightley and Charlesworth (2005) and may be outlined as follows. An isogenic line from the organism of interest is taken as the inbred progenitor, and is subdivided into several MA lines. Going through several generations of inbreeding (selfing or full-sib mating) these lines accumulate mutations at random, causing divergence at the phenotypic as well as the DNA level. One or a few individuals from each generation are isolated to form the next generation so that the potential role of selection is diminished. The inbred progenitor population is considered to be mutation-free and should be preserved in that state. The mean fitness of all MA lines is estimated at generation time t. Also, the mean fitness of the mutation-free population can be inferred. The first parameter of interest is the change in mean fitness per generation that is due to MA (ΔM). Secondly, we can estimate the increase in genetic variance in fitness among inbred lines per generation (V_m). If we assume that the average deleterious effects of mutation (s) are equal, then $\Delta M = U_d s$, where U_d is the genomic rate for mutations affecting fitness per generation. Furthermore, Keightley and Eyre-Walker (1999) showed that $V_m = U_d s^2$, so that U_d can be estimated from the formula: $U_d = \Delta M^2 / V_m$.

A lot of this work on MA has been performed using *D. melanogaster* as a model organism, but data are also available for *C. elegans*, *E. coli*, and humans. Keightley and Eyre Walker (1999) showed that U_d estimates vary greatly, ranging from 0.00017 in *E. coli* to 0.47 in humans. In *D. melanogaster* several U_d values have been estimated in different studies and these data vary over an order of magnitude (0.02–0.47), whereas the effect of deleterious mutations was estimated at 3%. We have to conclude that these figures are not very reliable and probably underestimate the actual mutation rate in a genome. The method only detects mutations of moderate effect that change phenotypic traits, whereas mutations with very small effects and mutations caused by transposon activity are not recognized.

The problems associated with mutation-rate estimates based on phenotypic traits were overcome in a study on *C. elegans* by Denver et al. (2004). These authors applied a genomics approach that provides a direct estimate of mutation rate in DNA sequences from a set of MA lines. Some 4 Mbp of randomly selected DNA segments was sequenced directly from a set of MA lines after $t = 280$, 353, and 396 generations. A per-nucleotide mutation rate can be determined from the sequence data, which can be used to calculate a haploid genomic mutation rate U_t per generation. Denver et al. (2004) detected 30 mutations, which translates to a mutation rate of 2.1×10^{-8}

mutations per site per generation and a U_t value of approx. 2.1 mutations per genome per generation. Surprisingly, the direct estimates of Denver et al. (2004) are at least 10 times higher than previous estimates (Drake et al. 1998). Furthermore, 17 of the 30 observed mutations were indel mutations, and 13 of these 17 indels were insertions. How can these high mutation rates and the predominance of insertions be explained? Denver et al. (2004) suggested that their study was less biased by genetic selection than previous studies. It seems that pseudogenes, on which earlier studies were based, do not evolve neutrally but may be under selection (Hirotsune et al. 2003). Alternatively, Rosenberg and Hastings (2004) propose that phenotypes are masking molecular variation in such a way that mutation rates are higher than predicted. They also propose that accumulation of harmful mutations in MA lines can induce increased mutability. Harmful mutations will cause cellular stress, and in response to that stress mutation rates may increase. However, Keightley and Charlesworth (2005) state that this mutator state is unlikely, because such a genetically unstable state would induce a rapid decline of fitness, and this was not observed.

It remains difficult to make generalizations about mutation rates in organisms. The genomic data on direct mutation-rate estimates generated by Denver et al. (2004) raise new questions, for instance on how their data on *C. elegans* relate to earlier estimates in *Drosophila*. As proposed by Keightley and Charlesworth (2005), it may be worthwhile to repeat the Denver approach in *Drosophila*.

If DNA sequences of the same gene from two individuals or from two species are aligned, differences are often observed. As discussed above most of these differences do not result in differences in the amino acid sequence of the encoded proteins, but some do. The extent to which differences between two sequences imply differences in amino acid composition is measured by the K_a / K_s ratio, defined by Hurst (2002) as the ratio between the number of nonsynonymous substitutions per synonymous site (K_a) to the number of synonymous substitutions per synonymous site

(K_s) in a specific segment (window) of DNA sequence. This ratio is also designated the d_n/d_s ratio, where d_n is comparable to K_a and d_s comparable to K_s. The rate of synonymous substitutions (K_s) is often equated to neutral mutation rate of the gene, whereas K_a indicates the amount of protein evolution (mostly functional). The K_a/K_s ratio of a certain sequence therefore tells us something about how that sequence has evolved.

Usually K_a is much smaller than K_s, because a change in the amino acid is less likely than a silent substitution. This is due to the fact that selection eliminates deleterious mutations to keep the function of a protein intact. Under purifying selection, $K_a/K_s < 1$, approaching 0 for complete conservation. However, sometimes K_a is greater than K_s, for instance in genes associated with the immune system that are co-evolving with parasites (Nei et al. 1997). $K_a/K_s > 1$ indicates selection acting positively on protein change in one sequence compared to the other.

Yang and Bielawski (2000) evaluated two ways for measuring molecular adaptation. The first class is based on intuition; the method of Nei and Gojobori (1986) is best known. Synonymous and nonsynonymous sites and synonymous and non-synonymous differences are counted between two sequences. Subsequently, the synonymous and nonsynonymous rates are corrected for multiple substitutions at the same site using simplistic assumptions to yield estimated K_a and K_s values. Nei and Gojobori (1986) assume equal rates of transition (T to C, A to G, and vice versa) and transversion (T or C to A or G and vice versa), and uniform codon usage. However, this method yields incorrect estimates of K_a and K_s because the assumptions have been proven to be unrealistic. The method tends to perform especially poorly when codon usage is biased due to differences in translation efficiency (genes that have a high transcription/translation rate are biased in their codon usage as compared to genes that are transcribed/translated at a low rate).

A more reliable class of methods was developed using the principle of maximum likelihood. The codon is considered to be the unit of selection, and a model is developed for substitutions in the codon. Parameters in the model are, for instance, sequence divergence, transition/transversion rate ratio, and K_a/K_s ratio; these parameters are estimated from sequence comparisons by applying the maximum-likelihood principle. Parameters also are corrected for biased codon usage. An important feature of these methods is that a statistical test can be applied to test whether the K_a/K_s ratio is significantly different from unity (neutral evolution). Several software packages have been developed to study molecular adaptation using the K_a/K_s ratio.

As an example, we discuss a study by Talbert et al. (2005), who analysed adaptive evolution in genes encoding centromere-binding protein C (CENP-C). Centromere-binding proteins play an important role in fixing microtubules to the centromeric region of a chromosome, which is essential in the formation of spindles during mitosis and meiosis. It is remarkable that the centromere has such a conserved function, although its DNA sequence is non-coding and highly variable. The centromere-binding proteins seem to co-evolve with the rapidly changing satellite DNA sequences in the centromere. Talbert et al. (2005) screened the sequences of *Cenpc* genes in pairs of related species and applied a K_a/K_s analysis to locate regions of selection (Fig. 3.9). Statistically significant selection was detected in the regions encoding DNA-binding domains. Mammals showed stretches of positive selection, whereas negative selection seemed to be more pronounced in maize and sorghum in the comparable region. Why the centromere and consequently the centromere-binding proteins would evolve so rapidly is not known. Talbert et al. (2005) propose a model in which there is competition between centromere variants during female meiosis. The female meiotic spindle has an asymmetric shape, such that the 'stronger' centromere variants may be better captured and included in the meiotic product that becomes the egg nucleus. Such biased inheritance has also been described for other loci, especially in *Drosophila*, and is known among geneticists as *segregation distortion* or *meiotic drive*. According to Talbert et al. (2005) meiotic drive can explain the apparent selection on DNA-binding centromere proteins.

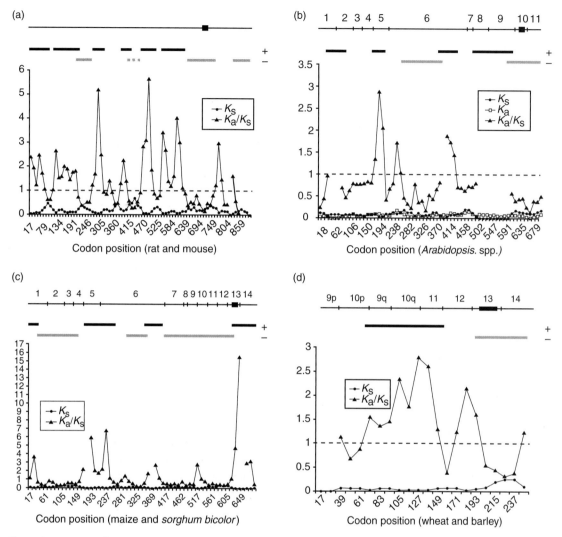

Figure 3.9 Patterns of variation between pairs of related species across the sequence of centromere-binding protein C (*Cenpc*) genes displayed using a sliding window analysis of K_a/K_s ratios. Each point represents the value of K_a/K_s for a 99-nucleotide (33-codon) window plotted against the codon position of the midpoint of the window. The aligned coding sequence is plotted at the top of each graph. On the sequences, the black rectangles indicate the locations of 24-amino acid CENP-C motifs, the defining structure of this type of protein. Exons are indicated by numbers for the plant sequences. Beneath the sequences regions of positive (black bars) and negative (grey bars) selection are indicated. (a) Rat and mouse, (b) *A. thaliana* and *Arabidopsis arenosa*, (c) maize and *Sorghum bicolor*, and (d) wheat and barley, exons 9p–14. From Talbert *et al.* (2004), by permission of BioMed Central.

Another example of the use of the K_a/K_s ratio is illustrated by Liberles *et al.* (2001), who studied adaptive evolution of amino acids at the genomic level. They calculated K_a/K_s values on nodes of branches within evolutionary lineages. The evolutionary lineages were taken from the Master

Catalog, a compilation of sequence alignments and evolutionary trees for all protein modules encoded by genes in Genbank, constructed by Benner *et al.* (2000). They focused on subtrees containing only chordates and Embryophyta (mosses, ferns, and higher plants) and could identify branches with high K_a/K_s values. These branches may be indicative of positive selection, where the mutated protein has a higher fitness than the ancestral form, probably associated with a change in function. The gene families that display high K_a/K_s values were stored in *The Adaptive Evolution Database* (TAED). Currently, TAED 2.1 (www.bioinfo.no/tools/TAED) contains 6657 families that are fully processed. In 10–20% of these families positive selection was determined in at least one branch. High K_a/K_s values on branch points in evolutionary protein trees may be caused by gene amplification followed by differentiation in function of the paralogues. Orthologous genes under different selective regimes in different species may also be found. However, the database cannot distinguish between paralogues and orthologues, so the results should be interpreted cautiously. Besides gene families that were identified previously to be under positive selection, such as the MHC proteins (proteins of the adaptive immune system), quite a number of families were newly identified in TAED to have undergone change in function. The authors conclude that TAED is a useful resource for biologists searching for potential examples of molecular adaptation as a starting point for further experimental study.

We can conclude that the use of K_a/K_s ratios for measuring molecular adaptation in coding sequences of the genome is a very effective approach. We expect that databases like TAED will become an important framework for ecological genomics to study adaptation at the molecular level. The challenge will be to integrate this information with gene-expression profiling and to link molecular variation to phenotypic differences between related species.

3.2 Prokaryotic genomes

The genomes of prokaryotes typically consist of one circular molecule, representing a haploid genome, characterized by an *origin of replication* and two semicircles of sequence that are read leftward and rightward, with the terminus of replication positioned on the opposite side of the molecule. The origin of replication is a stretch of

200–300 bp that has characteristic sequences allowing DNA unwinding and binding of the initiator protein DnaA. The genes in prokaryotic genomes are organized in *operons*, clusters of genes, the expressions of which are regulated jointly by interactions between operator and repressor proteins (see Fig. 4.9). The structural genes in an operon are transcribed into a single *polygenic mRNA*. The situation that an mRNA molecule contains protein-coding sequences from more than one gene (cistron) is called *polycistronic*. It was long thought that polycistronic transcription was limited to prokaryotes and that all eukaryotic mRNAs were monocistronic; however, it turned out in 1994 that the nematode *C. elegan*s has approximately 25% of its genes organized in polycistronic units of two or more members (Hodgkin *et al.* 1995), so polygenic transcription is not exclusive to prokaryotes.

The genes in a microbial genome are usually subdivided into two main classes: *informational genes* and *operational genes*. In the first category (also called *essential genes*) are genes related to information processing, such as transcription, translation, replication, etc. These genes define the 'essence' of the species. They are usually large, complex systems, with many interactions, located on the main chromosome of the cell and less prone to lateral gene transfer (see below). The second category includes genes that function in basic cell maintenance processes and metabolism, such as protein, energy metabolism, and phospholipid acid biosynthesis. These genes are usually members of small assemblies of a few gene products and more often found on plasmids.

3.2.1 Chromosomes and plasmids

As an example of the architecture of a prokaryote genome we reproduce in Fig. 3.10 the sequence of *H. influenzae*, published by Fleischmann *et al.* (1995). The figure summarizes some general aspects of the sequence and provides a great deal of information on restriction sites, arrangement of gene clusters, GC content, tandem repeats, etc. The relatively small prokaryotic genomes are the only ones for which this information can be reproduced

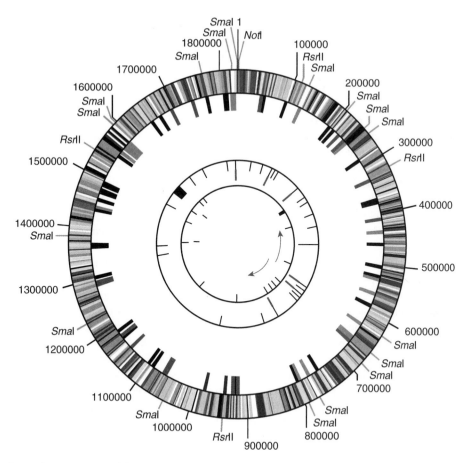

Figure 3.10   Circular representation of the annotated genome of *H. influenzae*, as published by Fleischmann *et al.* (1995). Its characteristics are indicated by four concentric circles and rings. Outer perimeter, restriction sites. The *Not*I restriction site occurs only once and was arbitrarily chosen as the site at which to begin the clockwise numbering of base pairs (top of the figure). Outer ring, locations of genes that could be assigned a role. In the original paper genes were classified according to 16 different functions, which in this figure are represented by different grey tones. Second ring, regions of high GC content (grey) and high AT content (black). Third concentric circle, locations of six ribosomal operons (extended markers), tRNAs (short markers), and a cryptic μ-like prophage (black box). Fourth (inner) circle, locations of simple tandem repeats. The origin of replication is indicated by two outward-pointing arrows near base pair no. 603 000. Two potential termination sites are indicated near the opposite midpoint of the inner circle. Reprinted with permission from Fleishmann *et al.* (1995). Copyright 1995 AAAS.

in a still more-or-less readable figure, using 16 different colours to indicate various functional details (colour-blind people are at a real disadvantage in genomics!). An analysis of the functional categories of these genes allows some first insight into the functions that can and cannot be performed by the organism. For example, the genome of *H. influenzae* has a complete metabolic machinery required for survival and reproduction as a free-living organism, but it lacks genes such as

the fimbrial gene cluster that encodes proteins allowing the bacterium to attach to host cells. This is explained by the fact that the sequenced strain is not pathogenic. The sequence also includes a cryptic μ-like prophage, which is a segment resembling the DNA of bacteriophage μ. We will see below that many bacterial and archaeal genomes contain genetic material from bacteriophage origin. A bacteriophage that is 'sitting' in its host genome, being replicated with the chromosome of the host and waiting to become active, is called a *prophage*. In some cases these prophage sequences become debilitated and are no longer functional but are still recognizable in the genome.

Although the single circular chromosome is the most common genome organization among prokaryotes, an increasing number of exceptions to this are being identified. In addition to circular chromosomes, many prokaryotes have smaller circular DNA molecules, called plasmids. The classical example of a bacterial plasmid is the *F-plasmid* of *E. coli*, which is transferred from a donor (F^+ cell) to a recipient (F^- cell) during *conjugation*. The F-plasmid carries genes associated with the formation of *F-pili*, hair-like surface structures which allow the physical union between the F^+ and F^- mating types. BACs, an extremely important class of cloning vector (see Section 2.2), were modelled after F-plasmids and use the F-plasmid's origin of replication.

Like the F-plasmid, many bacterial plasmids contain genes dedicated towards specific, so-called non-essential, functions. These functions may relate to antibiotic resistance, pathogenicity, or specialized metabolic pathways. In *Agrobacterium tumefaciens* (Alphaproteobacteria), one of the plasmids carries a discrete set of genes (*transforming DNA*, or T-DNA) that can be transferred to a plant host and, when expressed in the host, promote the synthesis of plant growth hormones that redirect the local development of plant tissue to form a gall. In the gall, the bacterium also directs the plant to produce specific amino acid derivatives (*opines*) that can be used as the sole carbon and nitrogen source for *Agrobacterium*. The transfer of the T-DNA and its integration in the chromosome of the host is supported by proteins encoded in virulence (*vir*) genes, which are also on the plasmid. The plasmid is thus specialized for gall-formation and is therefore called a *tumor-inducing plasmid* or Ti plasmid. The capacity of the Ti plasmid of *A. tumefaciens* to introduce foreign DNA into a host has been widely exploited by plant molecular biologists to artificially transform plants.

A similar specialized function located on a plasmid is present in the symbiotic nitrogen-fixing soil bacterium *Sinorhizobium meliloti*. This bacterium belongs to a group of three closely related Alphaproteobacteria of the family Rhizobiaceae that now have been sequenced completely (*S. meloti*, *Mesorhizobium loti*, and *A. tumefaciens*; Galibert et al. 2001; Goodner et al. 2001; Wood et al. 2001). *S. meloti* is a symbiont of alfalfa (*Medicago sativa*) and because of the profound ecological importance of symbiotic nitrogen fixation in the global nitrogen cycle, genomic studies of these types of bacteria are particularly relevant for ecological genomics. Fig. 3.11 shows the position of the three Rhizobiaceae in relation to some other bacteria from the protobacterial lineages with fully sequenced genomes.

The plasmid of *S. meliloti* which carries most of the symbiotic functions is called SymA or *symbiotic plasmid*. An extensive annotation of gene functions of this large plasmid (1354 kbp, comparable in size to an entire prokaryotic genome!) was published by Barnett et al. (2001). Specific nodulation (*nod*) genes on the plasmid encode biosynthetic enzymes for the so-called *Nod factors*, molecules which are active in very low concentrations (down to 10^{-12} M) in the signalling between plants and rhizobia in the early stages of root-nodule formation. The expression of Nod genes is triggered, via a signal transduction pathway, by phenolic compounds excreted by plant roots. The plasmid also carries several genes encoding enzymes necessary for nitrogen fixation, denitrification, and opine metabolism, and many genes involved with transport and osmotic stress. It is interesting to note that no informational genes, for example relating to DNA replication or cell division, were found on pSymA, demonstrating that despite its size, it is a true plasmid.

In addition to circular chromosomes and plasmids, some bacteria also have *linear chromosomes*. A particularly complicated situation is present in the genome of *A. tumefaciens*, which was published by Wood et al. (2001) and Goodner et al. (2001). It consists of one circular chromosome (2.8 Mbp), one linear chromosome (2.1 Mbp), and two plasmids, which in this case are called *megaplasmids* because they are relatively large (543 and 214 kbp). The smaller of the two plasmids is the Ti plasmid

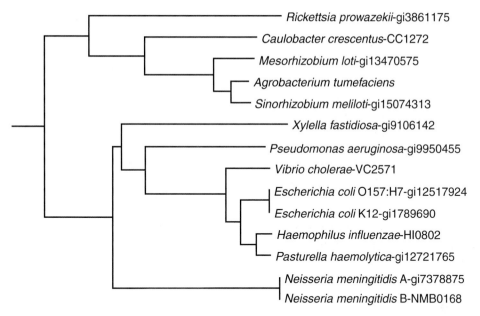

Figure 3.11 Phylogeny, based on the RpoA (RNA polymerase α) gene, for 14 bacteria with fully sequenced genomes from the Alpha-, Beta-, and Gammaproteobacteria phyla. Reprinted with permission from Wood *et al.* (2001). Copyright 2001 AAAS.

referred to above. The linear chromosome appeared to carry 35% of the genome's genes, including those encoding ribosomal and DNA-replication proteins, as well as 21 complete metabolic pathways. The presence of these genes confirmed the chromosome-like nature of the linear element; however, other genes on the same chromosome, especially those involved in conjugation, are reminiscent of a plasmid. Most interestingly, the sequence revealed that there was a *repABC* operon (a gene known to be associated with replication of circular chromosomes) near the centre of the chromosome, plus an inversion in the GC skew (see Section 3.1.3). This seems to indicate that the linear chromosome has a bidirectional, plasmid-like mode of replication. Why *A. tumefaciens* maintains such a complicated arrangement of its genome and the advantages of having multiple chromosomes remain a mystery. Multiple chromosomal elements are especially prominent among proteobacterial phyla and spirochaetes, and this in itself would suggest phylogenetic determination of the tendency to form extra-chromosomal elements.

3.2.2 Lateral gene transfer

For many years it has been known that micro-organisms can absorb DNA directly from the environment. The relative ease by which antibiotic resistance can be donated from one bacterium to another constitutes further proof that genetic information is not only transferred during cell division (vertical transmission), but also from one intact cell to another (*lateral* or *horizontal transmission*). Lateral gene transfer can occur via three mechanisms (Zhou and Thompson 2004), as follows.

Transformation. This is the uptake of DNA directly from the environment. If prokaryotic cells can do this they are called *competent*. Very few bacteria are competent during their whole life cycle, but some are during certain physiological stages.

Transduction. Bacteriophages can transfer DNA between species if two host species share similar bacteriophage receptors. Transduction may concern random pieces of the host DNA, packaged during phage assembly in the lytic cycle, or it may be limited to the sequences flanking the insertion site.

Conjugation. Lateral gene transfer can occur via specialized plasmids during physical contact between F^+ and F^- cells, as discussed above.

Lateral gene transfer is not limited to bacteria of the same species, it can also occur among species of widely different origin and even between Bacteria and Archaea. The latter phenomenon was discovered by Nelson *et al.* (1999), when they sequenced the genome of the thermophilic bacterium *Thermotoga maritima* (Fig. 3.12). *Thermotoga* derives its name from the sheath-like envelope that surrounds the cell (the 'toga') and the fact that it has a temperature optimum for growth of 80 °C. It is usually placed in a completely separate lineage of the Bacteria, called Thermotogales, because of some unique characteristics, including rRNA sequences that are unusual among the Bacteria and a set of fatty acids that is only found in this group. Phylogenetic analysis, aimed at enlightening the position of *T. maritima* within the Bacteria, resulted in a great lack of congruence when different genes were used as a basis for the comparison. Further analysis of the ORFs in the *T. maritima* genome showed that no less than 24% of the predicted genes were most similar to proteins in archaeal species, rather than to Bacteria. The Archaea-like genes were found to lie in clusters (islands) in the genome of *T. maritima*; in several of these islands even the archaeal gene order was conserved. These observations and those by Aravind *et al.* (1998) on another thermophilic bacterium, *Aquifex aeolicus*, provided great support to the theory that lateral gene transfer was not to be considered an oddity, but a very significant process for many micro-organisms.

Koonin *et al.* (2001) performed a quantitative analysis of the frequency of lateral gene transfer by analysing the genomes of eight archaeal and 22 bacterial species. They estimated that the percentage of new genes acquired from another domain (Bacteria, Archaea, or Eukarya) was on average 0.9% for the bacterial genomes and 3.4% for the archaeal genomes. When looking at inter-species transfers within the Bacteria, the percentages of acquisition of new genes varied considerably, from 0.4 to 19.8%, depending on the group. A particularly high frequency of foreign DNA is found in the genomes of the Spirochaetales. In general, it turned out that bacteria living at high temperatures had more archaeal genes in their genomes than mesophilic bacteria, and bacteria with a parasitic life style more often had eukaryotic genes in their genome than non-parasitic bacteria. It therefore seems that lateral gene transfer is especially common among organisms that live in close proximity to each other.

Several authors have pointed out that not all evidence for lateral gene transfer is equally reliable. If lateral gene transfer is inferred only from

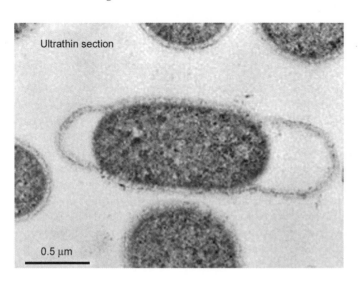
Ultrathin section

0.5 µm

Figure 3.12 Electron micrograph of *Thermotoga maritima*, a thermophilic bacterium belonging to the group Thermotogales, which was isolated originally from a geothermal heated marine sediment at Vulcano, Italy. Courtesy of K.O. Stetter, University of Regensburg.

the genome sequence, showing that certain ORFs have a BLAST match outside the group considered, this is not sufficient evidence, because alternative explanations may be given, such as the loss of genes in some lineages or widely diverging rates of evolution across the groups compared (Eisen 2000; Nesbø *et al.* 2001). To prove that lateral gene transfer has occurred, one needs to conduct a gene-by-gene phylogenetic analysis. As an example we discuss the work of Deppenheimer *et al.* (2002).

Deppenheimer *et al.* (2002) sequenced the genome of *Methanosarcina mazei* (Euryarchaeota), an archaeon of great ecological importance since it derives its energy from fermenting simple organic substrates to methane. Methane production in underwater sediments and inundated land (notably rice paddies) is an important link in the global carbon cycle. The genome of *M. mazei* (4.1 Mbp) was more than twice as large as other methanogenic Archaea that had been sequenced completely. Of the 3371 identified ORFs, no less than 1043 (31%) had their closest homologue not in an archaeal but in a bacterial species. Unlike the situation in *Thermotoga*, the bacterial genes in *Methanosarcina* did not cluster together in islands, but were found scattered in the genome. Gene phylogenies in which the laterally transferred genes were compared with orthologues in other Bacteria and Archaea showed that the foreign genes clustered with bacterial homologues, rather than with Archaea (Fig. 3.13).

Detailed analysis of the metabolic role of the bacterial genes in *Methanosarca* showed that the imported genes had considerably enlarged the metabolic spectrum of the archaeon. The suggestion from the data was that the metabolism of *M. mazei* has evolved from a simple methanogenic pathway (using hydrogen and carbon dioxide to produce methane), to a much more versatile substrate spectrum, allowing the use of acetate, methanol, and methylamine. Interestingly, most of the laterally transferred genes seemed to have been obtained from obligate and facultative anaerobic bacteria; that is, from organisms that live in the same microenvironment as the methanogenic archaeon (sediments, inundated land). This is in

line with observations on thermophilic communities showing that the greatest transfer is between organisms living close together.

As indicated by the examples above, lateral gene transfer is an important mechanism for recruiting new microbial functions. By lateral gene transfer, microorganisms may be able to exploit new ecological niches that were inaccessible prior to the event. However, not all lateral gene transfers are necessarily to be viewed within a purely adaptive framework. The presence of foreign genes in a genome might well be a consequence, rather than a cause, of adaptation (Nesbø *et al.* 2001). Mira *et al.* (2001) viewed the bacterial genome as resulting from an evolutionary balance between ongoing recruitment and removal processes. Lateral gene transfer is the most important recruitment process. If there is no pressure from natural selection to maintain a newly recruited gene, it will quickly be removed or inactivated. Assuming that in all bacterial genomes there is a bias towards a higher rate of deletions over insertions, this explains the small and streamlined genomes of many bacteria. According to G. Gottschalk (personal communication) the large genomes of methanogens may just be a consequence of the fact that removal of laterally transferred genes has not yet occurred. This argument is similar to the neutral theory of genome evolution by Lynch and Conery (2003), discussed in Section 3.1.

Are all genes equally subjected to lateral gene transfer? Jain *et al.* (1999) postulated the *complexity hypothesis*, which states that genes that have few interactions with other genes will integrate more easily into a new genomic background and are therefore more likely to be successful in lateral gene transfer, compared with genes that are part of a complicated network and dependent on many other genes. This hypothesis explains the greater tendency of operational genes to be transferred, compared to informational genes; however, there are many exceptions to the hypothesis (Zhou and Thompson 2004).

Although the actual rate of lateral gene transfer is considered to be low compared to the life cycle of microorganisms, lateral gene transfer may leave a permanent trace in the genome; the presence of

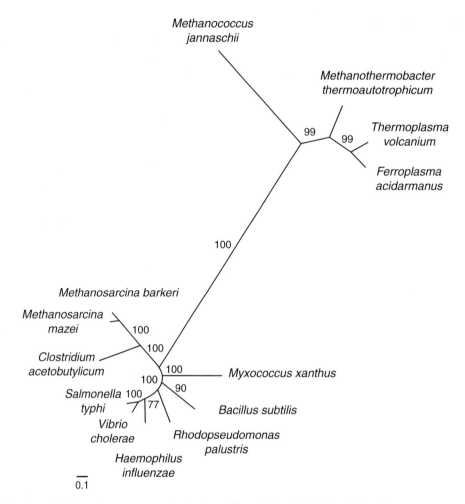

Figure 3.13 Unrooted phylogenetic tree, constructed using the neighbour-joining principle, for ATP-dependent Lon protease, a gene suspected of lateral transfer. The tree shows two clusters of archaeal (top right) and bacterial (bottom left) genes; however, the gene from the archaeon *M. mazei* clusters with the bacteria, rather than with the Archaea. This greatly supports the hypothesis that ATP Lon protease was acquired by *M. mazei* from a bacterial donor. Scale bar indicates the proportion of amino acid difference. After Deppenheimer *et al.* (2002), by permission of Horizon Press.

DNA of 'foreign' origin dating from a transfer event millions of years ago is still visible to the present-day genome investigator. At the same time, lateral gene transfer complicates the construction of phylogenetic trees, because the phylogenetic reconstruction of one gene may be different from the reconstruction of another gene if one of them underwent lateral gene transfer. Even the classical gene used for prokaryote phylogeny, the 16S rRNA gene, which is assumed to be less prone to lateral gene transfer than operational

genes, has caused some problems. Phylogenies based on 16S rRNA are not always consistent with those derived from another essential gene, RNA polymerase. Consequently, the phylogenetic history of a gene is not always a correct indicator of the phylogenetic history of the organisms themselves. In a much-discussed paper, Doolittle (1999) proposed a radical way out of this dilemma, which is to abolish the whole idea of a universal phylogenetic tree of life. Instead, Doolittle (1999) argued that the accepted taxonomic categories, for example Bacteria and Archaea, may be used as convenient descriptors of shared genes, but not as diagnostic indicators of common ancestry. Doolittle (1999) presented a sketch of early evolution in which the base of life is seen as a highly reticulate structure, a network of promiscous gene exchange, from which the three main domains of life finally rise (Fig. 3.14).

Despite the fact that the importance of lateral gene transfer is now well established, the argument that it is an impediment to classifying prokaryotes is not necessarily true. Snel *et al.* (1999) developed a genome-based phylogenetic approach that goes one step further than just comparing the sequences of genes. These authors considered the fraction of genes shared between genomes as a measure of distance between two species, as follows:

$$d_{AB} = 1 - \frac{n_{AB}}{n_A}$$

where d_{AB} is the distance between genomes A and B, n_{AB} is the number of genes shared (using an arbitrary threshold level for orthology), and n_A is the number of genes in the smallest genome of the two. So in this method the phylogenetic distance between two species is characterized by a single parameter, not by as many parameters as there are shared genes. The phylogeny based on a matrix of pairwise distances calculated in this way was

called a *genome phylogeny*. Snel *et al.* (1999) applied this approach to 12 fully sequenced prokaryotes and showed that the resulting tree was actually quite similar to the tree generated by the 16S rRNA gene. The authors concluded that, despite lateral gene transfer, there is still a very strong phylogenetic signature in the gene content of prokaryotic genomes. However, we should realize that a genome phylogeny captures the central trend of evolutionary history, but does not provide the complete picture (Wolf *et al.* 2002).

3.2.3 From bacteria to organelles

The haploid, circular structure of prokaryotic genomes extends to the genomes of mitochondria and chloroplasts, which have a similar arrangement but are smaller due to loss of many genes. Gray *et al.* (1999) indicated that the most probable ancestor of the mitochondrion is to be found in the order Rickettsiales of the Alphaproteobacteria. Members of this group include various obligate intracellular parasites such as *Rickettsia* and *Wolbachia*. The species *Rickettsia prowazekii*, the causative agent of a form of typhus transmitted by lice, was long considered to have the most mitochondrion-like genome. However, when more members of the group were sequenced, such as *Wolbachia pipientis*, an obligate intracellular parasite of Diptera (Wu *et al.* 2004), doubt was cast on the grouping of mitochondria within the Rickettsiales. Still, evolutionary analysis supports the hypothesis that mitochondria share a common ancestor with the Alphaproteobacteria. Also, the common view remains that the collective mitochondrial genomes are monophyletic; that is, they all originate from the same ancestor.

The most bacterium-like mitochondrial genome is found in the flagellate *Reclinomonas americana* (Sarcomastigophora, Histionida). When the mitochondrial DNA (mtDNA) of this species was sequenced (Lang *et al.* 1997) it was considered by some as a missing link between bacteria and mitochondria, because of its unusually large number of genes (97, including all the proteins found in other sequenced mtDNAs). The organelle genome database (GOBASE), coordinated by the

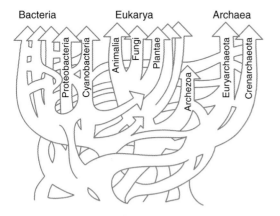

Figure 3.14 The network phylogeny of life which, according to Doolittle (1999), must replace the traditional phylogenetic tree to account for lateral gene transfer. Reprinted with permission from Doolittle (1999). Copyright 1999 AAAS.

University of Montreal in Canada (http://
megasun.bch.umontreal.ca/gobase), has now collected information on some 50 000 mitochondrial
sequences, including 429 complete organelle genomes. The mitochondrial genomes of protists
probably hold the key to the evolution of the group,
because they comprise most of the phylogenetic
diversity within the eukaryotes (Gray *et al.* 1999).

Mitochondria obey the rule that endosymbiosis
is accompanied by genome miniaturization (see
Section 3.1). In the course of evolution, most of the
genetic information for mitochondrial biogenesis
and function has moved to the nuclear genome;
the proteins needed by the mitochondrion are
synthesized in the cytoplasm and then transported
across the mitochondrial membrane. Still, the
mitochondrial genome encodes several RNAs
and proteins essential for mitochondrial function,
mostly respiratory complexes of the electron
transport chain such as NADH ubiquinone oxidoreductase, succinate ubiquinone oxidoreductase,
ubiquinol cytochrome *c* oxidoreductase, and
cytochrome *c* oxidase.

Comparisons among mitochondrial genomes are
troubled by the fact that loss of genes has occurred
many times independently. For example, genome
miniaturization in the bacterial rickettsias has
taken place independently of miniaturization in
the mitochondrial lineages of protists. To complicate things further, some plant mitochondria
contain genes that originate from chloroplasts;
this holds for two tRNA genes in the mtDNA of
Arabidopsis. The result is that the size and the gene
content of mitochondria are remarkably divergent
between the eukaryote lineages.

Gray *et al.* (1999) pointed out that mtDNAs
come in two basic types, designated as ancestral
and derived. *Ancestral mitochondrial genomes* (for
example, the one from *Reclinomonas americana*) have
retained clear vestiges of their eubacterial ancestry,
with many non-animal genes, tightly packed in a
genome with no or few introns. *Derived mitochondrial genomes* are characterized by substantial
reduction in genome size, marked divergence of
rRNA genes and adoption of biased codon-usage
patterns in protein genes. All metazoan and most
fungal mtDNAs fall into this category.

The size of mitochondrial genomes varies
between less than 6 kbp in *Plasmodium falciparum*
(the human malaria parasite, belonging to the
group Apicomplexa) to more than 200 kbp in land
plants. Gene content is similarly variable across
species. In angiosperms, the mitochondrial genome has evolved to become recombinationally
active, which has led to extensive rearrangements
of genes, breaking up bacterial gene clusters, and
loss of tRNA genes. The mitochondrial genome of
Arabidopsis is among the largest of the eukaryotes,
but it does not encode many more genes than
some of the protist mtDNAs. An overview of
the diversity of mitochondrial genome size and
content is given in Fig. 3.15.

Interestingly, mitochondrial genomes may fragment into different molecules. This is thought to be
due to recombination between repeat segments
in different parts of the genome. An example is
found in the potato cyst nematode, *Globodera
pallida* (Tylenchida, Heteroderidae), which has a
mitochondrial genome consisting of six different
circular small chromosomes each 6.3–9.5 kbp. The
12 mitochondrial genes are scattered over the six
units, but the ribosomal genes are all concentrated
on one of them (Armstrong *et al.* 2000). Even more
surprising is that the frequencies of these mitochondrial components differ between populations
of the nematode. Such small mitochondrial genomes are called *subgenomic-sized mtDNAs*; they are
also found in some green algae and higher plants.
Another peculiar situation is due to the presence
of plasmids inside mitochondria. Mitochondrial
plasmids are especially ubiquitous among filamentous fungi and some of them cause a syndrome of growth loss and early senescence when
inserted into ribosomal genes of the mitochondrial
genome (Maas *et al.* 2005).

In addition to mitochondria, the other main
eukaryotic organelle of bacterial origin is the
chloroplast. Comparative studies on bacterial and
chloroplast genomes have now demonstrated
convincingly that chloroplasts are derived from a
cyanobacterium related closely to the present
species *Nostoc punctiforme*. It is also evident that
chloroplast genomes jointly are one monophyletic
group; that is, they all descend from the same

(a)

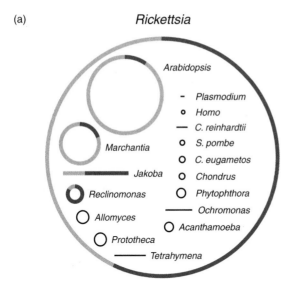

(b)

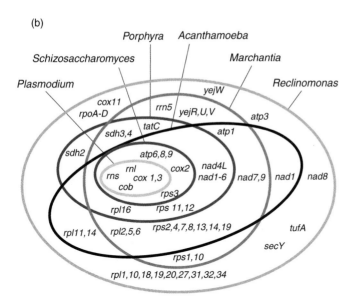

Figure 3.15 Genome size and gene content of mitochondrial DNAs across a wide range of species. (a) Circles and lines represent circular and linear chromosomes, with the ORFs of known function shown as dark lines. The major groups to which the species mentioned belong are as follows: *Rickettsia* (Alphaproteobacteria), *Arabidopsis* (angiosperm plant), *Marchantia* (liverwort), *Jakoba*, *Reclinomonas* (flagellates), *Allomyces* (fungus), *Prototheca* (green alga), *Tetrahymena* (ciliate), *Acanthamoeba* (amoeba), *Ochromonas* (golden alga), *Phytophthora* (oomycote), *Chondrus* (red alga), *Chlamydomonas eugamatos*, *Chlamydomonas reinhardtii* (green algae), *Schizosaccharomyces pombe* (yeast), *Homo* (human) and *Plasmodium* (malaria parasite). (b) Gene complement of mitochondrial genomes, showing the overlap between species. Each ellipse corresponds to one organism and includes all the mitochondrial genes of that organism. For example, the tiny mitochondrial genome of *Plasmodium* has four genes, which are also found in all other mitochondria, while all known mitochondrial genes are found in the mitochondrial genome of *Reclinomonas*. The genes are designated by code names. Reprinted with permission from Gray *et al.* (1999). Copyright 1999 AAAS.

ancestor. This is not to imply that the symbiotic event that gave rise to the chloroplast occurred only once. In fact, it is assumed that the initial inclusion of a cyanobacterium (which led to red algae, green algae, and higher plants) was followed by a second endosymbiotic event, probably involving a red alga, which produced cells in which the chloroplast was surrounded by four, rather than two, membranes. From this type of cell

three different evolutionary lineages are assumed to have originated (Tudge 2000; Raven and Allen 2003; Falkowksi *et al.* 2004):

• Brown algae (Phaeophyta), diatoms (Bacillariophyta), golden algae (Xanthophyta), and water molds (Oomycota), with the chloroplast secondarily lost in the Oomycota.

• Dinoflagellates (Dinoflagellata), Apicomplexa (*Plasmodium* and other parasites), and ciliates

(Ciliata), with a strong reduction of the chloroplast in apicomplexans and a complete loss of the chloroplast in ciliates.

• Kinetoplastids (parasites like *Trypanosoma*) and Euglenozoa, with secondary loss of the chloroplast in kinetoplastids and some euglenoids.

So, the present-day scattered distribution of photosynthetic capacity across the eukaryotes is explained not only by gains but also by losses of chloroplasts. Losses are especially prominent in the lines that obtained the chloroplast through double symbiosis. Different degrees of chloroplast reduction can be observed in the phylum Apicomplexa. These unicellular organisms are characterized by an organelle called an *apicoplast*, an assumed relict of a chloroplast, with a greatly reduced genome; however, not all species have this organelle. A recent genomic survey of the human parasite, *Cryptosporidium parvum*, which lacks an apicoplast, demonstrated the presence of 31 genes of likely cyanobacterial and other prokaryote origin in the genome, confirming the theory that apicomplexans evolved from a plastid-containing lineage (Huang *et al.* 2004).

The genome of a chloroplast typically encodes 60–200 proteins, which is more than an order of magnitude less than the genome of a cyano-bacterium, which encodes at least 1500 proteins. Genes in the preplastid were either lost or transferred to the nucleus. Studies on *Arabidopsis* nuclear and chloroplast genomes (Martin *et al.* 2002) have shown that thousands of genes have been transferred. Some 4500 genes in the nuclear genome of *Arabidopsis*, or 18% of the genes, appear to have a bacterial (chloroplast) origin. This is not to say that the products of these genes are all functional in the chloroplast; actually, more than half of the originally chloroplastic genes are now targeted to other cell compartments. To complicate things further, the protein products of many nuclear genes that were not acquired from the plastid ancestor are now targeted to the plastid! In this complicated interplay between genomes, many issues remain unresolved, including the fundamental question of why chloroplasts have a genome at all, if genes can be transferred so easily to the nucleus (Raven and Allen 2003). There must be some crucial selective advantage in retaining some genes in chloroplasts, but not others.

3.3 Eukaryotic genomes

3.3.1 Yeast and other fungi

Fungi have received a great deal of attention from genome researchers. The genomes of 14 species have been sequenced completely (Table 3.4). Our list does not include the microsporidian *Encephalitozoon cuniculi*, a human parasite that is mentioned as a fungus on several websites, but is not a proper fungus according to modern taxonomy (Tudge 2000). Not surprisingly, most of the sequencing effort has been directed towards fungi with nutritional or pathogenic importance and sequencing is concentrated in the ascomycete order Saccharomycetales, because the exploration of fungal genomes began in that group with the model species *S. cerevisiae*. The unicellular mode of life, ease of culturability, and the relatively small genome were important factors in the success of investigating baker's yeast. Probably more is known about yeast genomics, biochemistry, and physiological responses to environmental conditions than about any other species of eukaryote.

Although yeasts can be isolated from many natural habitats (fruits, leaf surfaces, decaying plants, soil), they are not common objects of study among ecologists. Spencer and Spencer (1997a, 1997b) provide a general overview of the biology of yeasts, with particular emphasis on their taxonomy and the type of natural habitat colonized by them. The designation yeast is not a proper taxon within the fungal kingdom, it just indicates that we are dealing with a unicellular fungus that can grow in colonies by budding or fission, like bacteria. Yeasts are contrasted with filamentous, mycelium-forming fungi. Although most yeasts are classified in the Ascomycota, they are also found in the Basidiomycota and Deuteromycota. *S. cerevisiae* is a yeast associated with bread preparation. The closely related *Saccharomyces* species used in beer and wine production are considered to be separate species (Spencer and Spencer 1997a).

The fungal species in Table 3.4 that is of greatest interest to ecologists is *Phanerochaete chrysosporium*,

Table 3.4 List of fungal species with completely sequenced genomes

Species	Genome size (kbp)	Growth form, habitat, importance
Ascomycota, Saccharomycetales		
Ashbya (= *Eremothecium*) *gossypii*	8743	Filamentous, pest of cotton, used in industry to produce vitamin B_2
Candida glabrata	12 281	Unicellular, causes variety of infections in humans (candidiasis)
Debaryomyces hansenii	12 221	Unicellular, abundant on fish and salted dairy products (cheese)
Kluyveromyces lactis	10 689	Unicellular, associated with various food items, biotechnologically important
Kluyveromyces (*Lachancea*) *waltii*	10 613	Occurrence similar to *K. lactis*, but evolutionarily closer to *S. cerevisiae*
Saccharomyces cerevisiae	12 069	Baker's yeast, important physiological model organism
Saccharomyces paradoxus	11 750	Sequenced to provide comparison with *S. cerevisiae*
Saccharomyces mikatae	12 120	Sequenced to provide comparison with *S. cerevisiae*
Saccharomyces bayanus	11 540	Sequenced to provide comparison with *S. cerevisiae*
Schizosaccharomyces pombe	12 534	Fission yeast, used in African millet beer
Yarrowia lipolytica	20 502	In various food media, can grow on alkanes, possibly important for biotechnology
Ascomycota, Sordaliales		
Magnaporthe grisea	40 000	Filamentous, causes blast disease in rice and other domesticated grasses
Neurospora crassa	43 000	Filamentous mould, important model for genetics and molecular biology
Basidiomycota, Aphyllophorales		
Phanerochaete chrysosporium	30 000	White rot, in fallen trees, on forest floor, degrades lignin

Sources: www.genomenewsnetwork.org, www.ebi.ac.uk/genomes/index.html, www.genomesonline.org, and Kellis *et al.* (2003, 2004).

or white rot fungus, which is found commonly on dead trees and wood fragments on the forest floor. To date, white rot fungus is the only basidomycete (mushroom-forming fungus) that has had its genome sequenced (Martinez *et al.* 2004). White rot owes its name to the fact that the fungus 'bleaches' wood by degrading the (brown-coloured) lignin, rendering the white cellulose visible. Lignin forms protective sheaths around cellulose fibrils in plant cell walls. The biodegradation of lignin by white rot is supported by unique extracellular oxidative enzymes (peroxidases and oxidases) that act non-specifically via the generation of free radicals attacking the lignin molecule. There are many other ecologically important fungi commonly found on dead organic material that contribute to processing of organic matter and cycling of nutrients in natural ecosystems (*Aspergillus*, *Trichoderma*, *Cladosporium*, *Morteriella*), but their genomes have yet to be sequenced.

The genome sizes of fungi are between those of bacteria and higher eukaryotes. *Neurospora crassa*, a mycelium-forming fungus, has one of the largest fungal genomes sequenced so far (Galagan *et al.* 2003). Comparing this genome with *S. cerevisiae*, Braun *et al.* (2000) suggested that the relatively small genome of baker's yeast is due to a process of 'streamlining' by loss of genes. Detailed comparisons of *S. cerevisiae* with the filamentous fungus *Ashbya gossypii* (Dietrich *et al.* 2004) and the unicellular *Kluyveromyces waltii* (Kellis *et al.* 2004) have confirmed that the evolution of *S. cerevisiae* has included a whole-genome duplication, followed by extensive rearrangements and loss of genes. These conclusions are in accordance with the idea that unicellularity in fungi is an apomorphic condition and that yeasts evolved independently several times from multicellular ancestors, not the other way around.

N. crassa (orange bread mould) has been a famous model organism for genetic and biochemical studies since the classical experiments of Beadle and Tatum in 1942, who proved that for every enzyme there is one gene (the *one gene/one enzyme hypothesis*). Although *Neurospora* is an ascomycete like *S. cerevisiae*, it is more similar to animals than to yeast in several ways. For example, unlike yeast but like mammals, it has a clearly discernable circadian rhythm, it methylates DNA to control gene expression, and it has complex I in the respiratory chain. With all this biochemical research, *Neurospora* is the best-characterized of the filamentous fungi, but its ecology remains relatively unexplored. Being moderately thermophilic, *Neurospora* was thought to occur mainly in moist tropical and subtropical regions, but recent surveys have also found *Neurospora* colonizing trees and shrubs killed by wildfires in temperate regions (Jacobson *et al.* 2004; Fig. 3.16). Mycological inventories showed that in North America isolates

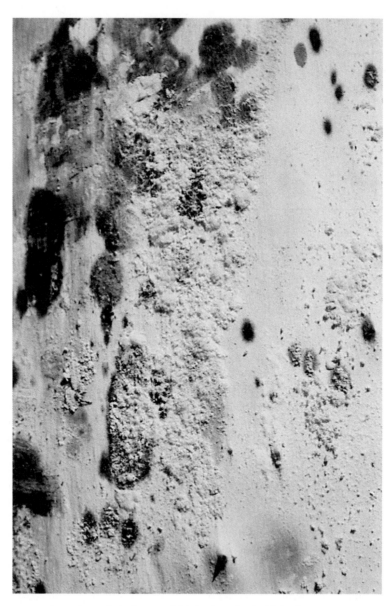

Figure 3.16 *Neurospora* growing under tree bark. Courtesy of D.J. Jacobson, Stanford University.

were comprised predominantly of a single species, *Neurospora discreta*, but in southern Europe species collected included *N. crassa*, *N. discreta*, *Neurospora sitophila*, and *Neurospora tetrasperma*.

The life cycle of *S. cerevisiae* is of the *diplobiontic* type; that is, it cycles through two distinct phases, one diploid and one haploid (Fig. 3.17). The diploid stage grows vegetatively by budding off new cells and forming colonies. Under certain conditions, such as deprivation of carbon or nitrogen, it can form stress-resistant ascospores. Exactly how *sporulation* is triggered is currently under investigation. Ascospores are of two types, called *a* and *α*. They sit together in the ascus and upon germination produce two so-called *mating types*. These haploid cells can grow vegetatively by budding, a property that provides unique opportunities to geneticists, because the expression of traits in this stage is not confounded by dominance: the phenotype is a direct result of the genotype. The two haploid mating types may interact with each other by means of hormones, which induce a characteristic change in shape, leading to

pear-shaped cells, called *shmoos* after the lovable creatures in Li'l Abner's comic strip from 1948. The process of *sexual conjugation* can occur only between opposite mating types. It involves a complicated series of cell-surface changes to facilitate fusion and is mediated by the hormones in a manner that is mating-type specific. The life cycle of *N. crassa* is similar, except that the diploid vegetative stage is suppressed and the zygote proceeds directly to form an ascus.

SGD™ is the Saccharomyces Genome Database (www.yeastgenome.org) where information about the molecular biology and genetics of baker's yeast is filed and presented to the world. The database includes a variety of search options that allow one to consult the genome sequence; analysis tools such as BLAST (see Section 2.4), programs for homology searches, and information about protein structure, as well as contact details for more than 1000 people in the yeast research community. The database also provides a list of recently published papers on all aspects of yeast molecular biology and links to databases of the other fungi listed in

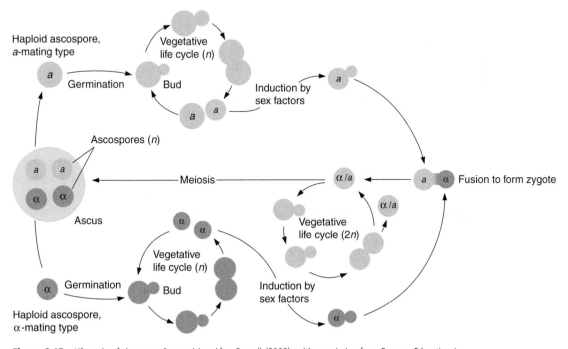

Figure 3.17 Life-cycle of the yeast *S. cerevisiae*. After Russell (2002), with permission from Pearson Education Inc.

Table 3.4. The organizational principles of the database are discussed by Dwight *et al.* (2004). Over the years the SGD has seen a dramatic increase in its usage, and has served as a template for other databases. The success of SGD, as measured by the numbers of pages viewed, user responses, and number of downloads, is due in large part to the network philosophy that has guided its mission and organization since it was established in 1993. The yeast genome was the first for which microarrays were developed (see Section 2.2). An oligonucleotide microarray (Affymetrix Yeast Genome 2.0 Array) is available that contains 10 765 probes, allowing one to address almost all genes of both *S. cerevisiae* and *Schizosaccharomyces pombe* at the same time.

A series of in-depth comparative genomic studies has recently been conducted in which the genome of *S. cerevisiae* was compared with other fungi in and outside of the order Saccharomycetales (Brachat *et al.* 2003; Cliften *et al.* 2003; Kellis *et al.* 2003, 2004; Dietrich *et al.* 2004; Dujon *et al.* 2004). An important argument for sequencing species that have a known phylogenetic relationship with *S. cerevisiae* is that it could help in the identification of genes and regulatory elements in the genome of *S. cerevisiae*. The idea is that orthologous sequences need to show a considerable degree of conservation before an ORF is considered a gene. True protein-encoding genes will typically be under selective pressure and show conservation, whereas ORFs that are not expressed will show mutations that are different for each species. When this approach is applied to regulatory elements it is called *phylogenetic footprinting*, referring to the fact that regulatory elements tend to be conserved across widely distant species and are recognizable in the genome as 'footprints' (Cliften *et al.* 2003). Comparative genomics applied to closely related rather than distant species has been called *phylogenetic shadowing*; this approach was first applied to sequences from 17 different primates as 'shadows' of the human genome (Bofelli *et al.* 2003).

As an example of a comparative genomics study in yeast, let us consider the work by Kellis *et al.* (2003). This author analysed the relationship among orthologous genes using a *reading frame conservation test*. This test classifies each ORF in *S. cerevisiae* as biologically meaningful or meaningless, depending on the proportion of the sequence over which conservation with other species is observed. Each of the other species was considered a 'voter', 'approving' or 'rejecting' the sequence in *S. cerevisiae*. Obviously, the procedure carries a risk that true genes under strong selective pressure in one of the species are rejected as biologically meaningful, but this was prevented by looking in detail at each rejection. Confidence in the method was increased when it appeared that only a few already annotated ORFs were rejected as genes. Inspection showed that in all of these possibly false-negative cases the annotated ORFs were indeed likely to be spurious. The analysis of Kellis *et al.* (2003) pruned the yeast gene catalogue of 503 genes, leaving only 20 ORFs in the database unresolved and decreasing the number of protein-encoding genes with more than 100 amino acids to 5538 (see also Table 2.1).

Further insight into the *S. cerevisae* genome has recently been obtained from comparisons with more distantly related species. Dujon *et al.* (2004) sequenced four species from the hemiascomycete group, *Candida glabrata*, *Kluyveromyces lactis*, *Debaryomyces hansenii*, and *Yarrowia lipolytica*, and compared their genomes with that of *S. cerevisae*. A total of approximately 24 200 novel genes was identified, and their translation products were classified into about 4700 families. Pairwise comparisons were made between the species to establish the degree of sequence divergence between orthologous genes. It appeared that the five yeast species together spanned a genetic diversity comparable to the entire phylum Chordata. For example, the average sequence identity between orthologous genes (translated into proteins) between *S. cerevisiae* and *C. glabrata* was 65%, between *S. cerevisiae* and *K. lactis* 60%, and between *S. cerevisiae* and *Y. lipolytica* 49%. This is less than the average sequence identity of proteins between mouse and fugu fish (70%) and comparable to that found between the urochordate sea squirt, *Ciona intestinalis*, and the mammals! The lesson of this large-scale comparative genomics study was that the evolutionary distance between yeasts, despite their very similar morphology, is extremely large.

3.3.2 Nematodes

Although molecular biologists have developed the habit (to the amusement of zoologists) of calling the nematode *C. elegans* a worm, the animal has nothing to do with the true worms, the Annelida, since it is classified in a completely separate phylum, Nematoda. Phylogenomic analysis has demonstrated that this phylum is related to the arthropods and belongs to the so-called moulting animals, the Ecdysozoa (Dopazo and Dopazo 2005); the annelids are classified with the molluscs in another superphylum, Lophotrochozoa.

C. elegans is one of the rhabditid nematodes, a group of tiny, free-living, bacteria-feeding animals, living in soils, dead organic material, or wherever there are bacteria. On the basis of rRNA gene sequences, 17 species are classified in the genus *Caenorhabditis*, including *C. briggsae*, the other nematode whose genome has been sequenced completely. Classification of nematodes is complicated by the fact that the external structure of the animals is not very diversified. The morphology of the most important diagnostic characters, the mouthparts, and other aspects of external morphology do not always fit with the molecular data and therefore the names assigned to higher-order categories in the classical taxonomy are sometimes illogical when arranging the species according to a molecular phylogeny. For example, the order Rhabditida does not indicate a monophyletic group, but appears to fall into at least two different phylogenetic lineages (see http://nematol.unh. edu/phylogeny.php, and Blaxter *et al.* 1998).

Despite their morphological uniformity, the phylum Nematoda is extremely diverse from a genetic point of view. Analysing a large collection of ESTs (>250 000) from 30 different nematode species, Parkinson *et al.* (2004) found that 30–50% of the transcriptome of each species was unique to that species. Consequently, a single nematode like *C. elegans* can reveal only a small fraction of the genomic diversity of even its own phylum. A phylogeny of 53 species of nematodes, based on small-subunit rRNA sequences, is given in Fig. 3.18 (Blaxter *et al.* 1998). The figure also provides information on feeding habits, which diverge widely within the nematodes as a whole; one can find bacteriovores (like *C. elegans*), fungivores, predators, omnivores, plant parasites, and animal parasites. Fig. 3.18 shows that there is no phylogenetic conservation of feeding habits; feeding modes are scattered throughout the tree. The great biodiversity of feeding habits, life-history patterns, and colonizing capacity makes the Nematoda a very suitable group for community bioindication. When each species is given a score on a scale of colonizers to persisters, the weighted sum of these scores for a given community (the *maturity index*) can be used as a indicator of habitat quality (Bongers and Ferris 1999).

The phylum Nematoda includes many parasites, such as the well-known intestinal roundworm of pigs, *Ascaris lumbricoides*, the small human pinworm *Enterobius vermicularis*, which infects 30–80% of schoolchildren in western countries, and various species causing serious diseases in tropical countries, such as *Onchocerca volvulus*, the causative agent of river blindness (onchocerciasis), which is spread by the bite of an infected blackfly. Experiments have shown that the inflammatory response in the human eye causing blindness, which is triggered by the presence of dying nematode microfilariae, is not only due to the worm itself, but also to toxins excreted by an endosymbiotic *Wolbachia* bacterium (Saint André *et al.* 2002). The genome of this *Wolbachia* is currently being sequenced. So the genomic studies on *C. elegans* have important ramifications for parasite research (Blaxter *et al.* 1998) and scientific networks are currently in development that address the field of nematode parasitomics: for example, the Filarial Genome Network (www.nematodes.org/fgn/ pnb/filbio.html), named after one of the parasitic species in the onchocercid group, *Filaria martis*. Although the interest in parasite genomics is exclusively medical at the moment, parasites are important agents in the population dynamics of many wild species and progress in the medical sector could well have a future spin-off to ecology. There are also several nematodes that form cysts

Figure 3.18 Phylogeny of 53 nematodes, based on sequences of the small-subunit rRNA gene. With each species an indication of the feeding habit is given (see key). The right-hand side indicates the orders of classical nematode taxonomy. After Blaxter *et al.* (1998), by permission of Nature Publishing Group.

in plant roots or otherwise damage below-ground plant tissues; however, none of these plant parasites is, as far as we know, on a list for a whole-genome sequencing project.

C. elegans is a 1 mm-long, transparant animal with *sequential hermaphoditism* and *self-fertilization*. Sperm cells are made first and stored in the spermathecae. Then the animal switches to the production of oocytes, which are fertilized by sperm from the same individual, mature partly while still in the body, and develop to the first larval stage, which emerges after the eggs have been laid. Around 10 eggs are in the body at any time, but the animal can produce more than 300 progeny during its adult lifetime, which depends on temperature (5.5 days at 15 °C, 3.5 days at 20 °C). This phenomenal reproductive capacity, within hours to days, was an important consideration when the species was chosen as a model. In addition to hermaphrodites, males sometimes occur. These males fertilize the hermaphrodites, as there are no gonochoristic females. The sex-determining system is chromosomal and the males lack one sex chromosome (hermaphrodites are XX, males are X0). The possibilities offered by this type of reproductive cycle are very convenient for genetic work, because clones can be made from hermaphroditic lines with no signs of inbreeding depression, and males can be used for cross-fertilization.

The life cycle includes four larval stages, each separated by a moult (Fig. 3.19). The development of gonads and the production of sperm are already taking place during the larval stage. In addition to the four normal larvae there is a resting stage, called the *dauer larva* (after the German word for endure), which is actually an arrested third larval stage, in which the animal goes into a state of torpor, and does not eat, although it can move slightly and may live for several months. The dauer larval stage is induced by adverse conditions such as crowding and food scarcety. When terminated, the dauer stage proceeds to the fourth larval stage. It is assumed that the dauer larva is the nematode's dispersal stage. Several features point towards an increased propensity to be transported, either by wind or by other animals. The dauer dries itself out, secretes an extra cucticle

and develops a behaviour known as nictation (winking); it tends to crawl up objects that protrude from the surface, stands on its tail and waves its head back and forth (Riddle 1988). An important aspect of the dauer is its extreme longevity, which may reach several months, rather than the normal 20 days. The dauer stage of *C. elegans* is an important model for investigating the genomic basis of longevity (see Chapter 5).

A very peculiar property of *C. elegans* is that it has a completely determinate developmental pattern, which is fixed for all 959 cells of the body. This was the reason why the animal was initially chosen as a model for developmental studies by Sydney Brenner at the beginning of the 1960s (Brown 2004). Brenner was inspired by earlier German work on the nervous system of the intestinal parasitic nematode *Ascaris suum*, which had shown that the fate and location of each cell could be traced, and was the same in all individuals. The original *C. elegans* strain on which the research in Cambridge was started by Brenner came from the laboratory of Ellsworth Dougherty in Berkeley, who had cultured *C. elegans* for several years. It is assumed that the culture actually originated from mushroom compost collected near Bristol, UK (Fitch and Thomas 1997).

C. elegans is a cosmopolitan species. More than 20 different strains have been isolated from soils of North America, Europe, and Australia (Fitch and Thomas 1997). Despite this broad distribution, the species is not a popular object of study among field ecologists, because it is very difficult to distinguish from other species in the same group and its distribution seems to be restricted to synanthropic habitats such as compost heaps and manure. For example, despite the fact that a good identification key of more than 600 species is available for the nematodes of the Netherlands, issued from an active university department specializing in nematology over many years (Bongers 1988), wild *C. elegans* have never been found in the Netherlands.

The complete genome sequence of *C. elegans* was the first to be published for a multicellular organism (the *C. elegans* Sequencing Consortium 1998). The WormBase consortium has continued to edit the sequence, brought the estimated error rate

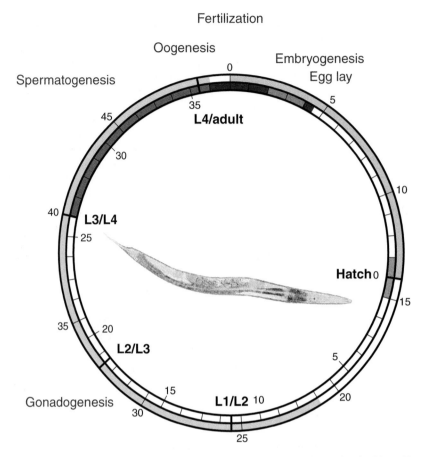

Figure 3.19 Life cycle (from egg to adult) of *C. elegans*, when cultured in the laboratory with abundant food (*E. coli*) at 25 °C. The outer scale is marked in hours since fertilization, the inner scale in hours since hatching. L1, L2, L3, and L4 are the first to fourth larval stages. The adult can live for several days more. After Wood *et al.* (1980), with permission from Elsevier.

down to 1 in 100 000, and closed the last gap in November 2002. This makes the *C. elegans* sequence the first and so far only metazoan genome database that has reached sequencing closure for all of the chromosomes. The interface on the World Wide Web (www.wormbase.org), described by Harris *et al.* (2004), offers a rich source of information, not only on the complete genome sequence but also on mutant phenotypes, genetic markers, developmental lineages of the worm, and bibliographic resources, including paper abstracts and author contact information. The genome sequence of the related species, *C. briggsae*, is now completely integrated into WormBase, which allows comparative analysis of orthologues

and synteny. WormBase also contains extensive information from large-scale genome analyses, microarray expression studies, and the assignment of gene ontology terms to gene products. New data releases are published regularly and from time to time a 'freeze' of the software and the database is deposited, which can be downloaded. For transcription profiling, commercial microarrays are available, such as the *C. elegans* whole-genome GeneChip® array (Affymetrix), which targets 22 500 transcripts.

The *C. elegans* sequence was announced in 1998 as a 'platform for investigating biology'. The consortium realized that the importance of the genome sequence of *C. elegans* extended beyond

Table 3.5 General features of the genome of *C. elegans* (*C. elegans* Sequencing Consortium 1998)

Category	Features
Protein-encoding genes	Many genes (25%) organized in cistronic units (Section 3.2); three times more genes than in yeast (see Table 3.1); more genes than was estimated from genetic studies
RNA genes	Many tRNAs on the X chromosome; several RNA genes in introns of protein-encoding genes; rRNA genes in long tandem arrays
Gene density	Uniform GC content (see Fig. 3.7); fairly constant gene density across chromosomes
Repetitive DNA	Tandem repeats account for 2.7% of the genome; inverted repeats account for 3.6% of the genome; repeat sequences overrepresented in introns; 38 different families of dispersed repetitive sequences associated with transposition; dispersed repetitive sequences abundant on the arms of the chromosomes

nematodes proper, and in fact could be considered the basic formula for constructing a multicellular animal, in the same way that the sequence of the *S. cerevisiae* genome contains all the information for making and maintaining a unicellular eukaryote. In addition, because nematodes branched off early in the evolutionary tree of life, the *C. elegans* sequence provides an outgroup for almost all other bilaterian animals (from Annelida to Chordata). For example, if a gene is identified in both *C. elegans* and a mollusc, it must also have been present in the ancestor of the Bilateria (Hodgkin *et al.* 1995). An overview of the specific features of the *C. elegans* genome is given in Table 3.5.

At the time of publication of the *C. elegans* genome sequence, information was available on only a few other eukaryote genomes. *C. elegans* could be compared with yeast and bacteria such as *E. coli*, as well as to the then-available gene content of the human genome. It turned out that 36% of the predicted *C. elegans* genes had a human homologue and that no less than 74% of the human genes had a homologue in *C. elegans* (Fig. 3.20). The similarity of *C. elegans* to *Homo sapiens* was greater than that to yeast or bacteria. This comparison demonstrated for the first time the striking unity that underlies the genomes of organisms as different as nematode and human. This tendency was reconfirmed many times when more eukaryotic genome sequences became available.

3.3.3 *Drosophila* and other arthropods

In terms of numbers of species, the arthropods as a group surpass any other phylum in the animal

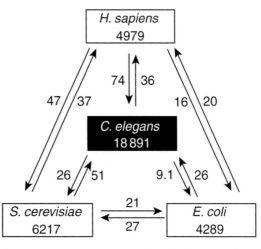

Figure 3.20 Pairwise comparison of predicted proteins in four species. The numbers adjacent to the arrows indicate the percentage of proteins in an organism that has a match in the organism indicated by the arrow. The numbers in the boxes indicate the actual number of proteins compared. Reprinted with permission from The *C. elegans* Sequencing Consortium (1998). Copyright 1998 AAAS.

kingdom. Of some 1.25 million known animal species, about four-fifths of these are arthropods; the class Insecta by itself represents almost three-quarters of all described animal species. Considering the small size of most members, it is probable that as many species again remain undescribed. However, the vast biodiversity of arthropods is not matched by proportional investment in genome sequencing (Heckel 2003). At the time of writing (June 2005), the genomes of just four species of arthropod have been sequenced completely; the fruit fly *D. melanogaster* (Adams

Table 3.6 List of arthropod species, other than *D. melanogaster*, *An. gambiae*, *B. mori*, and *A. mellifera* for which whole-genome sequencing projects have been initiated

Taxonomic group	Arguments, relevance
Insecta, Diptera	
Drosophila pseudobscura and other *Drosophila* species	Comparison with *D. melanogaster*, mechanisms of radiative speciation, phylogenetic shadowing
Aedes aegypti (yellow fever mosquito)	Disease transmission, comparative basis for haematophagy, together with *An. gambiae*, *C. pipiens*, and other mosquitoes
Culex pipiens (house mosquito)	Vector for Western Nile Virus, model for blood-feeding mosquitoes
Glossina morsitans (tsetse fly)	Transmitter of *Trypanosoma*, cause of sleeping sickness
Insecta, Hymenoptera	
Apis mellifera, other strains	Comparison with *A. mellifera* DH4, genetic basis of social behaviour
Insecta, Coleoptera	
Tribolium castaneum (red flour beetle)	Model organism for genetics and developmental biology
Insecta, Lepidoptera	
Heliothis virescens (tobacco budworm)	Larva attacks various crops such as alfalfa, clover, cotton, soybean, and tobacco
Acari, Ixodidae	
Amblyomma americanum (lone star tick)	Vector for transmission of bacteria causing spotted fever and rabbit fever in humans
Crustacea, Cladocera	
Daphnia pulex (water flea)	Important model species in aquatic ecology, widely accepted test organism in water-quality assessment

et al. 2000), the malaria mosquito *An. gambiae* (Holt *et al.* 2002), the silk worm *Bombyx mori* (Biology Analysis Group 2004), and the honey bee, *Apis mellifera* (sequence assembly Amel 2.0 released January 2005; see www.hgsc.bcm.tmc.edu/ projects/honeybee). Several other species are in the pipeline (Table 3.6).

As is evident from Table 3.6, many arthropods are considered for sequencing because of their medical (disease transmission) or agronomical (pest) importance. The species that raise the greatest expectations among ecologists are the water fleas *Daphnia pulex* and *Daphnia magna*, because water fleas (Crustacea, Cladocera) are extremely important organisms in aquatic food chains. On the one hand, they have a top-down effect by grazing and controlling algal populations, and on the other hand they have a bottom-up effect as food for fish. Numerous limnologists have studied population dynamics, life history, energy metabolism, and vertical migration of daphnids for nearly three centuries. The reproductive biology of *Daphnia* is very suitable for genomic studies because it

can reproduce by apomictic parthenogenesis under favourable conditions but also by sexual mechanisms under unfavourable conditions. The sexually produced resting stage is a saddle-shaped structure called an *ephippium*, which contains diapausing eggs; indeed, centuries-old *Daphnia* can be resurrected from ephippia recovered from natural sediments. Due to the combination of sexual reproduction and clonal propagation in localized populations, water fleas are differentiated in many ecotypes, showing diverging rates of genetic variation, depending on the temporal stability of their habitat and the rate of outbreeding (De Meester 1996). *D. magna* is an internationally accepted standard test organism in ecotoxicology and water-quality assessment in general.

The Daphnia Genomics Consortium (DGC) coordinates the release of the genome sequence (http://daphnia.cgb.indiana.edu) and a great variety of other genomic tools, such as nearly 2000 microsatellite markers (many linked to gene loci), a fine-scale genetic map, and no less than 50 000 cDNAs, expressed in a great variety of environments and developmental stages. A database, wFleaBase (http://wfleabase.org), provides the infrastructure to share genomics data and protocols via the World Wide Web. Techniques for cell culture and genetic transformation are in development. The genome sequencing effort is focused on *D. pulex*, but a considerable number of ESTs is also available for *D. magna*; microarrays with over 12 000 gene fragments have been developed for transcription-profiling studies in stress ecology (www.ams.rdg.ac.uk/zoology/daphnia). It is expected that, among animal models, *Daphnia* is one of the most promising species for realizing a true blend of ecology and genomics.

It is worthwhile probing the base of the phylum Arthropoda when evolutionary arguments are being advanced for future sequencing projects. How do insects relate to myriapods, arachnids, and crustaceans? This debate has occupied entomologists for a very long time and a convincing answer has not yet been given (Gillot 1980). Nardi *et al.* (2003) sequenced the mitochondrial genomes of two representatives of the class Collembola, a group of wingless hexapods, often called primitive insects and formerly classified as a subclass of the Insecta, the Apterygota. The sequences were compared with the mitochondrial genomes of a great number of insects and crustaceans and a phylogenetic tree produced (Fig. 3.21). The tree suggests strongly that the Collembola separated from the main line of the arthropods before the split between the crustaceans and insects. This would imply that the hexapod body plan, shown by insects as well as Collembola, evolved twice. Obviously, more genomic sequencing of representatives from the primitive hexapods, crustaceans, and myriapods is necessary to resolve this question.

Within the Diptera (the two-winged flies), *D. melanogaster* belongs to the family Drosophilidae, or fruit flies, which is one of the largest families of the Acalypterae, a group of flies belonging to the suborder Cyclorapha. Almost 3000 species of drosophilids in over 60 genera are described worldwide. Members of the family are found in association with fermenting substances, most notably fruit, but also in specialized habitats such as the sap of bleeding tree wounds, slime fluxes, rotting cacti, and flower heads as well as in general habitats such as on fungi and decaying leaves on forest floors. The adults (3–5 mm) are attracted by the alcoholic odours emanating from the activities of bacteria and yeasts colonizing a substrate rich in sugars. Eggs are driven into the substrate by means of the female's ovipositor and the maggot-like larvae emerging from the eggs develop inside the substrate through three stages followed by a pupa; at 25 °C new adults emerge within 10 days of hatching. *D. melanogaster* is believed to have evolved in Africa, but is now found worldwide in many synanthropic habitats, such as fruit markets and garbage cans. The classical booklet by Demerec and Kaufmann (1950) is still a valuable source of information about the biology of *Drosophila*.

The publication of the sequence of the *Drosophila melanogaster* genome by Adams *et al.* (2000) was a landmark achievement because it marked the end of nine decades of *Drosophila* research, starting with T.H. Morgan's discovery in 1910 of a mutant white-eyed fly and leading to major conceptual or technical breakthroughs in our understanding of animal genetics over the course of the century (Rubin and Lewis 2000). The *Drosophila* genome was the second and largest animal genome sequenced at the time and the short period (less than 1 year) in which the work was completed, as a combined academic and industry effort, was impressive. Interestingly, the superstar status that the fruit fly already enjoyed before the sequencing began was actually somewhat of an impediment to starting the project (Rubin and Lewis 2000). Over 1300 individual genes, nearly 10% of all the genes in *Drosophila*, had already been cloned and studied, and with such success a whole-genome sequence was considered unnecessary. Nevertheless, it was realized that the majority of genes eluded study by traditional methods because in *Drosophila*, as in *C. elegans* and *Arabidopsis*, many genes do not show obvious phenotypes when mutated.

The *Drosophila* genome-sequencing project was the first that had to deal with a substantial amount of heterochromatin, one-third (60 Mbp) of the genome. Attention was concentrated on the 120 Mbp of euchromatin, because the heterochromatic regions were intractable to the sequencing

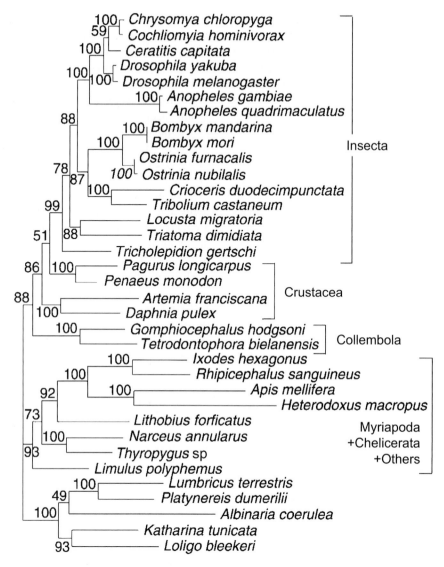

Figure 3.21 Phylogenetic tree of complete mitochondrial sequences of representative arthropod species. The position of two Collembola species (*Gomphiocephalus hodgsoni* and *Tetrodontophora bielanensis*), branching off before the Crustacea, suggests that the hexapod body plan arose twice in the evolution of the arthropods. Reprinted with permission from Nardi *et al.* (2003). Copyright 2003 AAAS.

method. Heterochromatin was present in the centromeres of the two autosomes, one half of the X chromosome, the complete Y chromosome, and the very small fourth chromosome. The heterochromatic, gene-poor, regions consisted primarily of simple sequence satellites and transposons. Interestingly, the transition zones between heterochromatin and euchromatin

regions appeared to contain many previously unknown genes.

The genome of *Drosophila* was found to contain around 13 600 genes; taking alternative splicing into account, 14 113 transcripts were postulated (Adams *et al.* 2000). Orthologues of these transcripts were sought in the databases of nematode, yeast, and mouse sequences to classify genes according to the gene ontology system. Rubin *et al.* (2000) compared the fruit fly genes with those of *C. elegans* and *S. cerevisiae*, the only two other eukaryote genomes sequenced at the time. A summary of these comparisons is shown in Fig. 3.22. Around 35% of the protein-encoding genes had a match in the nematode genome and 16.5% in the yeast genome. These relatively low values, which by themselves are in accordance with the phylogenetic relationships between the three species, showed that the genome of *Drosophila* was only remotely related to the other eukaryotes. In fact, further comparisons including a mammalian EST database demonstrated that half of the fly proteins showed similarity with mammalian proteins, which suggested that the *Drosophila* proteome is more similar to mammalian proteomes than are those of nematodes or yeast.

The basis for comparative genomics of arthropods was strengthened considerably when the genome of the malaria mosquito *An. gambiae* (Diptera, Nematocera, Culicidae) was published (Holt *et al.* 2002). This sequencing project posed another challenge that had not been encountered in any sequencing exercise so far, namely the high degree of genetic polymorphism in the clone library used. *An. gambiae* populations are highly structured into several genetic types, with different habitat preferences, but indistinguishable morphologies. The heterogeneous genetic background caused difficulties in genome assembly, because haplotype variation was difficult to distinguish from repeat sequences. The starting stock for the sequencing was a long-term culture maintained at the Institut Pasteur in Paris, which was established from a cross between a strain originating in Nigeria with a strain from Kenya, followed by three rounds of outbreeding. It was believed that the strain mostly represented the *Savanna* form of *An. gambiae*, presently found in

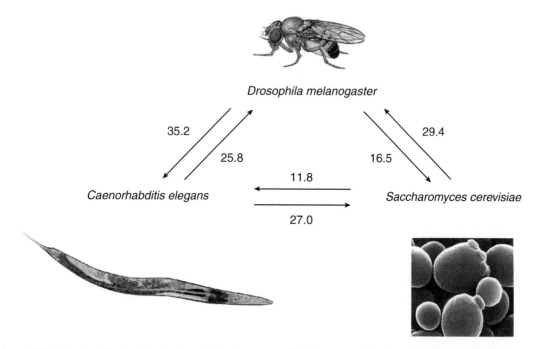

Figure 3.22 Pairwise similarities of predicted proteins in the genomes of fruit fly, nematode, and yeast. The numbers adjacent to the arrows indicate the percentage of proteins in an organism that has a match in the organism indicated by the arrow (at a BLAST *E* value of 10^{-10} or smaller), relative to the number of proteins in the first species. Each set of pairs was analysed without consideration of the third proteome. Data from Rubin *et al.* (2000).

western Kenya. The great degree of polymorphism was not found in earlier projects; the genomes of *Drosophila* or mouse were virtually entirely homozygous. Because the whole-genome shotgun sequencing approach is based on repeatedly sampling the genome, such that on average every region is covered a number of times (see Section 2.2), estimates of polymorphism could be made from the sequencing. Single-base-pair discrepancies between sequence reads were denoted as *single-nucleotide discrepancies*, but in regions where the assembly was supported strongly such variation was considered to be due to SNPs. Variability was not distributed uniformly throughout the genome; it was much higher in the automosomes than in the X chromosome. It was assumed that the genome of the strain used had resulted from a complex introgression between two different chromosomal forms (cytotypes); lack of introgression in the X chromosome (as a consequence of the hemizygous condition of the male) could explain the lower degree of polymorphism in this chromosome.

Compared with *Drosophila*, the genome of *An. gambiae* demonstrates several features that relate to its haematophagous mode of nutrition. Several gene families were found to be greatly expanded; that is, they appeared to consist of many more, and more diversified, members than the orthologous *Drosophila* family (Zdobnov *et al.* 2002). A prime example of gene-family expansion is in genes containing a fibrinogen domain (FBN genes; Christophides *et al.* 2002). The FBN genes of invertebrates are involved in the *innate immune response*. The innate or constitutive immune system is the system upon which all invertebrates rely and to which a newly evolved adaptive system was added later in the vertebrate lineage (see Section 6.4). FBN genes combat intruders by binding to foreign surfaces. The expansion of these genes in *Anopheles* might reflect selection pressure from growth of bacteria in the gut after a blood meal or from defence against *Plasmodium* and other parasites. Other genes of particular importance to *An. gambiae* are those related to antioxidant defence, which is a challenge to animals feeding on vertebrate blood, because the blood-meal-derived

haem group is a catalyst of free oxygen radicals. In summary, the genome of *Anopheles* is twice as large as the *Drosophila* genome, but this is not only due to expansion of specific gene families, but also to loss of non-coding DNA in *Drosophila*. The fraction of the genome which consists of introns and intergenic spacers is considerably larger in *Anopheles* (Table 3.7). The comparative analysis of Zdobnov *et al.* (2002) also showed that several genes known from yeasts, nematodes, and plants are absent from the two insect genomes. Examples are nine genes related to sterol metabolism, which is in accordance with the fact, noted by entomologists a long time ago, that insects cannot synthesize their own sterols. Table 3.7 summarizes some characteristic features of the genomes of *Anopheles* and *Drosophila*.

The genome sequence of *D. melanogaster* is curated by the FlyBase consortium (www.flybase.org). Along with other responsibilities, FlyBase is committed to maintaining up-to-date annotations of the genome (The FlyBase Consortium 2003). Although the genome sequence of the euchromatin is now essentially complete, predictions of transcripts and proteins will be ongoing for some time. A special challenge is to connect the genome annotations to the vast amount of information on phenotypic characters available for *Drosophila*. Since the original release, reannotations have changed the contents of the genetic database considerably, due to division of genes into two genes, merging previously separate genes, and addition or deletion of exons. FlyBase coordinates

Table 3.7 General features of the genomes of two insects, *D. melanogaster* (fruit fly) and *An. gambiae* (malaria mosquito)

	Anopheles	Drosophila
Total genome size	278 Mbp	123 Mbp
Total coding DNA	19.3 Mbp (7%)	23.8 Mbp (19%)
Total intron DNA	43.0 Mbp (15%)	27.6 Mbp (22%)
Total intergenic DNA	216.0 Mbp (78%)	71.3 (58%)
Number of genes	13 683	13 472
Number of exons	50 609	54 537
Genomic GC content	35.2%	41.1%

Sources: From Holt *et al.* (2002) and Zdobnov *et al.* (2002).

not only the sequence of *D. melanogaster* but also that of *Drosophila pseudoobscura*, and it will cover those of other Drosophilidae when these become available in the future. The web pages of FlyBase provide an enormous amount of information on genetic maps, cytological maps, genes, alleles, gene products, protein function, protein location, gene expression, transposons, transgene constructs, fruit fly stocks, collections, fly anatomy, literature, references, and fruit fly investigators. For transcription profiling the GeneChip® Drosophila Genome 2.0 Array from Affymetrix is often used. This microarray provides comprehensive coverage of the transcribed *Drosophila* genome using 18 880 probe sets, analyzing over 18 500 transcripts.

AnoBase is the *An. gambiae* genomic and biological database (www.anobase.org). The site also contains reference material and links to other mosquito resources, as well as current news and conference information. Anobase works in collaboration with Genoscope, the French Government's sequencing centre, who along with Celera Genomics were involved in the initial sequencing of the mosquito genome, and several organizations that specialize in research on tropical diseases, such as The Malaria Research and Reference Reagent Resource Center (MR4). Information on the *Aedes aegypti* sequencing project can be found on the mosquito genomics server through http://mosquito.colostate.edu/tikiwiki/tiki-index.php.

3.3.4 Plant genomes

The plant kingdom is a monophyletic evolutionary lineage, including green algae, mosses, ferns, and seed plants. Genome sequencing up to now has been completed for one representative from the green algae, *Chlamydomonas reinhardtii*, and three angiosperms, *A. thaliana* (thale cress), *Oryza sativa* (rice, two varieties), and *Populus trichocarpa* (black cottonwood). Many sequencing projects are being planned and genomic databases, for example collections of ESTs, are being developed worldwide. Researchers often organize themselves around a group of related plant species; for example the Legume Genomics network focuses on *Medicago truncatula* as a model, the Solanaceae Genomics

Network concerns tomato and potato, collaborators in TreeGenes are interested in forest genetics and the genome sequencing of forest trees, the Multinational Brassica Genome Project addresses the various *Brassica* species, SoyBase focuses on the soybean *Glycine max*, and BeanGenes addresses *Phaseolus* and *Vigna* species, among others.

As noted in Section 2.2 the very large genome sizes of some agriculturally important plants are a serious obstacle for full-genome sequencing (Table 3.8). This is especially valid for species from the family Poaceae (grasses), which includes the cereals (Triticaceae). Still, it is expected that even the genomes of these plants—despite their very large size—will be sequenced eventually. For some plant species commercial microarrays for gene-expression analysis are already available, even though a complete genome assembly and annotation has not yet been conducted. In these cases the microarrays were developed from publicly accessible databases (for example GenBank) complemented by EST sequences submitted by consortium members. Examples are the Affymetrix GeneChip® Soybean Genome Array, which can be used to study gene expression of over 37 500 soybean transcripts, and the Affymetrix GeneChip®

Table 3.8 Genome sizes of some agriculturally important crops, in comparison with *Arabidopsis*

Species	Scientific name	Genome size (Mbp)
Cabbages		
Thale cress	*Arabidopsis thaliana*	125
Oilseed rape	*Brassica napus*	1200
Cereals		
Rice	*Oryza sativa*	420
Barley	*Hordeum vulgare*	4800
Wheat	*Triticum aestivum*	16 000
Corn	*Zea mays*	2500
Legumes		
Garden pea	*Pisum sativum*	4100
Soybean	*Glycine max*	1100
Nightshades		
Potato	*Solanum tuberosum*	1800
Tomato	*Lycopersicon esculentum*	1000

Source: From Adam (2000). Reproduced by permission of Nature Publishing Group.

Barley Genome Array that was designed in collaboration with the international barley community.

Sequencing the genome of the green alga, *C. reinhardtii* (Chlorophyta, Volvocales), was inspired by the fact that this small unicellular organism served as a classical model for biochemical research into photosynthesis and cell motility (Harris 2001; Grossman *et al.* 2003; Gutman and Niyogi 2004). A key feature of its success was the availability of a mutation in a photosynthetic regulatory mechanism called the state transition. The state transition involves allocation of light-harvesting proteins from photosystem II to photosystem I in response to the wavelength of incident light. Because this shift is larger and more easily measured in *Chlamydomonas* than in higher plants the alga was a preferred organism for eukaryotic photosynthesis research. *Chlamydomonas* is also a very suitable organism for the study of experimental evolution. It can reproduce both sexually and asexually, and this property has been exploited by Kaltz and Bell (2002) to demonstrate that sexual, genetically diverse, populations were better able to adapt to a new hostile environment than equivalent asexual lines. From a comparative genomics point of view the genome of *C. reinhardtii* is interesting because together with *Arabidopsis* it forms a pair bracketing the entire lineage of green plants. *Chlamydomonas* species are also studied in freshwater algology and are commonly observed in plankton samples under the microscope as tiny, rapidly swimming flagellates. Thus a connection between the growing genomics insights and ecological studies seems very well possible, although a practical difficulty is the very large diversity in the genus *Chlamydomonas*, which has no less than 600 species worldwide. The Chlamy Center, found at www.chlamy.org, provides entry to a genome browser, an EST database, and various microarray projects, as well as a system for ordering specific strains. Another green alga, *Ostreococcus tauri* (Chlorophyta, Mamiellales) is also being considered for whole-genome sequencing. This species has a very densely packed genome of only 11.5 Mbp and may provide insight into the minimal number of genes necessary for a photosynthetic eukaryote.

The legume *M. truncatula* will probably be the next species of higher plant with a fully sequenced genome (www.medicago.org). *Medicago* is seen as a model for legumes in general, a family of plants that includes well-known genera such as *Lathyrus*, *Phaseolus*, *Glycine*, *Trifolium*, *Medicago*, *Pisum*, *Vicia*, and *Lotus*, among which are many agriculturally important species that provide a major source of proteins for the human population. Under nitrogen-limiting conditions, leguminous plants are able to establish a symbiotic relationship with bacteria from the family Rhizobiaceae, including *S. meliloti*, in itself a genomic model species (see Section 3.2). Symbiotic nitrogen fixation is a very important link in the global nitrogen cycle and so genomic studies of the legume–rhizobium relationship have great added value in comparison to *Arabidopsis*, which does not offer this possibility. In addition, *Medicago* can also be used to study endomycorrhizal symbiosis. *M. truncatula* is a close relative of alfalfa (*M. sativa*), but unlike the latter species, it has an annual life cycle and only half the genome size of the alfalfa genome. It exhibits simple genetics and a genome highly conserved with alfalfa and pea (*Phaseolus vulgaris*) and moderately conserved with soybean. The progress is supported by a dense physical map of nearly 200 000 ESTs, and a growing array of functional genomic tools. A consortium of laboratories in the USA and Europe will fully sequence the gene space of *M. truncatula* by 2006.

Populus trichocarpa (family Salicaceae) was the first tree that was considered for whole-genome sequencing. Forest biologists have developed strong justifications for why trees should be viewed as model systems in plant genomics (Bradshaw *et al.* 2000; Wullschleger *et al.* 2002). The physiology of trees includes a number of processes that cannot be understood from the model herbaceous plants, such as perennial growth, large size and complex crown structure, extensive secondary xylem, and bud dormancy (Taylor 2002; Brunner et al. 2004). Knowledge of the poplar genome will greatly contribute to the growth of a research area known as *forest genomics*, which focuses on questions related to mechanisms of wood formation, stress resistance, pathogen resistance, genetic diversity of trees, conservation, and tree breeding. Poplar is an important model species for forest genomics, along with spruce and to a lesser extent pine. With its genome consisting of 480 Mbp, more than four times larger than *Arabidopsis*, but still some 40 times smaller than pine, poplar has a relatively small genome among trees. The sequencing was done by the International Populus Genome Consortium (IPGC) in which US, Canadian, and Swedish scientists collaborated (www.ornl.gov/sci/ipgc). A recent series of papers on stress responses and developmental processes in poplar is testimony to the true beginning of genome-sequence-based poplar genomics (Strauss and Martin 2004).

The importance of poplar as a genomic model extends beyond the aims of forestry; poplar is also a very suitable species for fundamental ecological genomics, as illustrated by the following example. Cottonwoods (*Populus trichocarpa*) are central to the structure and ecosystem functioning of riparian forests of North American rivers, whereas black poplar (*Populus nigra*) fulfils the same role in European riparian systems. Ecological interactions in such ecosystems involve engineering by beavers (North American *Castor canadensis* and European *Castor fiber*), herbivory by the arthropod community on the leaves (caterpillars, beetles, aphids, etc.), and decomposition of fallen leaves by a diverse community of microorganisms, invertebrate shredders, and detritivores. Interestingly, the principal steering force in these interactions appears to emanate from the genetics of the tree. A number of hybrids from the genus *Populus* occur naturally and hybridization is associated with marked variations in leaf form and chemical composition of the leaves. Wimp *et al.* (2004) showed that the genetic variation of a cottonwood stand, estimated by AFLP fingerprints (see Section 2.1) was strongly correlated with the Shannon–Weaver index of the leaf arthropod community: genetically diverse stands had the highest species richness. The effect is most probably mediated by the chemical composition of the leaves, particularly the concentration of tannins. In the same system, leaf tannin was negatively correlated with decomposition and nitrogen mineralization of

poplar litter (Schweitzer *et al.* 2004). The fact that genetic diversity of a dominant species in an ecosystem pervades into the herbivorous and decomposing communities has important implications for conservation strategies. In addition, the system offers exciting possibilities for establishing a direct relationship between ecosystem functions and gene-expression profiles in the tannin-synthesis pathway.

Despite the promises of *Medicago* and *Populus*, at the moment *A. thaliana* is the best-characterized genomic model plant by far. Thale cress is a species of the Brassicaceae family, with a wide distribution in the morthern hemisphere (Fig. 3.23). The species is native to western Eurasia but is now found in the wild throughout Europe, the Mediterranean, the East African highlands, and eastern and central Asia (Hoffmann 2002). It has also been introduced into America and Australia (Fig. 3.23). Johannes Thal (hence, *thaliana*) first described *Arabidopsis* in the sixteenth century in the German Harz Mountains, although he called it *Pilosella siliquosa* at the time. The name underwent several changes before *A. thaliana* was settled upon in 1842.

A. thaliana is a small annual plant, 5–30 cm high, growing in open areas with sandy soil, along paths, and in agricultural fields. The life cycle is that of a winter annual, which germinates and grows in autumn, survives winter as a rosette, and flowers in early spring. The developmental switch from vegetative growth to reproduction, involving erection of the flower stem (*bolting*) is an important issue of research in ecological genomics (see Chapter 5). In fact the life history is more flexible than a typical winter annual, because there are also variants that germinate in spring and flower in July, and others that have more or less lost their phenological tuning (see Sections 2.1 and 5.3, and Fig. 2.7). *A. thaliana* has a high level of self-pollination. It does not cross-hybridize with its relatives, because the number of chromosomes is reduced to five, whereas all its closest relatives have a haploid chromosome number of eight. AFLP fingerprinting has shown that the population structure of *A. thaliana* over its native geographical range is shaped by postglacial

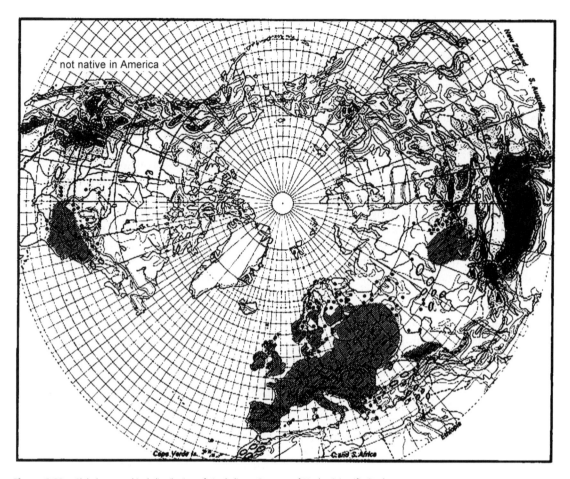

not native in America

Figure 3.23 Global geographical distribution of *A. thaliana*. Courtesy of Koeltz Scientific Books.

colonization from the Iberian peninsula and the near east, leading to a suture zone in central Europe (Sharbel *et al.* 2000).

A. thaliana is never a dominant species in wild vegetation. Its suitability as a model for ecological field work is limited by its sparse occurrence and its narrow phenological window. Molecular ecologists have been looking for related species that lend themselves better to ecological studies (Mitchell-Olds 2001). There are 10 species of *Arabidopsis* among the approximately 3000 species of Brassicaceae. *Arabidopsis halleri* and *Arabidopsis lyrata* are closely related species with a perennial life cycle. Genera within the same clade are *Cardamine, Rorippa, Barbarea, Arabis,* and *Thlaspi.*

Taken together these cruciferan species comprise a wide array of life cycles and ecological niches. Several of these species seem to be suitable as 'wild' counterparts of *A. thaliana*, and four of them are pictured in Fig. 3.24.

Scientific research on *A. thaliana* started in the beginning of the twentieth century with microscopic studies on the chromosomes, but it was not until the 1970s that molecular geneticists discovered its suitability as a model. A genetic map was developed by Maarten Koorneef and co-workers at the beginning of the 1980s and physical maps of the genome, based on RFLP and AFLP fingerprints, were developed thereafter. The possibilities of transforming *Arabidopsis*, first

Figure 3.24 Wild relatives of *A. thaliana*: (a) *Arabidopsis petraea*, (b) *Arabis alpina*, (c) *Boechera holboelii*, and (d) *Thlaspi caerulescens*. Courtesy of T. Mitchell-Olds, Max Planck Institute, Jena and C. Lefèbvre, Free University of Brussels.

using *Agrobacterium* and later more advanced genetic-engineering techniques, were also developed; extensive mutant collections were built concurrently. The Multinational Coordinated Arabidopsis thaliana Genome Research Project was launched in 1990 and the Arabidopsis Genome Initiative (AGI) started sequencing the genome in 1996 on a chromosome-by-chromosome basis. Chromosomes 2 and 4 were completed first and published in 1999. With the completion of chromosomes 1, 3, and 5 in 2000 the genome sequence was essentially complete and this was

considered a hallmark event for plant biology (Walbot 2000).

The first detailed analysis of the *Arabidopsis* genome content provided many surprises (Arabidopsis Genome Initiative 2000). From the total genome of 125 Mbp, 115.4 Mbp had been fully sequenced. This sequence appeared to contain 25 498 predicted genes, significantly more than *C. elegans* and *Drosophila* (see Table 3.2). When the protein-coding genes were compared with the genomes of other eukaryotes and prokaryotes, many matches were found (Fig. 3.25). It even

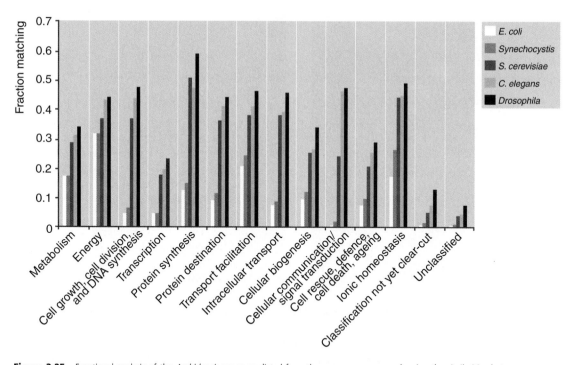

Figure 3.25 Functional analysis of the *Arabidopsis* genes predicted from the genome sequence, showing the similarities between *Arabidopsis* functional gene categories and bacterial genomes (*E. coli* and *Synechocystis*, a cyanobacterium) and those of yeast, nematode, and fruit fly. The *y* axis indicates the fraction of *Arabidopsis* genes in a functional category showing a BLAST match with the respective reference genome. The right to use this figure provided courtesy of members of the Arabidopsis Genome Initiative and Nature magazine. This figure first appeared in Nature 408, 796–815 (2000).

turned out that in the *Arabidopsis* genome genes may be found that share clear homology with human disease genes! However, the percentage of genes matching those of other species depended greatly on the functional category. Among the genes related to transcription only a small percentage (8–23%) had a match in another species, whereas among the genes related to protein synthesis up to 60% corresponded to a gene in another species. Overall, the similarity with prokaryote genomes was significantly less than with eukaryotes, but in the functional category of energy metabolism more than 30% of the plant genes were similar to a bacterial gene. This is obviously a consequence of the transfer of chloroplast genes to the nuclear genome. Maybe less surprisingly, in the category of cellular communication and signal transduction hardly any match was found between *Arabidopsis* genes and those of bacteria, but the correspondence with the

(unicellular) yeast genome was also relatively low (Fig. 3.25).

Why has the *Arabidopsis* genome 87% more genes than *Drosophila melanogaster*? Two explanations have been given (The Arabidopsis Genome Initiative 2000). First, individual genes have been subjected to wide-scale amplification events, generating large arrays of tandems and dispersed gene families; unequal crossing-over may be the predominant mechanism involved. Second, the genome of *A. thaliana* has undergone a whole-genome duplication after it diverged from most other dicotyledons (Bowers *et al.* 2003), classifying *A. thaliana* as a cryptotetraploid species (see Section 3.1). These two genome-enlargement mechanisms have led to a considerable degree of *genetic redundancy* in the genome; that is, more than one gene has the same function. This is consistent with observations from genetic engineering studies which show that many genes can be knocked-out

in *Arabidopsis* without any phenoypic consequences. The Arabidopsis Genome Initiative
speculated that such large-scale duplication events
may be needed to generate new functions, and that
creating new functions by duplication is more
common in plants than in animals, where novelties
are more often generated by rearrangements of
promoters and alternative splicing.

The possibility of ancient polyploidy in model
plants was analysed in more detail by Blanc and
Wolfe (2004), using whole-genome data and EST
sequences for 14 different species. The authors

estimated the sequence divergence between the
two genes of a paralogous pair by looking at the
average number of substitutions without amino
acid alteration (number of substitutions per synonymous site, K_s; see Section 3.1). The frequency
distribution of K_s values over all the genes is a cue
to the timing of the duplication process (Fig. 3.26).
Arabidopsis obviously has a peak in the frequency
distribution around a K_s value of 0.8, which is
indicative of synchronized duplication of many
genes together. The most likely explanation for
synchrony is a polyploidization of the whole

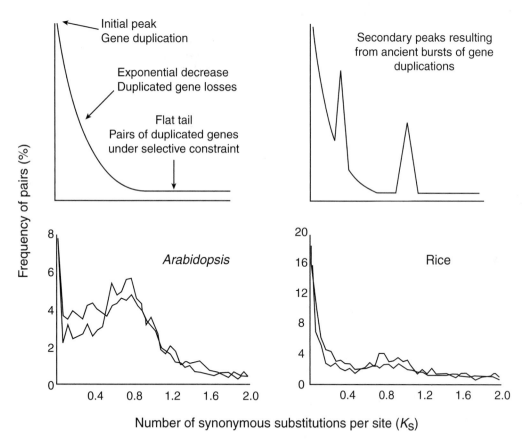

Figure 3.26 Top: theoretical age distributions of pairs of duplicated genes in a genome. The general decrease of the curve indicates that fewer and fewer genes remain as recognizable duplicate pairs with increasing time since duplication (measured by the number of substitutions per synonymous site, K_s). Peaks in the curve are indicative of 'cohorts' of synchronous duplications. Bottom: distribution of K_s values (a measure of divergence) of paralogous gene pairs in *A. thaliana* (left) and *O. sativa* (right). Distributions are shown for genomic gene sequences and for partial, sequenced cDNAs (ESTs). These two approaches result in practically the same pattern. The peak in the *Arabidopsis* curve around $K_s = 0.7–0.8$ is indicative of an ancient polyploidy event. In the rice genome the distribution conforms mostly to the theoretical prediction of the top-left panel. After Blanc and Wolfe (2004). Copyright American Society of Plant Biologists.

genome, dated around 25–26.7 million years ago. This was followed by extensive rearrangements and an accelerated loss of genes, with the consequence that the *Arabidopsis* genome is now relatively small among plant genomes (Table 3.8) and constitutes a complicated mosaic of duplicated genes. In rice, the distribution of K_s values is much more similar to the theoretical expectation following from a continuous process of individual duplications; however, there is a small elevation in the distribution, which according to Blanc and Wolfe (2004) is indicative of a partial chromosomal duplication dated at 70 million years (Fig. 3.26).

Rice was the second higher plant species with a completely sequenced genome. In fact, two different projects were conducted, one by Syngenta focusing on the *japonica* subspecies (Goff *et al.* 2002), and one by the Beijing Genomics Institute, focusing on the most widely cultivated subspecies in China, *O. sativa indica* (Yu *et al.* 2002). The *indica* genome was 466 Mbp in size, with the number of genes estimated to be between 46 022 and 55 615; the *japonica* data were similar. Again these counts show that the number of genes in plants can be much higher than in animals. Rice and *Arabidopsis* belong to two different lineages of angiosperm plants, the monocotyledons and dicotyledons, which diverged around 200 million years ago; however, despite this ancient evolutionary divergence, there appears to be a considerable degree of homology between individual genes. Goff *et al.* (2002) estimated that 85% of the *Arabidopsis* predicted proteins had a homologue in the rice genome and that 31% of the proteins shared between *Arabidopsis* and rice were not found in fruit fly, nematode, or yeast. Almost all genes related to disease resistance in *Arabidopsis* are also found in rice. These data show that the defence against pathogens is a very basic element of plant biology and is highly conserved between dicotyledons and monocotyledons.

Despite the large number of orthologues shared between *Arabidopsis* and rice, the degree of synteny between these two species is very limited. There is, however, a great deal of genome synteny (colinearity) between the species of the tribus Triticaceae, which in addition to rice includes

wheat, barley, rye, and some wild plants of the genus *Aegilops* (goatgrass). Analysing the genetic maps of the Triticaceae, Devos and Gale (2000) showed that only two chromosomal rearrangements need to be assumed to achieve colinearity between the genome of *Aegilops tauschii* (Tausch's goatgrass) and *Hordeum vulgare* (barley), whereas seven rearrangements can explain the relationship between *Ae. tauschii* and rye (*Secale cereale*). Similar syntenic relationships hold for the family Poaceae in general. So the sequence of the relatively small rice genome allows identification of chromosomal segments in other species. However, on a smaller scale (microsynteny), numerous discontinuities in gene order between wheat and rice were identified by Sorrells *et al.* (2003), so the use of rice as a model for cross-species gene isolation in other Triticaceae could prove to be limited.

The website for the Arabidopsis Information Resource (TAIR; http://arabidopsis.org) allows researchers to search for genes, proteins, alleles, markers, etc., and provides various analysis tools, such as sequence viewers, map viewers, BLAST protocols, and microarray analysis. There are also a great number of links, for example to the Arabidopsis Biological Resource Center, which has thousands of stocks in the form of clones or seeds, which are shipped around the world. The website includes a search engine for publications on *Arabidopsis* genomics in the widest sense. A frequently used platform for transcription profiling in *Arabidopsis* is the Affymetrix Arabidopsis Genome Array ATH1, which has probes for 24 000 genes. Specialized software packages for surveying and mining gene-expression data generated with the Affymetrix gene chips have also been developed (Zimmerman *et al.* 2004).

3.3.5 The deuterostome lineage

Genomes of higher animals are discussed jointly here with reference to the subkingdom Deuterostomia, which includes the phyla Pterobranchia, Echinodermata, Hemichordata, and Chordata. Nielsen (1995) also includes Phoronida and Branchipoda in the Deuterostomia, although most zoologists rank these two phyla with the protostome Lophotrochozoa group. Within the deuterostomes, genome sequencing is greatly biased towards mammals (see Table 1.1); however, in this section we will pay most attention to the genomes of the basal members of the Chordata. Two species of echinoderm, the California purple sea urchin, *Strongylocentrotus purpuratus*, and the green sea urchin, *Lytechinus variegatus*, are presently being sequenced, but an analysis of their genomes has not yet been made, so we will limit ourselves to the chordates.

An interesting view of the origin of chordates and vertebrates is obtained from the genome of a sea squirt, *Ci. intestinalis* (Dehal *et al.* 2002; Cañestro *et al.* 2003). This animal belongs to the chordate subphylum Urochordata, also called the Tunicata, after the tunic, a tough fibrous cover, excreted from the skin, in which the animal is contained. The Urochordata comprise three classes, one of them being the Ascidiacea or sea squirts. As an adult, *Ci. intestinalis* is sessile and attached to an underwater substrate where it filters food particles by pumping water through its elaborate pharynx, a basket-like structure, which fills most of the tunic. The name squirt is due to the regular pulses of water driven out of the exhalent siphon (Fig. 3.27a). Although aside from the gill slits in the pharynx no obvious characters indicate that the animal is closely related to vertebrates, the organization of the free-swimming ascidian larvae differs greatly from the adult and reveals it chordate body plan. In fact the larva of a sea squirt looks very much like a jawless fish, and is equipped with a chorda and a dorsal nerve cord, externally resembling a tadpole (Fig. 3.27b).

Ci. intestinalis is a solitary, small, and relatively short-lived marine animal that colonizes solid substrates in the sublittoral zone, such as protected rocky shores, ship wrecks, and buoys. Due to its rapid colonizing capacity, it is sometimes a conspicous and abundant representative of the 'fouling' community. With their large filtration capacity, the animals act as filters and so contribute to purification of coastal waters, although by the same mechanism they accumulate chemicals and are used for biomonitoring of coastal sea pollution. Ecological work on *Ciona* and other tunicates aims at answering questions about

(a)

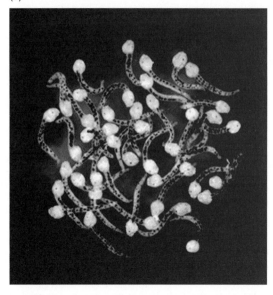

(b)

Figure 3.27 (a) Adult sea squirts. (b) A group of larvae. David Keys (photo) and Leila Hornick (artistic rendering), courtesy of the U.S. Department of Energy Joint Genome Institute. © 2005 The Regents of the University of California.

settlement in relation to density and intraspecific competition. Local populations seem to be highly dynamic and are characterized by cyclic retreat and recolonization events. Because of this type of population dynamics, ecologists are interested in geographical population genetic structure; microsatellite markers have been developed to support such analyses (Procaccini *et al.* 2000). The recently generated genomic information on

Ciona has, however, not yet penetrated into ecological studies.

The genomes of tunicates are considerably smaller than those of vertebrates, and *Ciona*'s genome measures about 160 Mbp (Dehal *et al.* 2002). The gene content represents an interesting blend between ancient protostome signatures and chordate innovations, with some tunicate autapomorphisms added. Dehal *et al.* (2002) found a total of 15 852 protein-encoding genes and these were compared with the gene complements of *Drosophila*, *C. elegans*, puffer fish, and mammals. It turned out that 60% of the genes shared homology with fruit flies and nematodes, so these represent the core physiological and developmental machinery that is common to all bilaterian animals. A few hundred of these genes even have a stronger similarity to fruit fly or nematode than to any vertebrate, and so these genes represent functions that were present in the invertebrates, but were lost in the vertebrate lineage. Examples are chitin synthase (there is no chitin exoskeleton in chordates), phytochelatin synthase (the role of the zinc-binding molecule phytochelatin was taken over by metallothionein), and haemocyanin (the copper-containing blood pigment of arthropods and bivalves, absent from vertebrates). Another 16% of the genes lacked a homologue in the protostome groups, but had a clear vertebrate counterpart. These genes apparently have arisen on the deuterostome branch before the split between tunicates and vertebrates. Then another 21% of the genes had no clear homologue in fruit fly, nematode, fish, or mammal and represent tunicate-specific genes.

Interestingly, *Ciona*'s genome has genes related to the synthesis and degradation of cellulose (cellulose synthase and several endoglucanases), genes that are never found in animals, only in plants, nitrogen-fixing bacteria, and bacteria living endosymbiotically with termites and wood-feeding cockroaches. The presence of these genes is related to the composition of the tunic, which is built largely of a cellulose-like carbohydrate called tunicin. How *Ciona* obtained these genes (a dramatic example of lateral gene transfer?) remains a mystery, but obviously it has been a very significant event in the evolution of this group (Matthysse *et al.* 2004). *Ciona*'s genome has all the genes related to the innate immune system, as in *Anopheles* and *Drosophila*, but genes implicated in adaptive immunity could not be found. This suggests that the adaptive immune system is an apomorphy of the vertebrates, not of the chordates as a whole. Ascidians are also known for their extremely high body concentration of vanadium, several orders of magnitude higher than any other animal. Vanadium is accumulated in specialized blood cells, vanadocytes, where it is localized in intracellular vacuoles, together with a similarly high concentration of sulphate. Three vanadium-binding proteins, *vanabins*, have been characterized recently in *Ascidia sydneiensis samea* (Ueki *et al.* 2003) and five vanabins are encoded in the genome of *Ci. intestinalis* (Trivedi *et al.* 2003). However, a genome-wide analysis of the peculiar vanadium metabolism of ascidians has not yet been conducted.

Turning our attention from urochordates to vertebrates, we note that three species of fish presently serve as genomic models, *Takifugu rubripes*, *Tetraodon nigroviridis* (both puffer fish, family Tetraodontidae), and *Danio rerio* (zebrafish, family Cyprinidae). *T. rubripes* (also known as *Fugu rubripes*) was proposed in 1993 by Sydney Brenner as a genomic model because with its small genome (470 Mbp) it would allow a cost-effective way of illuminating the human genome. In the far east, the fish is not only known for its small genome, but also for containing the extremely toxic compound tetrodotoxin, which, with an oral LD_{50} to mammals of 15 µg per kg of body weight is one of the most potent toxins known. Japanese men practice the habit of eating 'fugu' fish in restaurants that have obtained a special licence allowing the cook to separate the flesh from the hypertoxic liver and ovaria. The International Fugu Genome Consortium was formed in the year 2000, coordinated by the Institute of Molecular and Cell Biology in Singapore, in collaboration with groups in the UK and the USA (www.fugu-sg.org). The sequence was released less than 2 years later (Aparicio *et al.* 2002).

Because the *Takifugu* genome assembly remained highly fragmented, another team, coordinated by the French sequencing centre Genoscope, started on a related puffer fish, *Te. nigroviridis*. This species has an even smaller genome and it offered the additional advantage of being a popular aquarium fish, easily maintained in tap water. The name puffer is derived from the fish's habit of inflating itself when it is threathened. The analysis of the genome, published by Jaillon *et al.* (2004), revealed several interesting trends about gene duplications in the actinopterygian fish lineage (ray-finned fish, as opposed to lobe-finned fish, the Sarcopterygii, such as lung fish and coelocanths). The genome of *Tetraodon* measured 342 Mbp and had 28 918 putative protein-encoding genes, 1.8 times more than in *Ciona*, but somewhat less than in *Takifugu* (31 059). The slightly smaller genome size was ascribed to the absence of transposable elements, which are rather abundant in fugu fish. Careful analysis of

the content of each of the 21 *Tetraodon* chromosomes allowed reconstruction of a duplication event in the actinopterygian lineage (Fig. 3.28). Assuming that the original number of chromosomes of the ancestral gnathostome (jawed fish) was 12, a duplication event, followed by 10 different chromosomal rearrangements (fusions and translocations), can explain the present organization of the 21 chromosomes. The duplication is assumed to have taken place later in the evolution of the Actinopterygii, close to the origin of the Teleostei (modern bony fish), because some early-branching actinopterygian fish (bichirs, Polypteriformes) do not have the duplication. Similar conclusions were reached by Christoffels *et al.* (2004) in an analysis of the fugu genome.

The model of Jaillon *et al.* (2004) is consistent with an earlier analysis of the vertebrate *Hox* genes by Amores *et al.* (1998). These authors had

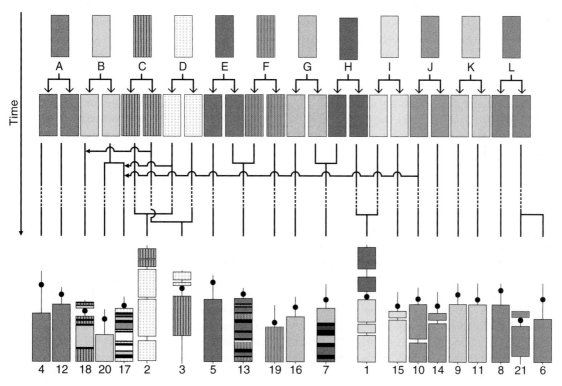

Figure 3.28 Model, inspired by detailed genomic analysis of the puffer fish *Te. nigroviridis*, showing how the present 21 chromosomes of teleost fish can be derived from an ancestral gnathostome karyotype with 12 chromosomes, by assuming a whole-genome duplication event followed by 10 different rearrangements of chromosomal segments (fusions and translocations). After Jaillon *et al.* (2004), by permission of Nature Publishing Group.

sequenced all 50 *Hox* genes of zebrafish and analysed paralogous and orthologous homologies across zebrafish, fugu fish, and mouse. The pattern of *Hox*-gene clustering could be explained by assuming that the ancestor of the Gnathostomata lineage had four clusters of *Hox* genes, each cluster consisting in principle of 13 genes. This system was continued in the Tetrapoda (amphibians, reptiles, birds, and mammals), but several losses led to a total of no more than 40 genes in the mouse, still arranged in four clusters. In the Actinopterygii lineage a duplication, identical to the one discussed by Jaillon *et al.* (2004), was assumed to have taken place, leading to eight clusters, followed by the loss of one cluster and several individual genes, leading to 50 *Hox* genes in zebrafish, arranged in seven clusters. How the ancestral gnathostome acquired its complement of four clusters, through two rounds of whole genome duplication (the 2R hypothesis), or through two local duplications, is not yet resolved (see the discussion in Section 3.1). The duplications of the *Hox* genes, both in the early evolution of the chordates and in the actinopterygian lineage, may have spurred innovation of the body plan and subsequent radiation of the highly successful vertebrate groups (Venkatesh 2003).

Although the two puffer fish had a head start as genomic models because of their unusually small genomes, the ultimate fish model is the zebrafish, *D. rerio*. This species has many experimental advantages, including ease of culture, a transparent embryo, and ample possibilities for manipulations such as cell labelling, transplantation, microinjection, and mutagenesis. Genome analysis of zebrafish started with the production of extensive genetic maps, to accelerate the molecular localization of mutations, and to allow comparisons of genome location with other vertebrates (Woods *et al.* 2000). The Zebrafish Information Network (ZFIN) now serves as the zebrafish model organism database. The design of the network (http://zfin.org) is described by Sprague *et al.* (2001); it aims to maintain the definitive reference data-sets of zebrafish research information, and to facilitate the use of zebrafish as a model for human biology. The zebrafish sequencing project

is conducted by the Sanger Institute and is expected to be completed by the end of 2005 (www.sanger.ac.uk).

To a certain extent, genomic information on zebrafish and fugu can be extrapolated to other fish species as long as genes are conserved. This cannot be stretched too far, probably not beyond the family Cyprinidae for the zebrafish. Ecologists working on other species will need to avail themselves of genomic sequences for their own model, but the community of fish biologists is rather fragmented (Clark *et al.* 2003). EST databases and microarrays are being developed for a considerable number of species including carp (*Cyprinus carpio*), large-mouth bass (*Micropterus salmoides*), fathead minnow (*Pimephales promelas*), killifish (*Fundulus heteroclitus*), and long-jawed mudsucker (*Gillichtys mirabilis*).

The three-spined stickleback, *Gasterosteus aculeatus* (Gasterosteiformes), seems to be the most promising ecological partner to zebrafish. The three-spined stickleback is an originally marine, anadromous fish species, but populations have permanently colonized a variety of freshwater habitats. Being reproductively isolated from each other, these populations have developed into a wide range of subspecies with different morphologies, habitat preferences, behaviours, and life cycles. For example, in lakes in British Columbia two ecotypes are present, one specializing in benthic habitats, with a larger body-size, reduced spines and armour plates, and fewer gill rakers, the other a pelagic form that is more streamlined and has larger eyes and well-developed spines, gill rakers, and armour plates. Despite reproductive isolation in the field the two forms can be crossed by artificial means in the laboratory. Peichel *et al.* (2001) developed a genetic map from such crosses after genotyping the animals with 438 microsatellite loci. QTLs (see Section 2.1) were identified for biometric characters of spines, armour plates, and rakers, and these loci were mapped into the linkage groups defined by the microsatellite markers (Fig. 3.29). The example shows a remarkable genetic flexibility for independent modification of the size and number of different structures related to feeding and armour. A full-length genome

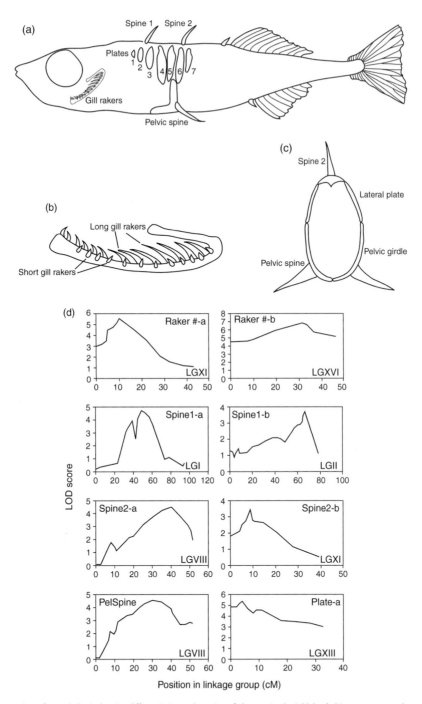

Figure 3.29 Mapping of morphological traits differentiating subspecies of three-spined stickleback (*Gasterosteus aculeatus*). (a, b, c) Schematic drawings of fish indicating the biometric characters investigated. (d) LOD (logarithm-of-the-odds, log-likelihood ratio) scores for various characters as a function of the position (in centiMorgans, cM) in the linkage group. The peaks in the LOD scores show that there are two QTLs for the number of short gill rakers, one in linkage group (LG) XI (Raker #-a) and one in linkage group XVI (Raker #-b), two QTLs for length of spine 1, two for length of spine 2, one for pelvic spine length, and one for lateral late size. From Peichel *et al.* (2001), by permission of Nature Publishing Group.

analysis will aid greatly in the further identification of these QTLs. The three-spined stickleback is a fascinating system for exploring the genomic basis of adaptive radiation and parallel evolution (Foster and Baker 2004).

The genomes of amphibians, reptiles, and birds are hardly explored at the moment, despite the fact that these animals are among the most popular ecological study objects. A genome sequence of the frog, *Xenopus tropicalis*, is presently being assembled (http://genome.jgi-psf.org), and is due for completion in 2005. A gene-expression tool is available for the African clawed frog, *Xenopus laevis* (Affymetrix GeneChip® Xenopus laevis Genome Array), which can be used to study gene expression of over 14 400 *X. laevis* transcripts. A community of *Xenopus* investigators has developed an information resource, Xenbase (www.xenbase.org), where data on the sequence are offered along with genomic tools, as well as an archive of basic biological information about clawed frogs, including animations showing anatomical features and developmental patterns. Sequencing *Xenopus* was mostly inspired by the eminent possibilities that the animal offers for studies into early embryonic development and cell biology;

however, it may also have some relevance to ecological studies in herpetology.

Regarding birds, the red jungle fowl (*Gallus gallus*) genome was recently assembled (International Chicken Genome Sequencing Consortium 2004; www.ncbi.nlm.nih.gov/genome/guide). This species, native to southeast Asia, is the ancestor of the various domestic breeds of chicken. The chicken genome will be an important resource for poultry science and applied avian studies, but its relevance to ecological studies of other birds is doubtful. As illustrated by a recent study (Postma and Van Noordwijk 2005), a tremendously rich system for investigating questions of population structure, life history, and behaviour is offered by species such as the great tit (*Parus major*) and so sequencing such species would be a true breakthrough for ecological genomics of birds.

This completes our overview of prokaryotic and eukaryotic model genomes and the promises that they hold for ecology. The field of comparative genomics is rapidly growing and we believe that many discoveries are still in store. Comparative genomics provides an indispensable evolutionary foundation for the still teneral state of ecological genomics.

CHAPTER 4

Structure and function in communities

In this chapter we will address one of the most fundamental issues of ecology: the relationship between ecosystem processes and species richness in communities, or as Lawton (1994) put it, 'What do species do in ecosystems?' Ecological genomics opens new avenues to explore this question. We will review scientific evidence concerning genome diversity in the environment and the function of genomes in nutrient cycles. Because microorganisms are in a key position at many crucial links of nutrient cycles, most of this chapter will deal with the ecological genomics of microorganisms.

4.1 The biodiversity and ecosystem functioning synthetic framework

A summary of the ecological framework that forms the background to this chapter is given by Naeem *et al.* (2002). At the beginning of the 1990s ecologists reformulated a question that had already existed for a long time in ecology; namely, what is the relation between structure and function in ecosystems? Structure includes all quantities that can still be observed in a snapshot of the system, at a particular moment in time. This includes things like species richness, biomass, dominance structure, and feeding groups. The functional aspects include the processes that cannot be observed in a snapshot, but need to be monitored in time, such as primary production, respiration, degradation of organic matter, and nitrification. With the staggering loss of biodiversity that we observe today, the question may be asked, how will ecological functions respond? Conversely, if we are interested in protecting functions, is it possible to achieve this aim through protecting the structure? An answer to these questions requires a scientific underpinning of the ecological importance of biodiversity.

Since 1993, when a group of scientists congregated in Bayreuth and the seed for the 'biodiversity and ecosystem functioning synthetic framework' was planted (Schulze and Mooney 1993), the role of biodiversity in maintaining ecological functions has been subject to intense theoretical and experimental analysis. These developments were also spurred by the Rio Convention on Biological Diversity held in 1992, followed by the spreading realization that global biodiversity is under serious threat. The issues were addressed by theoretical models, food-web analysis, microcosm experiments, and field-plot investigations.

In general it is assumed that there is an asymmetrical relationship between structure and function; that is, protection of functions does not require protection of all structures, whereas on the other hand protection of structures will always guarantee protection of functions. However, it is still a matter of debate what kind of form this asymmetrical relationship should take. Several alternative hypotheses have been formulated that differ from each other in the extent to which a decrease in the number of species endangers an important function of the system. The hypotheses are discussed in terms of graphs in which some ecosystem process is plotted as a function of the number of species in the ecosystem (Fig. 4.1). The argument is, what happens to ecosystem function (plotted on the vertical axis) when biodiversity

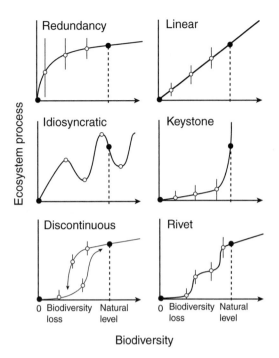

Figure 4.1 Graphical representation of six different hypotheses about the relationship between biodiversity and ecological processes. The central idea is that, commencing with the natural level of biodiversity and moving in the direction of a decrease (to the left), there are different ways in which the function of a system can change; in every case it ends at a zero level when no biodiversity is left. From Naeem *et al.* (2002), with permission from Oxford University Press.

(plotted on the horizontal axis) decreases or increases? Although Fig. 4.1 pictures six different relationships, other authors have distinguished no less than 50 different hypotheses, which can, however, be categorized in three main classes (Lawton 1994; Naeem *et al.* 2002), as follows.

The redundant species hypothesis. With a decrease in biodiversity, ecosystem functions are unaffected up to the point where only a small number of key species remains; if one of these species is removed, the system collapses. The idea is that many species in the ecosystem are redundant in the sense that they are at least partly substitutable while their contribution to the ecosystem process can be taken over by other, functionally similar, species

The rivet hypothesis. With a decrease in biodiversity, ecosystem functions decrease proportionally

(linearly or in steps). The idea is that every species makes a (smaller or larger) contribution to the process, so if it is removed, that contribution is subtracted from the process.

The idiosyncratic hypothesis. There is no universal relationship between structure and function; rather, the relationship is ecosystem-specific. In one case there may be a strong reduction of function with a loss of biodiversity, in another case there may hardly be an effect or maybe even an increase.

The general feeling among ecologists is that *functional redundancy* indeed plays an important role in many ecosystems. The argument is supported by observations on systems in which members of the community are suppressed by toxicants. In soil ecology it is well known that respiration is considerably less sensitive to the effects of toxic substances than nitrification, which is attributed to the fact that all heterotrophic organisms contribute to respiration, while only a few bacterial genera are responsible for nitrification (Domsch 1984). In studies of heavy-metal contamination in forest ecosystems, it has been demonstrated that in a gradient of pollution around a metal-smelting works a considerable loss of species of fungi can occur, whereas soil respiration is hardly affected and decreases only at the very high levels of pollution close to the source (Nordgren *et al.* 1983).

Still, after more than a decade of ecological research the central question of the synthetic framework cannot be answered in a simple way. It is still very difficult to refute any one of the six hypotheses of Fig. 4.1. In very general terms, one may conclude that a minimal number of species is necessary to allow a system to function, and a larger number of species is necessary to guarantee stability of the processes in a changing environment (Loreau *et al.* 2001). In addition, two points have emerged showing that analysis of the problem may benefit from narrowing down the scope of the question.

First, it appears that the way in which biodiversity influences ecological processes depends on the way in which these processes are limited. This issue is of evident importance in aquatic ecosystems, in which primary production can either be limited by the substrate (nutrient loading, such as phosphorus or carbon), or by the capacity of the organisms to process that substrate (the biomass and the number of producer species). In *capacity-limited systems* the substrate is supplied rapidly enough so that every functional unit is saturated and the rate of throughput is insensitive to changes in the rate of substrate supply. An example is nitrogen fixation in cyanobacteria, which is usually not limited by the availability of nitrogen gas but by the biomass of fixating organisms, which in turn are limited by other factors (e.g. zooplankton grazing, phosphate). A capacity-limited process is sensitive to a loss of biodiversity, because any reduction in the number of functional units will decrease the process rate. In a *substrate-limited system*, the capacities of the functional units are not fully deployed and if such a system loses biodiversity the overall throughput may remain unchanged because each functional unit can easily increase its share in the process (Levine 1989; Van Straalen 2002). Substrate limitation may be less important in terrestrial ecosystems and even less in below-ground systems, because of the abundance of dead organic matter as a food source for the decomposer community. However, even soil communities may be limited, for example by nutrient imbalance (Pokarzhevskii *et al.* 2003) and by microhabitat heterogeneity causing spatial dislocation between the food and the hungry.

Second, there is an increasing awareness that biodiversity as such is not as important as biodiversity in relation to the properties of the species. That is, to evaluate the effects of diminishing species richness on ecosystem processes we must look at the biodiversity of species attributes in a community, not only at species numbers. As an example, consider the work by Walker *et al.* (1999), who investigated vegetation structure in Australian savannahs. The authors noted that dominant plant species in the same community tend to be positioned apart from each other when classified according to species-specific attributes, such as height, biomass, specific leaf area, longevity, and leaf-litter quality. Rare species may contribute to resilience of the vegetation because they often have attributes similar to the dominant species and may act as a functional substitute. Another way to phrase the issue is the *principle of complementarity*: the stability of the system benefits if species complement each other in their function. The argument extends to communities of decomposer invertebrates. In microcosm experiments with earthworms, isopods, and millipedes Heemsbergen *et al.* (2004) demonstrated that the effect of detritivore invertebrates on soil respiration and litter breakdown depended not on the species composition per se, but on the *functional dissimilarity* within that community. The suggestion was that positive interactions in the community cause a functionally dissimilar assembly to have a larger effect on soil processes than a functionally similar assembly, independent of the number of species.

The biodiversity and ecosystem functioning synthetic framework has not yet been probed using the genomics approach. Yet there is a lot of mechanistic knowledge, especially in microbiology, about the actors behind biogeochemical cycles. This knowledge has increased tremendously with the large-scale sequencing and transcription profiling of microbial genomes. In the sections below we explore a possible link between the two fields of investigation; community ecology and microbial functional genomics.

4.2 Measurement of microbial biodiversity

To estimate the number of species that are present in a specific habitat is more difficult than it may seem. The situation is aptly described by the following phrase from the classical book by Charles Elton (Elton 1927), the founder of animal ecology:

Two boys of rather good powers of observation were sent into a wood in summer to discover as many animals as they could, returned after half an hour and reported that they had seen two birds, several spiders, and some flies—that was all. When asked how many species of all kinds of animals they thought there might be in the wood one replied after a little hesitation 'a hundred', while the other said 'twenty'. Actually there were probably over ten thousand.

Now it is obvious that if Elton's boys had been asked to include microorganisms, their estimate would have been even more inaccurate. For obvious reasons, estimating the biodiversity of microorganisms is more difficult than estimating species richness of plants or animals. In addition, microbiologists struggle with an even more fundamental question; that is, it is often not clear what constitutes a microbial species.

According to classical bacteriological taxonomy, an isolate is recognized as a proper species if its morphology is described plus some key aspects of its metabolism (trophic system, substrate use, etc.). Two isolates are considered to belong to the same species if their DNAs are similar by more than 70% or if there is less than a 5°C difference in the melting temperature of a DNA–DNA hybridization (Wayne *et al.* 1987). Obviously, species that cannot be put into pure culture cannot be characterized in this way. It is estimated that anything between 50 and 99% of microorganisms may belong to this group of *unculturables* and these remain undescribed as species, although parts of their genome may be sequenced from the environment. Why so many organisms cannot be cultured in the laboratory is unclear and probably there are many reasons, including specific growth conditions, unknown nutrient requirements, very slow growth, and special surfaces to which cells must attach. Recently, microbiologists have discovered that some 'uncultivable' bacteria can be brought into culture when placed in close proximity to other species, from which they are separated only by a membrane; apparently, chemical signals from other members of the community are sometimes crucial to induce growth (Kaeberlein *et al.* 2002). We will see later in this chapter that genomics approaches provide another solution to the problem: the DNA of species in the environment can be assembled and its functions characterized without even attempting to put them into a culture tube.

Microorganisms have been given little attention in ecological studies until recently. The last decade has produced a new awareness of microbial diversity and the suitability of microorganisms to address questions of fundamental ecological importance (Øvreås 2000; Horner-Devine *et al.* 2004; Kassen and Rainey 2004; Jessup *et al.* 2004). Microorganisms have been reported from extreme habitats in which they are the only type of organism surviving, such as hot springs, deep ocean vents, volcanic crater lakes, and sediments under permanent ice cover. Such extreme habitats hold many surprises in store. For example, a completely new phylum of Archaea, the Nanoarchaeota, was discovered in a hot submarine vent north of Iceland and a new division of Euryarchaeota was found in a hypersaline anoxic basin in the Mediterranean Sea (Huber *et al.* 2002; Van der Wielen *et al.* 2005). The development of universal phylogenetic trees on the basis of genes that are common to all life forms has demonstrated that the biodiversity of the Bacteria and Archaea is at least as large as that of the whole domain of the Eukarya (Fig. 4.2).

4.2.1 Diversity of small-subunit rRNA genes

In the so-called *polyphasic taxonomy* of current microbiology a species is differentiated on both genetic and phenotypic grounds. The genetic characterization is derived from the sequence of the small-subunit rRNA gene. From basic biochemistry we know that the size of ribosomes may be characterized by Svedberg units (S), a measure of sedimentation velocity during ultracentrifugation (1 S corresponds to 10^{-13} s). The prokaryotic ribosome measures 70 S and is made up of a *small subunit* (SSU) of 30 S, consisting of 21 proteins and an RNA molecule of 16 S, and a *large subunit* (LSU), measuring 50 S, consisting of 34 proteins and two RNA molecules, one 23 S and the other 5 S. In the prokaryotic genome, the genes encoding these RNAs are organized in an *rRNA transcription unit* (*rrn* region), with the 16, 23, and 5S rRNA genes lying behind each other, separated by spacers and being transcribed as one unit. The 16S rRNA gene (also called the *SSU rRNA gene*) has been chosen as the basic diagnostic instrument of prokaryote phylogeny and classification. The gene is assumed to fall into the category of essential genes, which are not, or at least infrequently, subjected to lateral transfer (see Section 3.2).

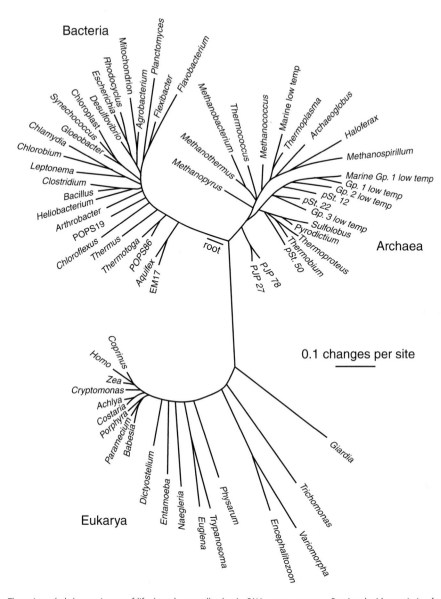

Figure 4.2 The universal phylogenetic tree of life, based on small-subunit rRNA gene sequences. Reprinted with permission from Pace (1997). Copyright 1997 AAAS.

The size of ribosomal components is slightly different in eukaryotes (Table 4.1). The SSU rRNA is 17–18 S in eukaryotes (18 S in vertebrates) and the LSU rRNA molecule, which is 23 S in prokaryotes, is enlarged to 28 S. In addition, eukaryotes have the prokaryote rRNA genes in their mitochondria and chloroplasts. The fact that all life forms have the same basic organization of rRNA

genes allows comparison across domains and the development of phylogenies such as that shown in Fig. 4.2.

The reason why the 16 S rRNA is particularly suitable as an anchor for prokaryote classification is that it shows a mosaic of conserved and variable regions. The molecule is shown in Fig. 4.3. A characteristic feature is that the RNA molecule

Table 4.1 Composition of ribosomes of prokaryotes and eukaryotes

	Prokaryotes	Eukaryotes
Overall size	70 S	80 S
Size of SSU	30 S	40 S
Proteins in SSU	~21	~30
RNA in SSU	16 S, 1500 bp	18 S, 2300 bp
Size of LSU	50 S	60 S
Proteins in LSU	~34	~50
RNAs in LSU	23 S, 2900 bp	28 S, 4200 bp
	5 S, 120 bp	5.8 S, 160 bp
		5 S, 120 bp

LSU, large subunit; SSU, small subunit. *Source*: From Madigan *et al.* (2003).

folds into a determinate structure with many short duplex regions as well as hairpin loops. The secondary structure of the molecule is crucial to its function in translation of mRNA to peptides, and so the nucleotide sequences of the duplex regions are highly conserved. Other parts of the molecule can undergo substitutions without change of function and so these parts provide an apomorphic signature, characterizing a certain prokaryotic lineage. There are nine variable regions in the molecule, numbered V1–V9 (Fig. 4.3).

Microbiologists have agreed that a 3% difference in the overall 16 S rRNA sequence is to be considered as the species boundary. Justification for this is obtained from the fact that if the genomic DNA of two bacteria hybridizes by more than 70% (see above), the 16 S rRNA always differs by less than 3% (Madigan *et al.* 2003). Since the genomic hybridization threshold is considered a valid species boundary, two organisms differing in their 16 S rRNA by more than 3% can always be considered different species. However, the converse is not true; there are many valid species differing by less than 3% in their 16 S rRNA. For example, in the genus *Bacillus* there is a high degree of similarity among species and some species (*Bacillus cereus* and *Bacillus anthracis*) even have completely identical 16 S rRNA sequences.

Because of the diagnostic value of the 16 S rRNA gene, an enormous number of sequences has accumulated in the international nucleotide

sequence databases (GenBank, EMBL). There is an internationally accepted system for numbering the base-pair positions, modelled on *E. coli*. The Ribosomal Database Project (RDP-II; http://rdp.cme.msu.edu) provides alignment and annotation of a large number of rRNA sequences to serve microbial taxonomy and biodiversity research. In the 9.0 release of RDP-II, an alignment is provided of 50 000 SSU rRNA sequences, prepared using a program that incorporates constraints from rRNA secondary structure (Cole *et al.* 2003). An annotated collection of more than 700 oligonucleotide probe sequences targeting rRNAs is available from probeBase (www.microbial-ecology.de/probebase; Loy *et al.* 2003). The use of databases for optimal probe design has become a crucial element in the development of oligonucleotide (synthetic) microarrays (DeSantis *et al.* 2003).

The 16 S rRNA gene is the basis for a popular method of community profiling called *denaturating-gradient gel electrophoresis* (DGGE). The principle of this method was borrowed from medical research, where it was applied for the detection of mutations. Muyzer *et al.* (1993) suggested that the same method could be applied to PCR-amplified 16 S rRNA genes. A PCR was designed that amplifies a segment of the gene, using primers targeting sites of the molecule in which the sequence is conserved across all bacteria (*universal bacterial 16 S primers*). The amplicon, however, is chosen to span a variable region, so that when the PCR is applied to an environmental sample containing a community of microorganisms, a mixture of fragments is produced with identical lengths but different sequences. These fragments are separated on the basis of differential sensitivity to denaturation in a gradient of urea (DGGE) or temperature (*temperature-gradient gel electrophoresis*, TGGE). The sensitivity to denaturation is determined mainly by the GC content of the sequence. Sequences with higher GC content will denature later and run further on the gel. To prevent complete separation of the two DNA strands, a so-called *GC clamp* is included in the amplicon by extension of the 5' primer with 40 bp of a GC-rich sequence; this ensures that the sequence is halted in the gel as a stable partially melted molecule. Preferably the profile is

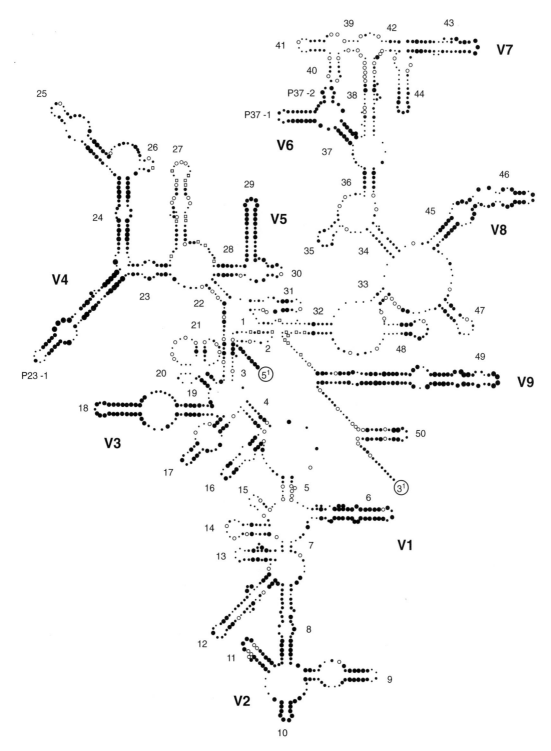

Figure 4.3 Model of a 16 S rRNA molecule, showing nine variable regions numbered V1–V9. Each dot represents a nucleotide and its variability across species is indicated by the size of the dot (in five classes). From Neefs *et al.* (1993) with permission from Oxford University Press.

calibrated by marker samples that contain a mixture of known sequences isolated from a clone library from the same environment.

The 16S rRNA-targeted DGGE approach has become one of the most popular methods with which to profile microbial communities in the

environment. An example of such a study is the work by Röling *et al.* (2001). These authors were interested in microbial communities of groundwater and their potential to degrade aromatic compounds (benzene, toluene, xylene), leaching from a landfill. Groundwater samples

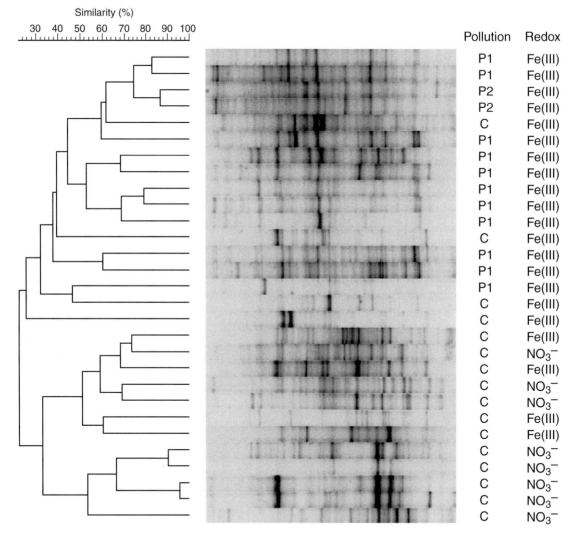

Figure 4.4 DGGE profiles of bacterial communities in an aquifer polluted by landfill leachate. Profiles are shown for samples at different distances and depths; for each lane the level of pollution is indicated (P, in the plume; C, outside the plume, below, above or remote). The right-hand column identifies the dominant redox process of the sample: Fe(III), iron reduction; NO$_3^-$, denitrification. The profiles were clustered on the basis of similarity measured by Pearson's correlation coefficient, using the unweighted pair-group method using an arithmetic average (UPGMA). There is a clear separation between clean samples and polluted samples, although the latter group comprises three clean samples (which were actually suspected of being influenced by the plume as well). Degradation of contaminants is taking place under iron-reducing conditions. From Röling *et al.* (2001) by permission of the American Society for Microbiology.

were taken from bore holes at different distances and microbial communities were profiled using DGGE. Profiles from samples taken in the plume were clearly different from those outside the plume (Fig. 4.4). Sequencing of cloned 16 S rRNA gene fragments confirmed the trend. In the uncontaminated area upstream of the landfill the community was dominated by Betaproteobacteria, but directly beneath the landfill Gram-positive bacteria dominated. At the end of the gradient the community was dominated by Deltaproteobacteria whereas Betaproteobacteria reappeared. The main determinant of community structure was the redox potential; under iron-reducing conditions in the plume the community was dominated by *Geobacter* species, which are known to be able to degrade organic compounds.

In addition to DGGE, variation in 16 S ribosomal genes can also be visualized by restriction polymorphisms. *Amplified ribosomal DNA restriction analysis* (ARDRA) applies a restriction to rRNA genes amplified by PCR using universal bacterial primers. Restriction of the PCR products will lead to a collection of differently sized DNA fragments and when these fragments are separated by electrophoresis a profile is seen that is characteristic of a community, due to differences between species in the restriction sites. A further development in this direction is *fluorescently tagged ARDRA* (FT-ARDRA), in which the terminal fragment of the restriction is marked by labelling one of the two PCR primers, thereby producing only one band per species. Following separation of the fluorescently labelled fragments in an automatic sequencer with laser detection, a species-composition profile of the community is obtained. FT-ARDRA is a form of *terminal restriction fragment length polymorphism* (T-RFLP); however, this more general term is often used interchangeably with FT-ARDRA.

4.2.2 Microarray-based biodiversity assessment

Because gel-based approaches such as DGGE have the disadvantage that only a limited number of species can be resolved, more sophisticated methods using genomics technology have been developed for profiling complex communities. 'A ray of hope' is expected to come from microarray-based hybridization approaches (Polz *et al.* 2003; Zhou 2003). In Section 2.3 microarrays were introduced as tools for transcription profiling and this is their most common use in ecological genomics studies of plants and animals. In microbial ecology, however, microarrays are mainly used for identification (detection) purposes. The principle is that DNAs from a microbial community, or PCR-amplified fragments of environmental DNA, are hybridized to a microarray which contains probes from a series of known species. Those species in the community whose probes are on the microarray will show their presence by hybridization. Unlike in the case of transcription profiling, hybridizations are usually not conducted with two pools of DNA in competition, unless two communities are being compared directly.

Using microarrays for detection, complex communities can be very rapidly profiled without the need for culturing. In principle, the technique can even be implemented in a portable system, allowing species identification in the field within 50 min (Bavykin *et al.* 2001). However, we need to realize that, unlike gel-based methods such as DGGE, microarray-based profiling can only be conducted with species for which genomic information is already available, because this information is required to develop the probes on the array. This implies that the only species that are detected are those that we already have knowledge of—although sometimes that knowledge is limited to the probe sequence alone. We will never find completely new species using microarray-based detection.

Four different types of microarray can be distinguished depending on the nature of the probes used (Zhou 2003; Zhou and Thompson 2004; see Table 4.2). One approach is to use the entire genome of a number of target species whose presence is expected (*community genome arrays*). This approach, based on the principle developed by Voordouw *et al.* (1991), but then called reverse sample genome probing, does not require prior knowledge of genome sequences, it just spots the whole bacterial genome on the array. This can be

Table 4.2 Overview of four types of microarray approach used in microbial ecology to detect microorganisms in the environment

	Community genome arrays	Genome fragment arrays	Phylogenetic arrays (phylochips)	Functional gene arrays
Source of probes	Entire genomic DNA of species	Random fragments of DNA from target species	Selected regions of 16 S rRNA genes	Genes with specific function in an ecological process
Type of information provided	Community composition	Relatedness between species	Taxonomy, species identification	Potential functions
Specificity	Species	Species, strains	Species, strains, mutants	All species within 80–90% sequence homology
Sensitivity (ng of DNA detectable)	0.2	Not determined	Not determined	1–8

Source: Adapted from Zhou (2003).

done with microorganisms that can be isolated in pure culture; the array then detects the species in a sample from the environment. Another, similar approach is to use random genome fragments rather than the entire genome (*random-genome fragment arrays*). The third type of microarray uses 16 S rRNA sequences. Because this gene is diagnostic for prokaryote classification, this type of microarray is called a *phylogenetic microarray* or *phylochip* (perhaps taxochip would be a better qualification, because the information retrieved is essentially taxonomic, not phylogenetic in the evolutionary sense). The final approach is to develop probes on the basis of specific functional genes, often genes related to nutrient cycles such as nitrification or sulphate reduction. These arrays are called *functional gene arrays*; however, it should be noted that they only detect the presence of genes, not their activity or the functions themselves. Depending on which parts of the gene are used as probes, functional gene arrays may detect the same gene in a wide variety of organisms, so the information retrieved is not of a taxonomic nature, unless species-specific parts of the gene are targeted. Functional gene arrays can be equipped with oligonucleotide probes (50–70 bp) and manufactured using photolithography, or they can be prepared by spotting longer gene fragments (200–1000 bp), which are amplified by PCR from a clone library.

The proof of concept for microarray-based detection of bacteria was given by Cho and Tiedje (2001). These authors used four species of

Pseudomonas as reference for the preparation of probes. The genome of each species was fragmented mechanically and fragments of 1–2 kbp (60–90 per species) were spotted on a coated slide (Fig. 4.5). The fragments represented 1–3% of the genome. The array was then tested by hybridization with labelled DNA from one of the species and with related species. It turned out that each species was recognized uniquely and other *Pseudomonas* species whose DNA was not on the array were detected with lower intensity. Comparing the hybridization patterns of related species and strains, a distance tree could be developed that was consistent with the phylogenetic tree obtained from 16 S rRNA sequences. Due to the use of many probes per species, the sensitivity of this approach is potentially very high; however, for profiling complex communities very large arrays would be necessary (e.g. for 1000 species, each represented by 100 probes, an array of 100 000 positions would be required). This is possible with oligonucleotide arrays prepared by photolithography, but more difficult with spotted arrays. Also, since the probes are a random selection from the genome, many of them may represent conserved genes that are not diagnostic of the species. This work nevertheless showed that the principle of microarray-based detection was valid and it has triggered further developments.

A strong argument in favour of the use of microarrays in microbial detection is that it can, at least in principle, replace PCR amplification of target sequences. PCRs applied to a heterogeneous

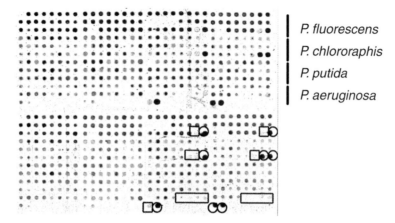

P. fluorescens

P. chlororaphis

P. putida

P. aeruginosa

Figure 4.5 Format of a microarray used by Cho and Tiedje (2001) for bacterial species determination. Random fragments of 1–2 kbp from the genome of four *Pseudomonas* species were spotted in duplicate on amino silane-coated glass slides. Duplicates were placed in the upper and lower halves. Positive and negative controls (genes from yeast and bacteriophage λ) are indicated by circles and rectangles, respectively. The different shades of grey indicate the ratio between the channels of test DNA (Cy3) and reference DNA (Cy5). From Cho & Tiedje (2001) by permission of the American Society for Microbiology.

collection of DNAs from the environment often suffer from preferential or aspecific amplification of certain sequences over others (*PCR bias*), especially when the PCR is addressed to a specific, limited group of organisms. This is due to the initially small number of template copies that must be found in a large background of DNA to be ignored. Biodiversity screens in which a PCR step is included are vulnerable to the criticism that the diversity in the amplified assemblage may be different from the diversity of the environment. With microarrays the PCR step can (in principle) be skipped altogether, as demonstrated by some authors (Small *et al.* 2001; Urakawa *et al.* 2002; El Fantroussi *et al.* 2003). In this strategy RNA extracted from environmental samples is fragmented, labelled, and hybridized directly with the microarray. The target RNA will consist mostly of rRNA that hybridizes with 16S rRNA probes on the array. A multiple melting-curve analysis is then performed to detect aspecific hybridizations. Although this strategy seems to be particularly applicable in an environmental context, most authors still prefer to start with a PCR to obtain sufficient target DNA for hybridization (e.g. Loy *et al.* 2002; Peplies *et al.* 2003; Taroncher-Oldenburg *et al.* 2003). Such a PCR uses universal bacterial primers (targeting all bacteria) to minimize

possible bias, after which specific groups are picked out by microarray hybridization.

Oligonucleotide phylochips are presently considered as the most promising instruments for microbial detection, because of the enormous capacity of such chips and the possibility of maximally exploiting the information in rRNA sequence databases. From the databases, 20-mer probe sequences can be developed that target different taxa in the hierarchical classification system of microorganisms. For example, in a microarray system targeting *E. coli* some probes are universal for all bacteria, others are specific for the phylum Proteobacteria, and still others for the γ subdivision of Proteobacteria; then some probes target only the group of enteric bacteria and the most specific probes address specific genera of Enterobacteria, such as *Escherichia* or *Salmonella*. So DNA from *E. coli* should hybridize with all these probes, and DNA from *Nitrosomonas*, a betaproteobacterium, should hybridize only with the first two. From the hybridization pattern of a complex community an indication can be obtained of the taxonomic composition; however, not all groups of bacteria can be resolved in this way. In most cases the discriminatory power is limited to genera; in several cases, when genera are very similar in their 16S rRNA gene, only the higher taxonomic levels are resolved. The taxa (be it genera, families, or divisions) that can be detected in this way are designated as *operational taxonomic units* (OTUs).

The strategy of hierarchical arrangement of SSU RNA probes was applied by Wilson *et al.* (2002) to develop a photolithography gene chip with 31 179 oligonucleotide probes. As in the case of gene chips used for transcription profiling, each 20-mer probe was paired with a control probe in which the eleventh nucleotide was replaced by a mismatch position; these control probes should not hybridize with the target and so serve to control for nonspecific effects. The number of probe pairs targeting a specific taxon ranged between 1 and 70, depending on the completness of sequence information of the rRNA region. In total there were 1945 prokaryotic and 431 eukaryotic sequences represented on the chip. By way of a test, the chip was used to detect bacteria and fungi recovered from filtered air samples. The results showed that the airborne microbial assembly was dominated by bacteria from the Gram-positive *Bacillus-Lactobacillus-Streptococcus* subdivision and the Gram-positive high-G+C group; in addition significant amounts of ascomycete and basidiomycete DNA (presumably spores) were found in the 'aeroplankton'. There was a good correspondence between chip-based detection and PCR amplification of 16S RNA genes followed by cloning and sequencing; however, cloning revealed 28 novel sequences that did not correspond well to any phylogenetic group in the RDP database and so were not detected by the microarray.

The principle of microbial detection in air was further developed with a gene chip of 500 000 probes allowing detection of 9121 OTUs (Andersen *et al.* 2004). A sampling campaign was conducted with this array to survey the air of two cities in the USA, Austin and San Antonio, over several months. The use of microarrays to gauge air biota had a mainly medical relevance, and was specifically motivated in the USA by the need for bioterrorism defence strategies against possible mischievous release of pathogenic agents. In the monitoring campaign, significant seasonal trends were observed; the abundance of bacteria was correlated negatively with air temperature and positively with the dew point. An interesting observation was that, although the time trends were similar for the two cities, the compositions of the microbial assemblies were different, indicating that in addition to climatic determinants there are local sources of airborne microorganisms. Low levels of *Bacillus anthracis* (causing anthrax) and *Clostridium botulinum* (causing botulism) were detected, because these bacteria are normally found associated with livestock and occur naturally in sediments. Inevitably, some of these bacteria are aerosolized and become airborne with the wind. It is therefore necessary to know the background levels of these pathogens so as to single out situations hazardous to human health.

Similar detection strategies applied in ecological settings, for example using DNA from soils, sediments, or water, meet more difficulties than air sampling, because the chemical matrix in which the environmental DNA is contained is much more complex. In particular, humic derivatives are notorious for causing disruption to DNA samples. Nevertheless, several successful attempts have been made. Loy *et al.* (2002) developed an oligonucleotide microarray called SRP-PhyloChip for cultivation-independent detection of sulphate-reducing prokaryotes (SRPs). Sulphate reducers form a highly heterogeneous, polyphyletic, group of bacteria. The capacity to gain energy from

Figure 4.6 Showing a phylogeny of sulphate-reducing bacteria in the orders Desulfobacterales and Syntrophobacterales of the Deltaproteobacteria. The tree was developed on the basis of 16S rRNA sequences using maximum parsimony and other methods. Non-sulphate reducers are underlined. The probes used on the array are indicated by short names such as DSTAL, DSRHP185, etc. Braces indicate the taxonomic span of the probes, e.g. DELTA495a, DELTA495b, and DELTA495c target all Deltaproteobacteria, DSB706 targets the upper clade of 11 species plus *T. norvegica*, DSB230 targets a group of eight species within the upper clade, and DSRHP185 targets two species of the genus *Desulforhopalus*. Numbers in parentheses next to a probe name indicate that more than one probe was used to target different parts of the rRNA molecule, e.g. five different DSTAL probes target the genus *Desulfotalea*. The codes after the species represent GenBank accession numbers of their 16S rRNA sequences. Adapted from Loy *et al.* (2002), by permission of the American Society for Microbiology.

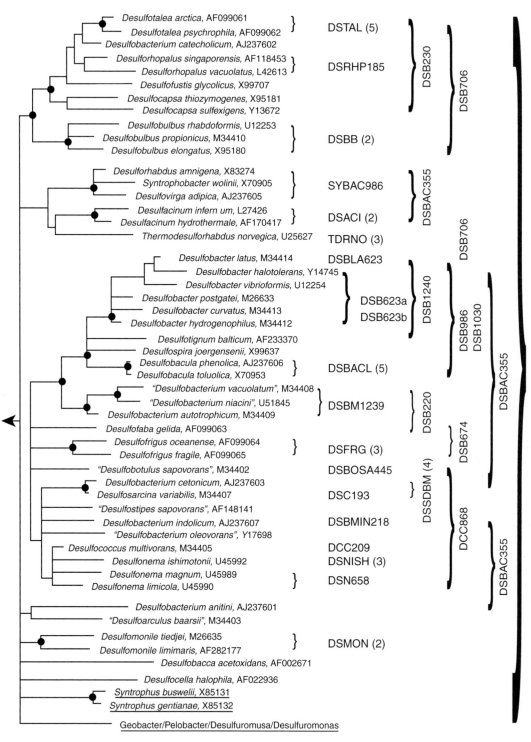

sulphate reduction has evolved independently several times in different lineages. In some cases species that can and cannot perform sulphate reduction are taxonomically closely related. This situation makes it impossible to target sulphate reducers on the basis of a PCR with a single 16S rRNA sequence; instead, several different PCRs would be needed. For microarrays, which can deal with many probes in parallel, the situation presents no problem. As in the example discussed above, Loy *et al.* (2002) based their probes on sequences available from public databases. Sulphate reducers (134 recognized species in total) are present in five different lineages of prokaryotes: Deltaproteobacteria, Firmicutes, Nitrospirae, Thermodesulfobacteria (all Bacteria), and Euryarchaeota (Archaea). Fig. 4.6 provides an example of the affiliations in two orders of the Deltaproteobacteria.

The alignments of 16S RNA sequences led to the development of 138 18-mer probes, which in conjunction are diagnostic for the community of sulphate reducers. For example, the genus *Desulfotalea* was specifically detected by five probes and was also targeted by three probes with broader specificities (Fig. 4.6). In some cases, however, the taxonomic resolution did not go further than a group of three genera (see SYBAC986 in Fig. 4.6). If the array is hybridized with a sample of PCR-amplified 16S rRNA genes, a read-out of hybridizations will produce a list of taxa present in the sample. This can be conveniently done using a software tool named ChipChecker (Loy *et al.* 2002). The SRP-PhyloChip was used to investigate the sulphate-reducing community of a hypersaline cyanobacterial mat from Solar Lake, Egypt. Bacteria from the genera *Desulfonema* and *Desulfomonile* were identified and this was confirmed by cloning and sequencing.

Another genomics survey of bacterial community composition was applied by Valinsky *et al.* (2002). These authors were interested in the *disease-suppressiveness* of agricultural soils. Agricultural experience shows that soils can become suppressive to soil-borne pathogens if they are managed in certain ways. One of the best-known examples is suppression of the fungus *Gaeumannomyces graminis* var. *tritici* (Ascomycota), which causes a root disease

of wheat worldwide (the syndrome is described as take-all). When wheat is cropped in continued monoculture, one or more severe outbreaks of the disease usually follow, but thereafter the soil spontaneously develops a suppressiveness and the fungus is unable to grow any more in such soils (take-all decline). This is correlated with an increase in the abundance of bacteria of the genus *Pseudomonas*, some of which are known to produce an antifungal compound, 2,4-diacetylphloroglucinol. This well-known example of soils developing suppressiveness is a model for research at the interface between plant pathology and soil microbiology (Weller *et al.* 2002; De Souza *et al.* 2003; Garbeva *et al.* 2004).

Valinsky *et al.* (2002) investigated a case of soil suppressiveness towards the plant-parasitic nematode *Heterodera schachtii* (sugar beet cyst nematode). Two adjacent agricultural fields, one suppressive and the other not, were sampled and microbial communities were screened using an array-based method. The authors used a membrane to which the soil-extracted and PCR-amplified 16S rRNA genes were fixed, while different radioactively labelled oligonucleotide probes were added to detect a 16S rRNA sequence of a specific species or group of species. A great number of bacterial clusters were identified, which were grouped into five major taxa (Table 4.3). There were obvious differences between the soils; the disease-suppressive soil had fewer *Bacillus* species, more Alphaproteobacteria, and fewer Enterobacteria. Interestingly, DGGE analysis of the same soils revealed only 13 bands and did not detect any bands differential between the two soils. This illustrates the enormous increase in resolution that can be obtained by genome-wide analyses. Further research is necessary to reveal the mechanistic relationships between community composition and suppressiveness to cyst nematodes in these soils.

4.2.3 Statistical approaches to prokaryote diversity

The recurrent question of how many species of prokaryote there are has also been addressed using

Table 4.3 Taxonomic composition of bacterial rRNA clones obtained from two agricultural soils using an array-based screening method

	Number of clones Reference soil	Soil suppressive to sugar beet cyst nematode
Bacillus	405	35
Cytophaga–Flexibacter–Bacteroides group	5	25
Actinobacteria	130	185
Alphaproteobacteria	10	142
Beta- and gammaproteobacteria	162	87
Enterobacteria	127	8

Source: From Valinsky *et al.* (2002).

statistical approaches applied to genome data (Hughes *et al.* 2001; Curtis *et al.* 2002; Ward 2002). When a community is sampled and the number of species is counted in successive samples, the total number of species retrieved will show a curve of diminishing returns. A plot of the cumulative number of observed species versus sample size produces a *species-accumulation curve*, also called collector's curve, which tends to level off to a plateau beyond a certain sample size (Krebs 1999). Assuming a suitable model (e.g. a hyperbola or a negative exponential) the total number of species in the community may be estimated by extrapolation, even though a fraction of species remains unnoticed. Such methods are commonly applied to field inventories of plants and insects, but they are rarely found in microbial ecology.

Hughes *et al.* (2001) reviewed various microbial data-sets and showed how the ecological methodology can be applied to estimate species richness in a microbial setting. One of the data-sets analysed was due to McCaig *et al.* (1999). These authors had sequenced 16 S ribosomal genes from clones isolated from soils of two upland pastures in Scotland, one natural but grazed by sheep in summer, the other reseeded, fertilized and grazed during the whole season ('improved'). Applying the 97% similarity criterion as a species boundary, there were no differences in observed species richness between the two sites (113 in the fertilized habitat and 114 in the natural grassland); however, there were differences in species composition. Hughes *et al.* (2001) re-analysed the data and

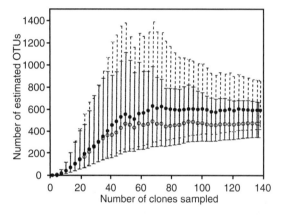

Figure 4.7 Estimates of species richness according to Chao applied to 16 S rRNA sequences differing more than 3% from each other (OTUs) in two upland pastures in Scotland, one natural (•), the other fertilized (○), with 95% confidence intervals (dashed lines for the natural grassland, solid lines for the fertilized habitat). From Hughes *et al.* (2001), by permission from the American Society for Microbiology.

applied Chao's estimator for total species richness, S_T, which is

$$S_T = S_{obs} + \frac{n_1^2}{2n_2}$$

where S_{obs} is the observed number of species, n_1 is the number of singletons (species captured once) and n_2 is the number of doubletons (species captured twice). The theory also accounts for a standard error of this estimate. Applying this equation to the Scottish grassland data produced the species-accumulation curve shown in Fig. 4.7. Total species richness S_T was was estimated as

590 OTUs in the natural grassland and 467 OTUs in the improved grassland. Due to the significant number of rare species (singletons relative to doubletons), this estimated total number of species in the community was considerably higher than the actual number observed. There seemed to be more species in the natural grassland soil than in the fertilized soil, although the difference was not statistically significant.

Another approach to estimation of microbial diversity was applied by Curtis *et al.* (2002). These authors used the distribution of species over abundance classes as a basis for extrapolation. We know from basic ecology that in any community there are a few species with very high abundance, whereas most species have intermediate abundance. The distribution of species over abundances can be characterized by a function f, where

$$S_T = \int_{N_{min}}^{N_{max}} f(N)dN$$

in which S_T is the total number of species in the community, N_{min} the abundance of the least abundant species, and N_{max} the abundance of the most abundant species. Integrating f over all abundances is equivalent to cumulating species over all abundance classes, from rare to dominant. Community ecologists have argued that f often takes a bell-shaped symmetrical form if abundance is grouped into geometric-scale units, e.g. in classes whose widths increase by a factor of 2 with every successive group (octave scale; Krebs 1999). The function f then approaches a normal curve on a logarithmic scale. *Lognormal species-abundance curves* are assumed to hold particularly well for microorganisms, which exhibit highly dynamic and random growth, influenced by many independent factors. The lognormal distribution is defined by only two parameters, allowing S_T to be estimated from the number of species in one of the classes (e.g. the largest class) plus the spread of the distribution (Krebs 1999). However, these two quantities are still difficult to measure in the case of microorganisms, therefore Curtis *et al.* (2002) developed an expression allowing S_T to be estimated from two quantities easily accessible to measurement: the total number of individuals in the community (N_T)

and the abundance of the most abundant member (N_{max}). Applying the formula to data for microbial clone abundance reported in the literature, the authors were able to estimate prokaryote species diversity for several ecosystems. In addition, they also estimated the global diversity of Bacteria and Archaea in the sea (Table 4.4).

The analysis of Curtis *et al.* (2002) depends crucially on abundance data, which, even for the most abundant organisms—prokaryotes—are not very reliable. Nevertheless, these estimates show that prokaryote species richness may be orders of magnitude greater than can ever be achieved with clone libraries; even the most extensive clone libraries do not reach beyond a few hundred species. When better quantitative data become available in the future the estimates can be refined; however, to develop a reliable empirical species-abundance curve for a microbial community will be a serious experimental challenge. For the moment the theoretical predictions seem to stand on firmer ground. There are also some intriguing questions inherent to the estimates in Table 4.4; for example, why the global species richness of the Archaea is so much lower than that of the Bacteria and why the biodiversity of soil ecosystems increases less with increasing scale than the biodiversity in the sea.

Table 4.4 Estimates of local and global biodiversity of prokaryotes, derived from abundance data assuming a lognormal distribution of species over abundance classes

Environmental compartment	Estimated number of species
Bacteria in 1 ml of ocean water	163
Bacteria in the entire sea (pelagic)	2 000 000
Archaea in the entire sea (pelagic)	20 000
Bacteria in an average lake	8000
Bacteria in 1 g of soil	6300–38 000
Bacteria in 100 g of soil	100 000–1 000 000
Bacteria in 1000 kg of soil	4 000 000
All soil bacteria in the world	4 500 000
Bacteria in the global atmosphere	4 000 000
Bacteria in 1 ml of sewage sludge	70
All bacteria in a sewage works	500

Source: From Curtis *et al.* (2002).

Estimates of species richness can also be derived from computer-aided assemblies obtained by applying the WGS approach (see Section 2.2) to communities. This approach was applied in a large-scale sequencing campaign examining marine biodiversity (Venter *et al.* 2004). The Sorcerer II expedition is an undertaking of the Venter Institute aimed at analysing the vast untapped and unseen world of oceanic biota around the world (www.sorcerer2expedition.org). In the first phase of the project, samples were collected from the Sargasso Sea, off the coast of Bermuda, and that DNA was cloned into plasmid vectors and sequenced using the WGS approach. Traditionally, WGS sequencing has been applied to identify the genome sequence of one particular organism, but the approach taken by Venter *et al.* (2004) was to capture representative sequences from as many organisms as possible (see also Section 4.4 below). Obviously, when sampling the community at random, the most abundant species will be sequenced many times and the rare species only occasionally. Using an adapted assembly algorithm with enormous computational power, scaffolds were constructed that could be assigned to species. A 'genomic species' was defined as a clustering of assemblies or unassembled reads with more than 94% sequence similarity.

Venter *et al.* (2004) developed a new method for estimating species-richness from the data obtained. In Section 2.3 we saw that in a WGS assembly the probability of a base position being sequenced r times, $P(r)$, can be modelled as a Poisson distribution, for which the mean equals the average depth of coverage. Similarly, for a mixture of rare and abundant genomes, different species have different depths of coverage and $P(r)$ is a composite Poisson distribution. By fitting the theoretical distribution to the data, the number of genomes at each depth of coverage could be estimated (Table 4.5), and the sum of these numbers is an estimate of the number of species in the community. For a 2001 sample from the Sargasso Sea, species richness was estimated as at least 1800, which is consistent with the extrapolations from the model of Curtis *et al.* (2002) discussed above. However, since 80% of the assembled sequence is contributed by organisms of very low abundance, the actual species richness could be much greater, perhaps an order of magnitude more.

Our *tour d'horizon* of prokaryote biodiversity screening has demonstrated an inherent weakness

Table 4.5 Illustrating a model to estimate genome (species) diversity in microplankton communities analysed by WGS sequencing

Sequencing coverage depth	Fraction of assembly consensus organisms	Expected fraction of genome sequenced	Genomes
25	0.0055	1.0	2.5
21	0.0050	1.0	2.3
13	0.0035	1.0	1.6
9	0.0040	1.0	1.8
7	0.0080	0.999	3.6
6	0.0047	0.998	2.1
4	0.0100	0.982	4.6
2.4	0.0258	0.909	12.8
2	0.0700	0.865	36.4
0.25	0.8635	0.221	1756.7
Total	1.0		1824.4

Notes: Assembly consensus sequences are classified by the sequencing depth coverage. For example, 0.55% of the organisms have been sequenced 25 times and such sequencing covered the whole genome. This results in an estimated number of 2.5 genomes falling into this class, assuming a mixture of Poisson distributions. The total number of species estimated for a 200-l sample from the Sargasso Sea is 1824. Reprinted with permission from Venter *et al.* (2004). Copyright 2004 AAAS.

of microbial ecology compared to plant and animal ecology: the difficulty of obtaining reliable estimates for abundance and species richness. It is expected that genomics approaches will improve on this situation in the near future. Still, sequences cloned from environmental samples are rarely identical to sequences of cultured bacteria in gene databases. Several investigations report on completely new bacterial lineages. For example, only a few years ago it was realized that the phylum Acidobacteria, previously known from only three described species, is as diverse as the whole superphylum of Proteobacteria. DNA sequences of this group are found everywhere in clone libraries; often no less than 30% or even 50% of the 16S rRNA gene inserts in clone libraries from soil sample belong to Acidobacteria, which is indicative of a very important (but unknown) ecological role (Quaiser *et al.* 2003). Obviously, a levelling-off in the collector's curve of global microorganisms is not even in sight.

4.3 Microbial genomics of biogeochemical cycles

The crucial role of microorganisms in biogeochemical cycles has been known about for a long time, but only recently have we gained insight into the architecture and diversity of the genetic systems responsible for these ecological functions. In addition, new and unexpected links between element cycles have been discovered that were not taken into account before. The new evidence is coming from two sources: detection of functional genes using microarrays and cloning of environmental genomes. The second topic will be discussed in Section 4.4, and this section is devoted to an overview of the genes and genomic background associated with conversions in the major element cycles, as well as examples of studies using microbial functional genomics in the field.

All elements in the Earth's crust and the atmosphere interact with the biosphere and for some elements, especially those that are used as essential building blocks or catalysts in cells, the interaction is particularly strong. Many elements undergo transitions from one chemical compound to

another, from one redox state to another, or from one environmental compartment to another. Many of these transitions are catalysed by enzymes in specific biological entities. To understand and predict biogeochemical cycles it is necessary to know the mechanisms by which organisms direct the transitions in these cycles. In this section we will therefore focus on the genomic determinants of such transitions and conversions. Knowledge of the responsible genes and the organisms in which these are expressed can be used to develop screening instruments by which biogeochemical functions in environmental samples may be assessed. Table 4.6 is a guide for the discussion. We focus on key links in the carbon, nitrogen, sulphur, phosphorus, iron, calcium, and silicium cycles. We do not discuss all the biochemical aspects of nutrient cycles; for a full treatment of this the reader is referred to a microbiology textbook such as Madigan *et al.* (2003).

4.3.1 Key genes in the carbon cycle

Starting with carbon, we note that photosynthesis is the most important link in the cycling of nutrients through the biosphere, because it is the process upon which almost all other life depends. Most organisms that can use sunlight for the generation of metabolic energy (phototrophs) divert that energy to the synthesis of organic molecules from CO_2 and so are *photoautrophic*. There are also phototrophs that rely on organic carbon for their growth, although they use light as a source of energy; these are called *photoheterotrophs*. Another distinction relates to the source of reducing power for the conversion of CO_2 to organic carbon. This can come from water, producing O_2 as a by-product, from reduced sulphur compounds in the environment such as H_2S, or from molecular hydrogen, H_2. The first type of photosynthesis, which is practised by cyanobacteria, green algae, and land plants is called *oxygenic* (producing oxygen), whereas the second type, practised by purple bacteria and other prokaryotes, is called *anoxygenic*. The use of H_2S or H_2 as a source of reduction equivalents is not always driven by light, it can also be coupled

Table 4.6   Overview of genes associated with key links in biogeochemical cycles, classified by element and process

Element	General process	Specific process	Gene products, enzymes	Gene clusters
Carbon	Anoxygenic photosynthesis	Photopigment biosynthesis	Enzymes of bacteriochlorophyll and carotenoid biosynthesis	*bch, crt*
		Assembly of photosynthetic complex	Light-harvesting and reaction-centre complex subunits and assembly factors	*puf, puh, puc*
	Oxygenic photosynthesis	Photopigment biosynthesis	Enzymes of chlorophyll, phycobilin, and carotenoid pathways	*apc, cpc*
		Assembly of photosynthetic complex	Photosystem I and II subunits	*psa, psb*
	Carbon fixation	Calvin cycle	Ribulose bisphosphate carboxylase complex (Rubisco) subunits, phosphoribulokinase	*rbc, cbb, prk*
		Reverse citric acid cycle	ATP citrate lyase	*acl*
	Non-chlorophyll phototrophy	Light-driven energy generation	Proteorhodopsin	*gpr*
	Decomposition	Polysaccharide catabolism	Amylase, cellulase, pectinase, chitinase	*amy, cel, pem pel, chi,*
		Lipid catabolism	Phospholipase, acyl-CoA dehydrogenase	*phl*
		Lignin degradation	Peroxidase	*mnp*
	Methanotrophy	Methane oxidation	Methane monooxygenase	*pmo*
	Methanogenesis		Methylreductase, heterodisulphide reductase	*mrc, hdr*
Nitrogen	Nitrogen fixation	Nitrogenase	Dinitrogenase α and β, dinitrogenase reductase, electron-transport flavoproteins	*nif, fix*
	Denitrification		Nitrate reductase, nitrite reductase, NO reductase, N_2O reductase	*nap, nar, nir, nor, nos*
	Anammox	Anaerobic ammonia oxidation	Nitrite oxidoreductase	*nor*
	Nitrification	Aerobic ammonia oxidation	Ammonia monooxygenases α and β	*amo*
		Nitrite oxidation	Nitrite oxidoreductase	*nor*
Sulphur	Sulphate reduction		Adenosine phosphosulphate reductase, sulphite reductase	*aps, dsr*
		Dimethyl sulphide generation	Catabolism of dimethylsulphoniopropionate	Dimethylsulphonio propionate lyase
	Sulphur oxidation		Sulphite oxidase, sulphur dehydrogenase	*sox*
Phosphorus	Polyphosphate hydrolysis		Exopolyphosphatase	*pps*
Iron	Iron reduction	Periplasmic electron shuttling	c-Type cytochromes	*omc, mtr, cct, cym*
	Iron oxidation	Electron shuttling	Rusticyanin, c-type cytochromes	*rus, cyc, cox*
Calcium	Formation of calcareous skeletons	Calcium carbonate precipitation	Vacuolar-type ATPase proton pump	V-ATPase
Silicium	Formation of frustules, spicules	Silica uptake, silica precipitation	Silaffin, frustulin, silicatein	*Sil*

to chemoautotrophic energy generation, although the use of water and the production of O_2 is always driven by light.

Interception of light by photoautotrophs relies on chlorophyll, a porphyrin molecule with a magnesium atom in the centre of the ring structure. Several chlorophylls are known, which differ from each other in substituents on the porphyrin ring and side chains attached to it. These substituents influence the spectroscopic properties of the Mg-porphyrin, and so the different chlorophylls have different absorption maxima. Cyanobacteria, green algae, and land plants have chlorophyll *a* and *b*, several algal groups (Phaeophyta, Chrysophyta, Dinoflagellata) have chlorophyll *a* and *c*, and red algae have a type of chlorophyll *a* called bacteriochlorophyll *a*. In addition, most phototrophic organisms have various accessory pigments that support the capture of light (phycobilins) or protect against damage from radicals (carotenoids).

The main genomic models for investigating bacterial photosynthesis are purple bacteria of the genera *Rhodobacter* (Alphaproteobacteria, Rhodobacterales) and *Rhodopseudomonas* (Alphaproteobacteria, Rhizobiales). These bacteria are anoxygenic phototrophs, which generate sulphur from H_2S rather than oxygen from water. *Rhodobacter sphaeroides* is readily amenable to genetic manipulation, allowing the development of mutants that are deficient in some aspect of the process and this has greatly helped in reconstructing the genomic organization of the photosynthetic apparatus. In *Rhodobacter*, the genes involved in photosynthesis are clustered in operons in a 50 kb region of the chromosome called the *photosynthetic cluster*. The operons are transcribed in a highly coordinated manner to allow the correct assembly of new photosynthetic complexes. The cluster involves *bch* genes, which encode proteins involved in the synthesis of bacteriochlorophyll, *crt* genes, which encode proteins involved with carotenoid synthesis, *puh* and *puf* genes, which encode pigment-binding peptides of the reaction center, and *puc* genes, encoding assembly factors. These genes may be used to develop probes for functional gene microarrays (Table 4.6).

The main model for oxygenic photosynthesis in bacteria is *Synechocystis*, a cyanobacterium from the order Chroococcales, the fourth microorganism to have its genome sequenced completely and the first photosynthetically active species to do so (Kaneko *et al.* 1996). We know from plant physiology that cyanobacteria, green algae, and plants have a more complicated system for the flow of electrons than the anoxygenic phototrophs; namely, they use two interconnected photosystems. In the first step, conducted by *photosystem II*, water is split into oxygen and hydrogen, whereas the second step, conducted by *photosystem I*, produces NADPH. The double system is also called *noncyclic photophosphorylation* because the electrons liberated from chlorophyll by light excitation are transferred to $NADP^+$, rather than back to chlorophyll as in anoxygenic or cyclical photophosphorylation. The evolution of photosystem II in Cyanobacteria around 3.5 billion years ago changed the face of the Earth because it allowed the use of water as an electron donor and the production of oxygen, which came to accumulate in the atmosphere. The subunits of the two photosystems are encoded by genes of the *psb* cluster (for photosystem I) and *psa* (for photosystem II). Other genes of the photosynthetic complex are the enzymes of the pathways for chlorophyll synthesis and accessory photopigments (Table 4.6).

Several mechanisms are known for the fixation of CO_2 in organic molecules, but the *Calvin cycle* is the most widespread. Both photoautotrophs and chemoautotrophs use the generation of ATP and NADPH to fuel CO_2 fixation. The Calvin cycle uses two key enzymes, ribulose bisphosphate carboxylase (Rubisco) and phosphoribulokinase. Rubisco is the most abundant enzyme of the biosphere and it shows a remarkable constancy across all autotrophic organisms, from bacteria to flowering plants. The complex consists of several subunits, encoded by *rbc* and *cbb* genes. In plants some of the subunits are encoded in the chloroplast, others in the nuclear genome. The second enzyme that is unique to the Calvin cycle is phosphoribulokinase, which is encoded by genes from the *prk* cluster. In green sulphur bacteria (*Chlorobium* is the best-investigated genus), and in some chemoautotrophic bacteria and Archaea, an alternative CO_2-fixation pathway is active, the so-called *reverse citric acid cycle*. Most of the enzymes in this cycle are the same as in the normal Krebs cycle, except for the enzyme ATP citrate lyase, which is specific to this pathway and is used as a genetic marker for reverse citric acid carbon fixation. In *Chlorobium* the enzyme is encoded by two adjacent ORFs, *aclB* and *aclA* (Kanao *et al.* 2001).

Only recently, thanks to a genomics approach, has another mechanism of phototrophy been implicated in the carbon cycle. In this pathway energy is generated from light, not through chlorophyll but through a retinal-dependent light-activated proton pump, *proteorhodopsin*. This mechanism was known previously only from some Archaea (in which the protein was called bacteriorhodopsin) until Béjà *et al.* (2000a) discovered that rhodopsin-like genes were associated with uncultured Gammaproteobacteria of the SAR86 clade in the sea. Further research has shown that this protein is very common in marine bacterioplankton and is present well outside of the phylum Proteobacteria (Béjà *et al.* 2001; De la Torre *et al.* 2003; Venter *et al.* 2004). Proteorhodopsin most probably supports a photoheterotrophic lifestyle in which light is used to generate energy, relieving respiratory costs in bacteria that do not fix CO_2. However, proteorhodopsin may be coupled to an as-yet-unknown photoautotrophic pathway.

Turning to the catabolic part of the carbon cycle we note that a great variety of enzymes are used by heterotrophs to degrade polymers from photosynthetically fixed cabon. These are all part of the *decomposition* subsystem, which returns CO_2 or CH_4 to the atmosphere. Decomposition is a crucial process ocurring in soils and sediments. It actually involves more than just breakdown of organic matter, it includes mineralization of nutrients and synthesis of humus, leading to humus–clay complexes that stabilize the soil. All organic matter eventually ends in the decomposer network, to which an enormous variety of heterotrophic organisms contribute. The number of genes that can be used as indicators of the decomposition process is potentially very large; we have mentioned in Table 4.6 only a few enzymes associated

with carbohydrate and lipid catabolism. Many of these enzymes are present in bacteria, fungi, and animals; however, cellulase (more correctly called β-1–4-glucanase) has a limited distribution. Most fungi can degrade cellulose, but only few groups of bacteria; no animal, except the tunicates (see Section 3.3), has cellulase. Consequently all animals depend on microorganisms for the degradation of this abundant polysaccharide. Many animals have recruited a specialized microflora in one of their gut compartments to do the job of cellulose degradation for them. The capacity for lignine degradation is even more limited in nature. Certain basidiomycetes called wood-rotted fungi are responsible for this important link in the carbon cycle and they do this by means of aspecific peroxidases. We have seen in Section 3.3 that the white rot fungus *Phanearochaete chrysosporium* was selected as a genomic model for lignin degradation.

In addition to the polymeric substrates attacked by decomposition reactions, a great variety of simpler organic compounds may be used by heterotrophic microorganisms as a source of carbon and to generate energy (e.g. disaccharides, organic acids, phospholipids). A special case is C_1 metabolism, which is conducted by some Gamma- and Alphaproteobacteria (*Methylomonas*, *Methylosinus*) that are able to grow on compounds with only one carbon atom and thus have to synthesize all carbon–carbon bonds themselves. Methane is the most common one-carbon substrate; the microorganisms that use methane as their sole source of carbon are called *methanotrophs*. These organisms are found whereever a stable source of methane is combined with availability of oxygen; for example in the transition zone between oxic and anoxic strata of lakes and soils. Methane oxidation is an important link in the carbon cycle, because it converts methane back into cell material and CO_2. A key enzyme in the methane oxidation pathway is methane monooxygenase, a haem protein of the same family as ammonium monooxygenase (see below), which is encoded by the *pmo* gene cluster (Table 4.6).

The last link in the carbon cycle to be discussed is the process of *methanogenesis*. This is an extremely important process in the climate-change issue because methane is a very effective greenhouse gas. Methane is generated in marshes, swamps, lake sediments, paddy fields, and landfill sites under anoxic conditions by a group of strictly anaerobic Archaea, called methanogens. These organisms are found in five orders of the phylum Euryarchaeota and include the genomic model species *Methanococcus jannaschiii*, a hyperthermophile from the ocean floor (mentioned in Section 1.2 as the first fully sequenced archaeon), as well as many mesophilic species such as *Methanosarcina mazei*, whose genome was discussed in Section 3.2. The most important substrate used in methanogenesis is CO_2, which is reduced to CH_4 using H_2 as an electron donor in the overall reaction:

$$CO_2 + 4H_2 \rightarrow CH_4 + 2H_2O$$

In addition to CO_2, many methanogens can use simple alcohols and organic acids as well—for example methanol, methylamines, formate, and acetate—but not larger organic molecules such as sugars. This implies that methanogens must rely on decomposing (CO_2 producing) and fermenting (formate- and acetate-producing) microorganisms in their immediate surroundings for the supply of necessary substrates. This trophic interdependence within microbial communities is called *syntrophy*. The reactions leading to the reduction of CO_2 to CH_4 are quite complex and involve some unique enzymes and coenzymes. The terminal step is conducted by the methyl reductase complex, which is encoded by the *mrc* gene cluster. The reaction catalysed by methyl reductase produces not only free methane but also leads to the formation of a complex of two coenzymes (coenzyme B and coenzyme M), linked via a disulphide bridge. This disulphide is then reduced to produce the free coenzymes by the enzyme heterodisulphide reductase, another unique enzyme found only in methanogens; its three subunits are encoded by genes of the *hdr* cluster (Table 4.6).

4.3.2 Key genes in the nitrogen cycle

Figure 4.8 shows an overview of the nitrogen cycle. Most of the inorganic nitrogen on Earth is present

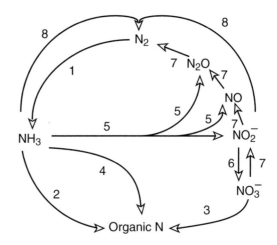

Figure 4.8 Overview of the nitrogen cycle. 1, Nitrogen fixation; 2, ammonium assimilation (uptake by microorganisms and plants); 3, reductive nitrate assimilation (uptake); 4, ammonification (deamination, mineralization, decomposition); 5, aerobic ammonia oxidation; 6, nitrite oxidation; 5 + 6, nitrification; 7, denitrification; 8, anaerobic ammonia oxidation. Modified after Kowalchuk and Stephen (2001), reproduced with permission from Annual Reviews.

as inert dinitrogen gas, N_2, which surrounds all plants and animals, but is unavailable to them. Only certain prokaryotes can utilize dinitrogen gas and reduce it to ammonia, which is then taken up into the biosphere. *Nitrogen fixation* is therefore a key link in the nitrogen cycle (process 1 in Fig. 4.8), although on a global scale it is now surpassed by industrial nitrogen fixation in behalf of the production of mineral fertilizers (the Haber process). Among the nitrogen-fixing organisms are freeliving aerobic bacteria such as Cyanobacteria, free-living anaerobic bacteria, such as *Clostridium*, symbiotic bacteria of the family Rhizobiaceae, living in association with leguminous plants, and symbiotic bacteria of the actinomycete genus *Frankia*, living in association with nonleguminous plants such as alder trees (*Alnus*) and buckthorn (*Hippophae*). The ability to fix nitrogen is not associated with a monophyletic group of prokaryotes, but is found scattered among several bacterial and archaeal lineages, phototrophs, chemo-organotrophs, and chemolithotrophs. Nitrogen fixers are also called *diazotrophs* (deriving nutrition from dinitrogen).

The reduction of N_2 to NH_3 is a highly energy-demanding process, catalysed by the enzyme complex *nitrogenase*, which consists of two units, dinitrogenase and dinitrogenase reductase Dinitrogenase contains iron and molybdenum in its active centre, whereas dinitrogenase reductase contains only iron. Because the fixation of N_2 requires such a large amount of energy, the activity of these enzyme complexes is highly regulated. The genetic architecture of nitrogen fixation has been studied in detail in the model species *Klebsiella pneumoniae* (Gammaproteobacteria, Enterobacteria), a species normally living in soil or water but occasionally causing pneumonia in humans, hence its name. The structural and regulatory genes associated with nitrogen fixation are organized in a complex network of operons, the *nif* regulon, encoding no less than 20 different proteins, all dealing with regulation, maturation, and assembly of the nitrogenase complex (Fig. 4.9). The various *nif* genes are indicated with letters; for example, *nifK* encodes the β-subunit of denitrogenase, *nifD* encodes the α subunit of the same enzyme, and *nifH* encodes the enzyme dinitrogenase reductase. The regulon also includes several proteins that support the

processing of Mo, its insertion in the apo-enzyme, and expression regulators and inhibitors, etc. Similar *nif* regulons are present in other nitrogen fixers such as cyanobacteria. Some of the *nif* genes have been used succesfully to develop probes for assessing diazotroph microbial communities in the field (see below). In addition to the *nif* cluster, symbiotic nitrogen fixers have a *fixABCX* gene cluster, which encodes *electron-transport flavoproteins* (ETFs), which presumably support the electron transport to nitrogenase. This *fix* gene cluster is under transcriptional control by one of the regulatory proteins from *nif*, NifA.

Nitrogen can assume six different oxidation states, with the valency of the nitrogen atom varying from +5 in nitrate to −3 in organic N (amino groups and heterocyclic compounds). Nitrogen fixation is a reduction process because it changes N from oxidation state 0 to −3. Another reduction takes place in the process of *denitrification*, where N changes from +5 to 0. Denitrification (process 7 in Fig. 4.8) is the biological conversion of nitrate to nitrogen gas, a process that is detrimental in agriculture, because it diminishes

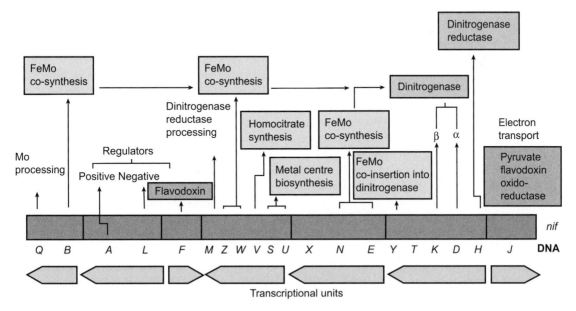

Figure 4.9 Schematic representation of the *nif* regulon in *Klebsiella pneumoniae*. The 20 *nif* genes are indicated by letters (*Q, B, A,* etc.). They are transcribed as seven polycistronic messengers indicated below the genes; the direction of transcription differs between operons. The functions of the various proteins are indicated in the boxes and by arrows. From Madigan *et al.* (2003), with permission from Pearson Education.

the effect of fertilization, but favourable in other situations, such as sewage treatment, because it removes eutrophicating nitrate from the effluent. Most denitrifying prokaryotes are members of the Proteobacteria and they represent a metabolically versatile group, growing both aerobically and anaerobically. Some denitrifying bacteria will also use other electron acceptors such as ferric ion (Fe^{3+}). Four consecutive enzyme systems are involved with denitrification, converting nitrate subsequently to nitrite, NO, N_2O, and finally N_2. The enzymes are referred to as nitrate reductase, nitrite reductase, NO reductase, and N_2O reductase, respectively. Some bacteria, such as *E. coli*, can only carry out the first two steps and therefore generate NO, not N_2. Natural denitrification, although considered beneficial in the combat against eutrophication, has negative side effects because it is always accompanied by the emission of NO and N_2O, which in themselves are greenhouse gases and by reaction with water in the atmosphere contribute to the acidity of rain. The reduction of nitrate in the environment is also called *dissimilative nitrate reduction*, to distinguish it from nitrate reduction following uptake in plants, which is called *assimilative nitrate reduction* (process 3 in Fig. 4.8).

The biochemistry of denitrification has been studied in detail in *Paracoccus denitrificans*, which can conduct all four denitrification steps. This organism can express both a periplasmic nitrate reductase (between cell membrane and outer membrane) and a membrane-bound nitrate reductase, where the latter is formed only under anaerobic conditions and also functions exclusively under these conditions. A third nitrate reductase is active in the cytoplasm and associated with assimilatory nitrate reduction. The membrane-bound reductase is encoded by a *nar* gene cluster and the periplasmatic nitrate reductase by the *nap* operon. As in the case of nitrogen fixation, these clusters include both structural genes, encoding subunits of the enzymes, and regulatory proteins. Another gene cluster, *nir*, encodes nitrite reductase, whereas *nor* genes encode NO reductase and a *nos* genes encode N_2O reductase. Because of the toxicity of nitrite and nitric oxide, the expression of

these genes must be highly co-ordinated; a comprehensive set of regulators (including nitrite sensors) is involved, as well as transporters that may prevent intracellular accumulation of nitrite. Van Spanning *et al.* (2005) provided a recent review of the distribution of denitrification-associated genes in Bacteria and Archaea.

The third link in the nitrogen cycle that we discuss here is the process of *nitrification*, of which the first and rate-limiting step is aerobic ammonia oxidation (process 5 in Fig. 4.8). It is commonly assumed that, unlike nitrogen fixation and denitrification, the capacity for ammonia oxidation has a limited distribution within the bacterial kingdom. However, recent metagenomic screening has suggested that ammonia oxidation may be more widespread (see Section 4.4). Up until now, all terrestrial ammonia-oxidizing bacteria have been found in the Betaproteobacteria, involving the well-known genera *Nitrosomonas* and *Nitrosospira*. Another group of ammonia oxidizers is present in the Gammaproteobacteria (*Nitrosococcus*), but these organisms seem to have a limited distribution, being mainly marine. Nitrite oxidizers are found in the Alphaproteobacteria (*Nitrobacter*), the Deltaproteobacteria (*Nitrospina*), and the phylum Nitrospirae (Kowalchuk and Stephen 2001). The β-ammonia-oxidizing bacteria *Nitrosomonas*/*Nitrosospira* group forms a monophyletic lineage and can be probed with a single set of 16 S rRNA primers. Most ammonia-oxidizing bacteria are autotrophic; in addition to the system for nitrate oxidation they have the Calvin cycle for CO_2 fixation. The ATP and reducing power requirements for this process, added to the relatively limited amount of ATP generated by nitrate oxidation, may explain the slow growth of these bacteria in laboratory cultures.

The key enzyme in aerobic ammonia oxidation is *ammonia monooxygenase*, which oxidizes ammonia to hydroxylamine (NH_2OH), which is then oxidized further to nitrite by hydroxylamine oxidoreductase. Ammonia monooxygenase splits O_2, and incorporates one of the oxygen atoms in NH_3, while the other reacts with H^+ to form water. Similar reaction schemes using monooxygenase enzymes apply to the oxidation of methane and organic compounds such as benzene. The genome of *Nitrosomonas europaea* has been completely sequenced and serves as the main genetic model for nitrification research (Chain *et al.* 2003). Two genes in the *amo* cluster of *Nitrosomonas*, *amoA* and *amoB*, encode the two subunits of monooxygenase, whereas a third gene, *amoC*, encodes a supporting membrane protein. *AmoA* is one of the genes that is used in microarray-based profiling of environmental samples (see below). The enzyme nitrite oxidoreductase, which is used by nitrite oxidizers to produce nitrate from nitrite, is encoded in the *nor* operon, which is essentially the same gene cluster as used by denitrifiers to reduce NO to N_2O.

In addition to being oxidized under aerobic conditions, ammonia can also be oxidized under anoxic conditions by means of a process called *anammox*. This involves the joint use of ammonia and nitrite in the overall reaction:

$$NH_4^+ + NO_2^- \rightarrow N_2 + 2H_2O$$

indicated by process 8 in Fig. 4.8. The first organism that was identified as responsible for anoxic ammonia oxidation, *Brocadia anammoxidans*, is a member of the bacterial phylum Planctomycetes, a somewhat unusual group of prokaryotes because they have membrane-enclosed compartments inside the cell. One such compartment, the anammoxosome, is geared specifically towards the anammox process. *Br. anammoxidans* is an autotroph and uses nitrite as an electron donor to fix CO_2. This is the same principle as used by nitrite oxidizers (e.g. *Nitrobacter*). In addition, the same enzyme, nitrite oxidoreductase, is used, encoded in the *nor* operon (Table 4.6).

Finally, the mineralization of organic nitrogen to ammonia (*ammonification*) takes place during the decomposition of organic material and involves *deaminase* reactions. Many heterotrophic organisms can do this and there are many different deaminases, so no key enzyme for ammonification can be indicated. Under neutral to acid pH conditions ammonia from decomposition reacts with water and is present in the environment as an NH_4^+ ion, adsorbed to the sediment or soil-exchange complex. At low pH very little NH_4^+ dissociates to form NH_3 and because NH_3, not the ammonium ion, is required for ammonia oxidation, low pH was long thought to be a limiting factor for nitrification; however, even at pH 3 in nitrogen-saturated coniferous forest soil a substantial rate of nitrification by acid-tolerant *Nitrosopira* has been measured (Laverman *et al.* 2001).

This overview of the nitrogen cycle shows that it is accompanied by a great variety of redox reactions and conducted by many different microorganisms. For most of the transitions the key enzymes have been well characterized. It is evident that in some cases similar enzymes conduct different reactions in different organisms and apparently their action depends on the context of the organism in which they are expressed. Several genes are indicative of key steps in the nitrogen cycle and these can be used in environmental genome-profiling studies.

4.3.3 Other nutrient cycles

Sulphur, like nitrogen, can take different oxidation states and the transitions between these are exploited by many microorganisms. One of the best known reactions is *dissimilative sulphate reduction*, the conversion of sulphate to sulphide, which can be liberated as smelly H_2S gas or precipitate with cations. Sulphate-reducing bacteria are found in many different lineages of Bacteria and Archaea, as we have seen above (Fig. 4.6). The species *Desulfovibrio vulgaris* (Deltaproteobacteria), whose genome has been sequenced recently (Heidelberg *et al.* 2004), is a model species for the study of sulphate reduction. Most sulphate-reduction processes take place under anoxic conditions and use electron donors from organic compounds. Intertidal sediments are prime examples of sulphate-reducing environments, because of the combination of high sulphate availability in seawater with abundant organic material from salt-marsh vegetation and anoxic conditions due to regular flooding. The black colour of intertidal sediments is due to the precipitation of sulphide with iron to form FeS. In the sea itself, sulphate reduction is limited by the availability of carbon sources. In addition to sulphate reduction in the environment, sulphate is also reduced after uptake by microorganisms, plants, and animals (*assimilative sulphate reduction*), because the major form of sulphur in organic molecules is the reduced form of the thiol (SH) group in proteins.

The reduction of sulphate proceeds through a series of steps. Sulphate is first activated by binding to ATP to form adenosine phosphosulphate (APS) before it can be reduced to sulphite by the enzyme APS reductase. In the next step sulphite is reduced to sulphide by sulphite reductase. The two reactions require electrons from a suitable donor, either hydrogen or acetate, which is oxidized to water or CO_2. The *aps* locus of sulphate reducers encodes the key enzyme APS reductase, and *dsrA* and *dsrB* encode the α and β subunits of sulphite reductase. The three genes are highly conserved in several deep-branching phyla of the Bacteria and Archaea and seem to have been subject to lateral gene transfer. Sulphate reducers often occur in close association with methanogenic Archaea and with methanotrophic (methane-oxidizing) Archaea and they use each other's products (another case of syntrophy). Taken together, the only oxidants that this small community requires are CO_2 and sulphate, compounds which existed before the Earth became oxygenated by photosynthesis. Therefore, communities consisting of sulphate reducers, methanogens, and methanotrophs, found in salt marsh sediment and in association with hydrothermal vents on the ocean floor, could represent the modern descendents of extremely ancient communities dating back to the beginning of the Proterozoic era (Teske *et al.* 2003).

In addition to H_2S, a major source of sulphur emission to the atmosphere is due to the volatile compound dimethyl sulphide (DMS; CH_3–S–CH_3). Microbial mats of salt marshes are well-known emission sources for DMS. This compound is a degradation product of β-dimethylsulphonio-propionate (DMSP), which is produced in large amounts by many marine algae, cyanobacteria, and salt-marsh plants, probably to protect them against osmotic shock (Yoch 2002). DMS is assumed to have an anti-greenhouse effect, because sulphate aerosols generated by DMS act

as cloud-condensing nuclei, reducing the amount of sunlight reaching the Earth's surface. However, by generating sulphate, DMS may contribute significantly to acid rain in places where there is no air pollution and the concentration of aerosols is low. When the DMSP-containing organisms die off or are grazed by zooplankton, DMSP becomes available in the environment and is converted to DMS by means of an enzyme, DMSP lyase, which is expressed by sulphate-reducing bacteria such as *Desulfovibrio* and *Roseobacter*. Some *Roseobacter* live in close association with dinoflagellates, which are major producers of DMSP, and it is assumed that the bacteria may benefit from the DMSP produced by the dinoflagellates (Miller and Belas 2004).

Reduced sulphur compounds such as elemental sulphur and sulphide can also be used as electron donors in the process of *sulphur oxidation*. Sulphur oxidizers often live attached to a surface, because elemental sulphur does not dissolve in water. Oxidation, from either S^0 or S^{2-}, leads to the production of sulphite, sulphate and H^+ and may lower the ambient pH considerably. Sulphur oxidizers are autotrophs that fix CO_2 by means of the Calvin cycle. The sulphur-oxidizing system was investigated in detail in the Gram-negative lithoautotrophic species *Paracoccus pantotrophus* (Friedrich *et al.* 2001). The *sox* gene cluster of *P. pantotrophus* comprises at least 15 different protein-encoding genes. Seven of these genes encode four proteins with sulphite oxidase and sulphur dehydrogenase activities. The cluster also contains transcriptional regulators and transport regulators. Evidence has emerged that similar proteins are present in other bacteria, but not in the archaeon *Sulfolobus solfataricus*. Homologies can be drawn between the genes of the various species (Fig. 4.10), and a common mechanism for all sulphur-oxidizing bacteria is emerging (Friedrich *et al.* 2001).

Regarding the *phosphorus cycle*, we note that it is dominated by physical processes (erosion, precipitation) more than by biological influences. Phosphorus undergoes hardly any microbe-mediated redox reactions in the environment; an exception is the process of phosphite oxidation conducted by

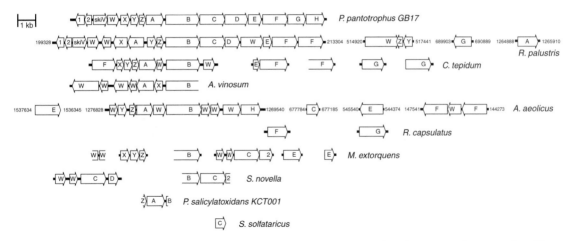

Figure 4.10 Map of the *sox* gene cluster of *Paracoccus pantotrophus* and other sulphur-oxidizing bacteria. Homologies are indicated by common letters. Explanation of gene codes: 1, ArsR-type transcriptional regulator; 2, periplasmic thioredoxin; skiV, membrane-bound transporter involved with cytochrome *c* biogenesis; W, periplasmic thioredoxin; X, Y, Z, A, B, C, and D, structural genes encoding four proteins—SoxXA, SoxYZ, SoxB, and SoxCD; E–H, proteins of unknown function. *P. pantotrophus, Paracoccus pantotrophus; R. palustris, Rhodopseudomonas palustris; C. tepidum, Chlorobium tepidum; A. vinosum, Allochromatium vinosum; A. aeolicus, Aquifex aeolicus; R. capsulatus, Rhodobacter capsulatus; M. extorquens, Methylobacterium extorquens; S. novella, Starkeya novella; P. salicylatoxidans, Pseudaminobacter salicylatoxidans; S. solfataricus, Sulfolobus solfataricus.* After Friedrich *et al.* (2001), by permission of the American Society for Microbiology.

lithoautotrophic sulphate reducers (Schink and Friedrich 2000), in which the redox state of phosphorus changes from +3 to +5. All organisms need phosphate, which they take up using membrane-bound transport systems, of which there is a great variety. Phosphorus may be present in the environment as *polyphosphates* (long chain-polymers of phosphate, contrasted with ortho-phosphate, PO_4^{3-}) and these need to be hydrolysed before take-up. Microorganisms use exopoly-phosphatase to hydrolyse polyphosphates. The *ppx* operon of *E. coli* K12 encodes this enzyme, and presumably many microorganisms have similar loci. In addition, all cells, prokaryotic and eukaryotic, accumulate polyphosphate internally, where it functions as a phosphate reserve and as a sequestration mechanism for otherwise toxic cations, for example Ni and Pb (Kulaev and Kulakovskaya 2000). Another major source of phosphate in the environment is organic phosphorus in dead organic matter, from which phosphate is released during decomposition by phosphatase activity of heterotrophic microorganisms. The fact that biota do not directly alter the chemical appearance of phosphorus in the environment to a great extent does not diminish the important growth-limiting effect that phosphate has in most aquatic systems and to an extreme degree in the sea.

Iron has two oxidation states, ferric (III) and ferrous (II), and undergoes redox reactions in the environment catalysed by microorganisms. Since the valency of the iron atom is correlated with differences in solubility and stability of its compounds, iron redox reactions have important consequences for the biogeochemical cycles of other elements, especially sulphur and phosphorus. For example, Fe^{2+} may precipitate with S^{2-} to form insoluble FeS, preventing the release of gaseous H_2S, and Fe^{3+} oxides may bind phosphate, preventing the dissolution of phosphate in water. The latter phenomenon is well-known among limnologists who are familiar with the fact that release of phosphate from sediment into the overlying water is much greater under anaerobic conditions than from the same sediment under aerobic conditions.

The most important bacteria conducting *iron reduction* in the environment are from the genus *Geobacter* (Deltaproteobacteria). Other iron reducers are *Shewanella* (mainly marine), *Geothrix*, and *Anaeromyxobacter* (North *et al.* 2004). Most iron reducers can also use other metal ions as electron acceptors, such as uranium (VI) and manganese (IV). Various organic molecules (e.g. acetate, benzene and toluene) can act as electron donors in Fe^{3+}-dependent anaerobic growth. By oxidizing organic molecules in conjunction with iron reduction, iron-reducing bacteria may contribute significantly to the natural purifying capacity of subsoils, for example when ferric-rich aquifers are polluted by leachate from landfills or oil-storage tanks (Lovley 2003; see also Fig. 4.4). The genomes of *Shewanella oneidensis* and *Geobacter sulfurreducens* have been sequenced completely (Heidelberg *et al.* 2002; Methé *et al.* 2003). The genomes of these species contain an unprecedented number of *c*-type cytochromes, some with two or more haem groups. The abundance of cytochromes highlights the importance of electron transport to these organisms. Three different strategies are followed by Fe^{3+}-reducing bacteria (Lovley *et al.* 2004): they can attach to an iron-rich surface and transfer electrons directly to the mineral, they can release iron chelators to bring iron to the cell where it can be reduced, or they can use extracellular *electron shuttles* to transfer electrons back and forth between the mineral and the cell. A model for the latter mechanism, iron reduction at a distance, was proposed by Croal *et al.* (2004) and Lovley *et al.* (2004) (see Fig. 4.11a).

Different iron-reducing bacteria seem to have different sets of cytochromes, although there are also some genes that are common among the three species *G. sulfurreducens*, *Sh. oneidensis*, and *Desulfovibrio vulgaris*. These genes are potential markers for iron reduction activity in the environment. Recently, Chin *et al.* (2004) showed that expression of *omcB*, a gene encoding an outer membrane *c*-type cytochrome (Fig. 4.11a), was very well correlated with Fe-reduction rates in *G. sulfurreducens*. It is not yet certain that this gene can be used as a universal indicator of Fe-reducing capacity in the environment; the relatively small overlap between the derived gene complements of different iron-reducing bacteria suggests that their metal-reducing capabilities are not simply related to their sharing an exclusive set of genes.

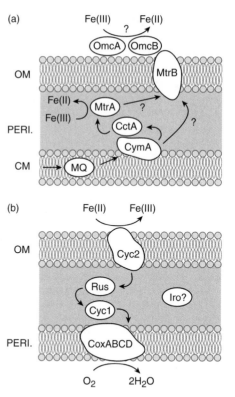

Figure 4.11 Models for electron-shuttling processes involved in (a) Fe^{2+} reduction by *Shewanella oneidensis* MR-1 and (b) Fe^{3+} oxidation by *Acidithiobacillus ferrooxidans*. OM, outer membrane; PERI., periplasmatic space; CM, cell membrane; OmcA and OmcB, outer-membrane cytochromes A and B; MtrA and MtrB, metal-reduction proteins A and B; CctA, small tetrahaem *c*-type cytochrome A; CymA, inner-membrane cytochrome A; MQ, menaquinone; Cyc1 and Cyc2, *c*-type cytochromes 1 and 2; Rus, rusticyanin; Iro, ferredoxin; CoxABCD, cytochrome oxidase complex. After Croal *et al.* (2004) with permission from Annual Reviews.

In addition to metal reductase activity associated with anaerobic respiration, iron is also reduced during uptake into the cell (assimilative iron reduction). Fe is often scarce in the environment because of the insolubility of $Fe(OH)_3$. Therefore many organisms have developed mechanisms of scavenging Fe^{3+} by excreting compounds in the medium with an extremely high iron affinity. Citrate is often used for this purpose, but also molecules completely geared to Fe chelation, *siderophores*, are used. These compounds may either donate Fe to a membrane-bound transporter or be taken up entirely by pinocytosis after binding to a membrane receptor. Genes encoding ferric citrate transporters and iron siderophore receptor proteins are found in many bacteria. Expression of siderophore receptor proteins is regulated by membrane-bound sensor systems encoded by *fecIR* genes.

Few bacteria use *iron oxidation* as an energy-yielding reaction; large amounts of iron need to be oxidized in order to deliver sufficient energy for growth. Iron oxidation can take place under anaerobic conditions by nitrate reducers and photoautotrophs (Straub *et al.* 2004), but most iron-oxidizing bacteria are aerobic acidophiles. At neutral pH hardly any energy can be gained from iron oxidation under aerobic conditions because ferrous ion then oxidizes spontaneously to ferric iron. Iron oxidation leads to the formation of $Fe(OH)_3$ precipitates; this process is recognizable in many places with a boreal climate, where iron-rich groundwater from acidic marshes comes into contact with aerobic sandy soils. Another environment where iron oxidation is important is acid-mine drainage; a very acid leachate develops under the influence of iron oxidizers and sulphur oxidizers when iron sulphide comes into contact with the air in coal mines (see Section 4.4).

Model species for iron oxidation are *Acidithiobacillus* (*Thiobacillus*) *ferrooxidans* (Gammaproteobacteria) and *Leptospirillum ferrooxidans* (Nitrospirae). A crucial role in iron oxidation by *Ac. ferrooxidans* is played by a periplasmic copper-containing protein, *rusticyanin* (Fig. 4.11b). This protein oxidizes ferrous ion in the periplasm and then reduces a cytochrome in the cell membrane. After a very short electron-transport chain O_2 is reduced to water under consumption of H^+, allowing the generation of ATP. The *rus* operon of *Ac. ferrooxidans* encodes the rusticyanin protein as well as two *c*-type cytochromes (*cyc1* and *cyc2*) and four subunits of a cytochrome *c* oxidase (*coxABCD*). Recent work by Yarzábal *et al.* (2004) has shown that expression of this operon is induced by ferrous ions and so *rus* seems an excellent candidate for indicating the capacity of iron oxidation in the environment (Table 4.6).

Regarding the *calcium cycle*, a significant influence, at least in oceanic systems, is due to coccolithophorid algae (Coccolithophorida,

Chrysophyta). *Emiliania huxleyi* is a well-known representative from this group, recognized by its beautiful platelets of calcite, *coccoliths*, which form a skeleton structure around the cell (Fig. 4.12). *Em. huxleyi*, although very small, is extremely common in the ocean, such that extensive fields of *Em. huxleyi* blooms can be observed even from space. Coccolithophorids are favourite study objects among palaeontologists because they are abundant in marine sediments extending back to the Cambrian era, and may be used as indicators for climate reconstruction. The ecological relevance of these organisms is due to the fact that they fix carbon not only for synthesis of organic molecules but also for coccolith formation. By sedimentation of cells to the bottom of the ocean significant amounts of carbon and calcium are withdrawn from the biological sphere of influence. The coccoliths are formed inside the cell in specialized calcifying vesicles of the Golgi apparatus, and later pushed outwards. The formation of calcifying vesicles is an interesting model for *biomineralization*. It is assumed to involve a vacuolar-type ATPase (V-ATPase), a proton pump that is known from many other eukaryotes as well. Ca^{2+} and CO_3^{2-} ions are assumed to precipitate as $CaCO_3$ on a macromolecular template inside the vesicle.

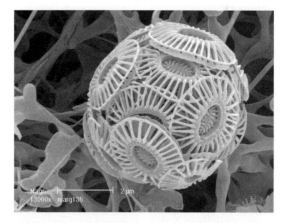

Figure 4.12 Scanning electron micrograph of *Emiliania huxleyi*, a representative of the Coccolithophorida (Chrysophyta, golden algae), well-known for its beautiful external platelets, coccoliths, made of calcium carbonate. Coccolithophorids are a key link in the oceanic carbon and calcium cycles. Courtesy of Jeremy Young and Markus Geisen, Palaeontology Dept., The Natural History Museum, London.

Coccolithophores have very large genomes, exceeding 200 Mbp, and there is hardly any genomic information about *Em. huxleyi*, although Wahlund *et al.* (2004) developed a cDNA library and sequenced 3000 ESTs. The library was found to contain multiple copies of genes for calnexin and calreticulin, two chaperones that play a key role in calcium homeostasis, but the authors did not find a V-ATPase. Further research is needed before a definitive marker gene for calcification can be identified.

The silicium cycle in the ocean is greatly influenced by diatoms (Bacillariophyta) who incorporate silicium in their silica cell wall or *frustule*. Like coccolithophorids, diatoms withdraw carbon from the atmosphere by sinking to the bottom of the sea, and in doing so also remove silicium from the sea water. The delicate structure of diatom frustules and their bewildering array of forms (Fig. 4.13) have fascinated biologists ever since the invention of the microscope and their reproducible fine structure presently also raises interest among nanotechnologists (Bradbury 2004). The genome of *Thalassiosira pseudonana*, a diatom with a centric cell shape (resembling a Petri dish, in contrast with elongate, pennate cells), has recently been

sequenced (Armbrust *et al.* 2004). Not surprisingly, the genome contains many genes with a function in silicium metabolism. The authors identified three genes that encode transporters for active uptake of silicic acid. Silica precipitates in a silica-deposition vesicle, in a process controlled by a family of phosphoproteins called silaffins, that are embedded in the frustule while it is being formed. Five such silaffins could be identified in the genome of *Th. pseudonana*, but surprisingly these did not match proteins of similar function in other diatoms and sponges. Another group of proteins associated with frustule formation are the frustu-lins (casing glycoproteins), which protect the frustule at the outside.

4.3.4 Microarray-based screening of functional genes

Not all the genes discussed above and mentioned in Table 4.6 have been exploited for environmental screening with genomics technology. Until now, the nitrogen cycle has attracted most attention, primarily because it offers such a rich variety of enzymatic systems, many of which have well-described biochemistries (Ye and Thomas 2001). We saw in Table 4.2 that *functional gene microarrays* are designed to detect the presence of genes with specific functions in the environment, irrespective of the species. Depending on hybridization conditions, all genes showing a sequence similarity of more than 80% with the probes will be detected. Microarray screening of functional genes provides an indication of the capacity to conduct the function, it does not demonstrate the function itself. Assessment of the function itself requires isolation of RNA and reverse transcription to cDNA, followed by competitive microarray hybridization, as in the case of transcription profiling in model organisms (Section 2.6). Such studies are more difficult in environmental microbial communities than in model eukaryotes, because of the instability of microbial mRNA, which is characterized by long transcription products without poly(A) tails covering several genes, and the ensuing difficulty of obtaining reliable RNA samples other than rRNA from environmental matrices (Ye *et al.* 2001).

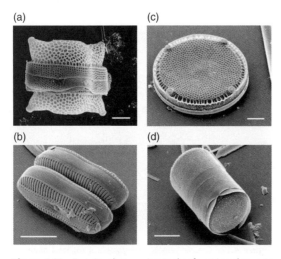

(a)　　　　　　(c)

(b)　　　　　　(d)

Figure 4.13 Scanning electron micrographs of centric and pennate species of diatom. (a) *Biddulphia reticulata*, (b) *Diploneis* sp., (c) *Eupodiscus radiatus* (single valve), (d) *Melosira varians*. Scale bars, 10 μm. Courtesy of Mary Ann Tiffany, San Diego State University.

One of the first attempts to apply DNA array technology for assessing functional gene diversity was made by Wu *et al.* (2001). These authors developed a prototype microarray with 104 probes representing PCR-amplified genes from bacteria involved in nitrogen cycling: the nitrate reductase genes *nirS* and *nirK*, and the ammonium mono-oxygenase gene *amoA*. Strong hybridization signals were obtained with DNA extracted from a marine sediment and from a surface soil. The diversity of *nirS* and *nirK* genes was similar in the two environments, indicating comparable capacities for denitrification. Similar results were obtained by Tiquia *et al.* (2004), who used an oligonucleotide microarray with several hundred 50-mer probes developed from genes in the cycles of nitrogen, carbon, and sulphur (*nirS, nirK, amoA, nifH, pmoA,* and *dsrAB*). Although the potential for detection was clearly demonstrated, sensitivity is still a critical issue in these hybridization experiments, because they use DNA extracted directly from the environment without prior PCR amplification. It is estimated that only genes from populations contributing more than 5% to the environmental genomes can be detected in this way (Cho and Tiedje 2002); however, this figure can be brought down to 1% by using modified 70-mer probes printed on special substrates (Denef *et al.* 2003).

Bodrossy *et al.* (2003) developed a functional gene microarray for methanotrophs. A survey of published *pmoA* and *amoA* gene sequences allowed the construction of a nested set of 68 probes that together could diagnose almost the entire known diversity of methanotrophs. A microarray with these probes was fabricated and used to survey methanotroph diversity in soils and microcosms. The study confirmed that all cells representing more than 5% of the targeted community could be detected.

As an example of successful use of functional gene microarrays (with prior PCR amplification) in an ecological context we will discuss the work by Taroncher-Oldenburg *et al.* (2003). These authors developed a microarray with 70-mer oligonucleotides targeting genes encoding the functions of nitrification (*amoA*), nitrogen fixation (*nifH*), and denitrification (*nirK, nirS*). Samples were taken from sediments in the Choptank river (in Maryland, USA) and extracted DNA was amplified with gene-specific primers. Data for *nirS* diversity are shown in Fig. 4.14. Among 64 *nirS* probes, 29 were detected in a sample from the upstream area, and 11 in the downstream area. Hybridizations were distributed over a great variety of sequences; that is, they were not limited to a distinct cluster in the dendrogram of probe sequences. The data suggest that the composition of denitrification genes is quite different between the two sampling sites. It is also evident that the diversity of denitrification genes is lower in the downstream location. Denitrification rates measured in sediment cores from the downstream river station were also lower than in the upstream location. So, this study suggests a positive correlation between community biodiversity and ecological function, in accordance with the 'rivet hypothesis' mentioned at the beginning of this chapter. The differences in the composition of the denitrifier community were possibly related to an environmental gradient of salinity and dissolved organic carbon in the river.

Studies like this are still relatively rare. This may be taken to illustrate that the declared promises of microarray-based microbial detection have not yet been realized. Most studies reported in the literature are still in the development and testing phase; more work needs to be done to increase sensitivity and specificity (Cho and Tiedje 2002; Zhou 2003). In addition to solid-support microarrays, the use of membranes for probe immobilization (macroarrays) is another attractive option (Jenkins *et al.* 2004; Steward *et al.* 2004). It is to be expected that technology development will not halt and that the coming years will see more definite tests of the diversity-function hypothesis with the aid of genomics technology.

An interesting new development is the combination of *stable-isotope probing* (SIP) with microarray analysis (Dumont and Murrell 2005). In SIP, a substrate (e.g. methanol), enriched with a stable isotope (e.g. ^{13}C) is added to an environmental sample and after a designated incubation time DNA is extracted. DNA from microorganisms that have taken up the substrate can be separated from other DNA by using CsCl density-gradient

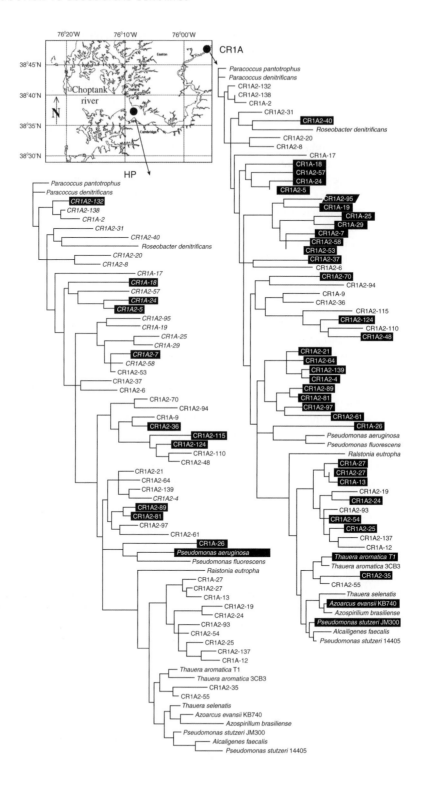

centrifugation. The diversity of this DNA can then be screened with a microarray, usually targeting the 16S rRNA gene. By focusing only on DNA in which ^{13}C is incorporated, a profile is obtained of a single functional guild, defined by the capacity to metabolize the substrate (DeLong 2001; Gray and Head 2001; Radajewski *et al.* 2003; Wellington *et al.* 2003).

Another promising, newly developed, technique goes under the name *isotope array* (Adamczyk *et al.* 2003). As in SIP, a labelled substrate is added to a community, but the label is a radioisotope such as ^{14}C. rRNA is then extracted, labelled with a fluorescent label, and screened for 16S rRNA diversity using a phylochip (see Section 4.2). By scanning for both fluorescence and radioactivity one can single out species from a designated functional guild that have proven to be active because they have taken up the substrate. Adamczyk *et al.* (2003) used this approach to profile nitrifying communities in activated sludge from two different wastewater-treatment plants. Diversity of ammonia-oxidizing bacteria in the sludge was assessed from the incorporation of added bicarbonate (leading to radioactively labelled rRNA) combined with hybridization to ammonia-oxidizing bacterium-specific 16S rRNA probes (which is possible since ammonia-oxidizing bacteria constitute a monophyletic lineage). Such a combination of different molecular tools seems to be particularly promising.

4.4 Reconstruction of functions from environmental genomes

An exciting new development in microbial genomics is the exploration of communities by isolating large fragments of DNA directly from the environment and cloning them into vectors such as BACs, followed by probing, sequencing, or screening for functions (Ball and Trevors 2002;

Handelsman 2004; Riesenfeld *et al.* 2004a; Tiedje and Zhou 2004; Allen and Banfield 2005; DeLong 2005; Schleper *et al.* 2005). This approach, designated *community genomics* or *metagenomics*, allows insight into microbial diversity, and it may lead to the discovery of new genes and functions, novel metabolic pathways, and previously unknown properties of microorganisms that cannot be cultured. Most importantly, metagenomic analysis may lead to reconstruction of functions from genome sequences of organisms that have never been cultured. Comparative analysis of different environments has demonstrated that metagenomes contain habitat-specific signatures that can be used for environmental diagnosis (Tringe *et al.* 2005). Sometimes genes found in metagenomes may be 'brought to life' in the laboratory by expressing the DNA segment in a suitable host.

A characteristic property of the metagenomics approach is that genes and functions are studied without consideration of the species from which the DNA derives. The metagenome of a habitat thus consists of the collective genomes of all organisms together. Of course, such an approach has its limitations, because a specific cell environment and the joint expression of several genes together in a delimited volume are often crucial to the function. In addition, the genomes of different species will differ in dynamics and responses to environmental change, and so the composition of a metagenome could be highly variable in time (DeLong 2001). Nevertheless, several important discoveries have been made by probing communities in this way, as we will see in the examples discussed below.

Two different approaches may be discerned for screening environmental genome libraries: *function-driven screening* and *sequence-driven screening* (Schloss and Handelsman 2003). In the first approach, the aim is to identify clones in the library that express a certain function, often one that

---

Figure 4.14 Comparison of *nirS* (cytochrome cd_1-containing nitrite reductase) gene diversity in two samples taken from sediments of the Choptank river, Maryland, USA. A total of 64 *nirS* probes (each 70 bp) were spotted on a microarray and the similarity between these sequences is shown as a dendrogram. Probes were developed from environmental *nirS* sequences isolated earlier (indicated by codes) and from pure cultures of bacteria (indicated by species names). Positive hybridizations are shown in white type on a black background. The diversity of *nirS* sequences is greater in the upstream location (CR1A) than in the downstream location (HP). After Taroncher-Oldenburg *et al.* (2003) by permission of the American Society for Microbiology.

has potential applications in medicine, agriculture, or ecology. For example, one may be interested in genes from biosynthetic pathways for antibiotics or genes associated with crucial links in biogeochemical cycles. Usually the frequency of active clones is quite low, so one needs a simple assay by which large numbers of clones in a library can be tested quickly. A clever, high-throughput, method was developed for this purpose by Uchiyama et al. (2005). The authors made use of the fact that many genes are induced by the substrate that they catabolize. A vector containing green fluorescent protein, suitable for shotgun cloning, was used with the effect that host cells with an insert carrying the promoter of the target gene expressed green fluorescent protein in the presence of the target substrate; these cells could then be sorted by an automated cell-sorting system.

In the second type of screening, sequence-driven screening, one uses hybridization probes to detect clones containing a desired known sequence. The probe can be a phylogenetic anchor, such as a 16S rRNA sequence, or a specific functional gene. A variant to sequence-driven screening was proposed recently by Sebat et al. (2003), who screened a metagenomic library with a microarray. In their study, the microarray consisted of probes from a stable reference community, cloned in a cosmid library. Hybridizations with the metagenomic library were evidence of the presence of certain species. A great advantage of microarray-based screening of libraries is that many probes are used in parallel.

Library screening is usually followed by partially or completely sequencing the clones of interest. In addition, the library may also be sequenced from the start, without prior screening, if one is interested in all sequences. In this case the library is not prepared with large-insert cloning vectors such as BACs, but with small-insert plasmid vectors that are suitable for direct sequencing. By picking out clones at random and following the WGS philosophy (see Section 2.2), assembly of multiple complete genomes is then attempted. We will discuss examples of all these approaches below.

4.4.1 Marine community genomics

One of the most appealing results of community genomics is the discovery by Béjà et al. (2000a) of proteorhodopsin in the ocean. As we have seen above, proteorhodopsin is a retinal-dependent light-driven proton pump, which may support a photoheterotrophic lifestyle of marine bacteria. The widespread presence of this pathway in the carbon cycle was an unexpected outcome of genomics exploration. Crucial to the approach applied by Béjà et al. (2000a) was the construction of a large-insert BAC library after preparation of DNA using a special type of electrophoresis, *pulsed-field gel electrophoresis* (Béjà et al. 2000b). With this technique, applied to environmental DNA digests, it was possible to isolate high-molecular-mass DNA fragments up to several hundred kilobase pairs. The library was screened with 16S rRNA probes to survey the taxonomic diversity and to find clones with new species of Archaea and Bacteria. Béjà et al. (2000a) decided to sequence a 130 kbp genomic fragment from a clone in which the 16S probe had detected an rRNA sequence of an uncultivated member of marine Gammaproteobacteria, the SAR86 group. Sequencing the rest of the clone revealed an ORF for a rhodopsin-like protein called proteorhodopsin, which showed similarity with rhodopsin genes from extreme halophilic Archaea and the fungus *Neurospora crassa* (Fig. 4.15).

Rhodopsins act as transmembrane channels that can bind the chromophore retinal (a derivative of vitamin A) to become sensitive to light. Absorption of light energy by the protein–retinal complex leads to a series of conformational shifts, promoting the transport of ions across the cell membrane. In the case of proton transporters, the outside surface of the cell membrane will become charged with protons and the resulting electrochemical membrane potential creates a motive force for another membrane-bound molecule, H^+-ATPase, to drive ATP synthesis. Three groups of rhodopsins are present in Archaea, one group acting as chloride pumps (halorhodopsins), another as proton pumps (bacteriorhodopsins), and the third

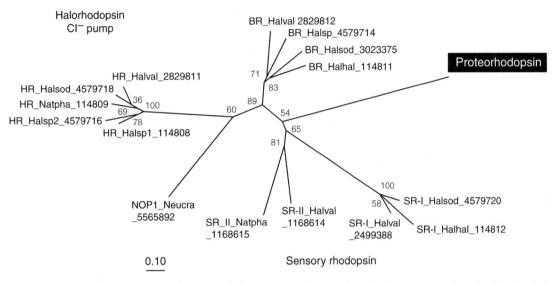

Figure 4.15 Unrooted phylogenetic tree of the proteorhodopsin sequence of an uncultured marine gammaproteobacterium found by Béjà *et al.* (2000a), aligned with rhodopsins in Archaea and the fungus *Neurospora crassa*. HR, halothodopsin, light-driven chloride pumps; BR, bacteriorhodopsin, light-driven proton pumps; SR, sensory rhodopsin; Halsod, *Halorubrum sodomense*; Halhal, *Halobacterium salinarum*; Halval, *Haloarcula vallismortis*; Natpha, *Natronomonas pharaonis*; Halsp, *Halobacterium* sp. (all Archaea); Neucra, *N. crassa* (Ascomycota). The scale bar indicates the proportion of amino acid difference. Reprinted with permission from Béjà *et al.* (2000a). Copyright 2000 AAAS.

as photosensory receptors (sensory rhodopsins). The last group of molecules is related to the opsin proteins found in eyes throughout the animal kingdom. In *N. crassa* a related opsin protein acts in the maintenance of circadian rhythmicity.

That the sequence found in the marine BAC clone represented a functional protein and not a pseudogene of some sort was proven by recombinant expression. *E. coli* cells, transfected with the rhodopsin sequence, expressed the protein and it was shown that a combination of retinal and yellow light triggered cross-membrane proton transport in *E. coli* cell suspensions (Fig. 4.16). Subsequent research using membrane preparations collected directly from seawater exposed to laser-flash photolysis demonstrated that similar photoactive molecules were very common in the environment (Béjà *et al.* 2001).

The widespread occurrence of proteorhodopsins in the sea was confirmed by Venter *et al.* (2004). As

mentioned in Section 4.2, these authors applied the WGS sequencing approach to microbial communities collected from the Sargasso Sea off the coast of Bermuda. The intention was to collect representative sequences from many diverse organisms simultaneously. Whereas the WGS approach is normally used to assemble a genome sequence for an individual species, the community WGS approach aimed to reconstruct as many genomes as possible from the mixture of genomes of varying abundance. In total 1.36 Gbp of microbial DNA sequence was generated, from at least 1800 genomic species, including 148 previously unknown bacterial phylotypes and 1.2 million previously unknown genes. Some distinct groups of sequence scaffolds could be distinguished, one clearly related to a *Burkholderia* species, others to *Shewanella, Prochlorococcus*, and a SAR86 gammaproteobacterium. The presence of *Burkholderia*, a nutritionally versatile genus of Betaproteobacteria,

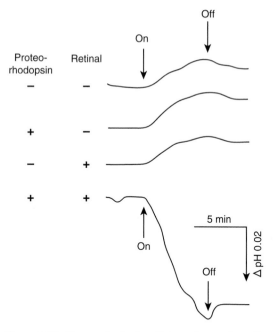

Figure 4.16 Diagram showing the pH change in a medium with cell suspensions of *E. coli* expressing a proteorhodopsin. In the presence of both the protein and retinal, an outward transport of protons occurs when the cells are exposed to yellow light (>485 nm, indicated by On/Off), leading to a decrease in the pH of the medium. Reprinted with permission from Béjà *et al.* (2000a). Copyright 2000 AAAS.

was unexpected, because this genus was considered typical for terrestrial environments. Similarly, *Shewanella* is an abundant genus in aquatic, nutrient-rich environments. The presence of these organisms in the open ocean shows that they have a wider ecological amplitude than thought previously, or that there are nutrient-rich microhabitats (possibly associated with marine animals or anthropogenic waste) in which they may survive.

In the metagenome of the Sargasso Sea Venter *et al.* (2004) found 782 different rhodopsin-like genes, which were classified into 13 distinct subfamilies. Four of these families consisted of the archaeal, fungal, and sensory rhodopsins mentioned above, but nine families were related to sequences from uncultured species, including seven only known from the Sargasso Sea samples. Analysis of scaffolds containing both a taxonomic marker (σ subunit of RNA polymerase) and a

rhodopsin gene demonstrated that rhodopsins are not limited to the Gammaproteobacteria in which Béjà *et al.* (2000a) had first discovered them. For example, in one scaffold a rhodopsin was found together with a σ subunit RNA polymerase from the phylum Flavobacteria.

Bioinformatic analysis of the massive sequence data from the Sargasso Sea is by no means exhausted. It is expected that many more functional aspects of marine bacterial communities can be recovered from these data, while still more sequence information is expected to come from the Sorcerer II expedition. Table 4.7 provides an overview of some functional aspects of the Sargasso Sea sequence data. As an example, consider the presence of an ammonia monooxygenase gene sequence associated with an archaeal taxonomic marker, which indicates scope for archaeal nitrification in this environment. Previously, marine biologists had argued that nitrification in the ocean was hardly possible due to the sensitivity of chemoautotrophic bacteria to high levels of UV irradiation. Nitrification by Archaea would not be inhibited by UV light and this activity would be in accordance with the relatively high nitrite concentrations that are seen along with nitrate at certain times of the year in the Sargasso Sea.

In a commentary to the paper by Venter *et al.* (2004), Falkowksi and De Vargas (2004) remarked that the massive sequencing approach is reaching its limits when applied to community genomes. For example, despite the huge sequencing effort, only two nearly complete genomes (those of *Burkholderia* and *Shewanella*) could be reconstructed, and this could only be achieved by using already existing databases as a reference to support the assembly. The major part of the community is represented by rare organisms, and to obtain 95% coverage of these more than an order of magnitude of sequencing depth would be needed. Furthermore, if the approach was extended from prokaryote to eukaryote DNA the project would become much more problematic, because some dominant eukaryotes in seawater plankton (dinoflagellates, coccolithophorids) have extremely large genomes. Despite these obvious limitations, further WGS sequencing of marine communities

Table 4.7 List of some remarkable functional insights reconstructed from WGS sequencing of Sargasso Sea microbial DNA by Venter *et al.* (2004)

Genomic property	Functional relevance
782 different rhodopsin genes belonging to 13 protein families	Rhodopsin-mediated phototrophy is very common in oceanic bacterial plankton
Rhodopsin gene in scaffold bearing a taxonomic anchor from the Flavobacteria/Cytophaga group	Rhodopsin-mediated phototrophy distributed well outside the Proteobacteria
Ammonia monooxygenase in archaeal-associated scaffold	Oceanic nitrification not limited to the Bacteria; there are nitrifiers among the Archaea, and ammonia oxidation is not inhibited by UV light
Genes encoding phosphonate and high-affinity phosphate transporters; many genes responsible for utilization of pyrophosphates and polyphosphates	Versatile use of phosphorus compounds in oceanic environment to deal with severe phosphorus limitation
Gene homologous to *umuCD* DNA damage-induced DNA polymerase of *E. coli* found on plasmid	Resistance against UV damage by allowing DNA replication even when damaged by UV irradiation
Genes for arsenate, mercury, copper, and cadmium resistance found on plasmids	Possible role of oceanic microorganisms in trace-metal cycling in an oligotrophic environment
At least 50 bacteriophage gene groupings in scaffolds and 150 in singletons	High diversity of phages in oceanic bacterial community; significant fraction of bacteria infected

will undoubtedly lead to new surprises and possibly new insights in oceanic functions, even though complete reconstruction of communities may remain beyond reach.

Another marine metagenomic study focused on viral communities (Breithart *et al.* 2002). Viruses represent a very important factor in biogeochemical cycles and microbial biodiversity; by means of transduction they interfere with the genomes of their hosts, which for the marine ecosystem are mostly bacteria and algae. Viral activities are also considered important drivers of microbial community diversity by 'killing the winner' and promoting growth conditions of species with low abundance (Weinbauer and Rassoulzadegan 2004). Obtaining an overview of the biodiversity of viruses is difficult, because these organisms do not possess universal taxonomic anchors like the 16S rRNA gene in prokaryotes. The DNA polymerase gene *pol* can be used as a taxonomic marker for a subset of viruses (Short and Suttle 2002). New taxonomic systems are being developed that use all the genes in a viral genome to determine distances between species. This is elaborated in the Phage Proteomic Tree, a database and taxonomic algorithm for classifying bacteriophages

(Edwards and Rohwer 2005). However, not many viral genomes have actually been sequenced completely (Paul *et al.* 2002). Another issue is that viruses cannot be cultured outside their hosts, which in the case of marine bacteria are themselves mostly uncultured. So, a metagenomic approach to surveying viral biodiversity seems very appropriate.

In the study of Breithart *et al.* (2002) free-living viruses were collected by differential filtration and density-gradient centrifugation from surface sea water at two sites—Scripps Pier and Mission Bay—along the coast of California, USA. Special precautions must be taken when cloning viral DNA in a bacterial host, due to the presence of modified nucleotides and genes that could lyse the host. A total of 1934 sequences was obtained in a WGS approach, of which 70% showed no significant hits on sequences reported previously in GenBank. Among the remaining sequences no more than 34% were annotated in GenBank as viral sequences, and the rest were sequences of Archaea, Bacteria, and Eucarya, as well as mobile elements and repeat sequences. It appears that viral genomes carry a significant amount of DNA that originates from their hosts. About 83% of the viral sequences were related to bacteriophages, and these were classified further over the major groups of phages (Fig. 4.17). The viral community seemed to differ between the two sampling sites. Viral genomes at Scripps Pier were more 'bacterial' in origin, whereas viral genomes of Mission Bay had a more eukaryotic signature. Among the phage types, the Siphoviridae (λ-type phages) were more dominant at Mission Bay than at Scripps Pier.

Why the viral community should differ between two sampling stations and whether there is any ecological relevance in such differences remains uncertain. A possibility could be that viral community composition is a reflection of eukaryotic versus prokaryotic dominance of the plankton, for example due to algal blooms of variable composition. Given the fact that the majority of the two marine communities appears to be uncharacterized, any conclusions in this direction stand on shaky ground. Like the Sargasso Sea study illustrated above, the community analysis of marine genomes is still in the exploratory stage.

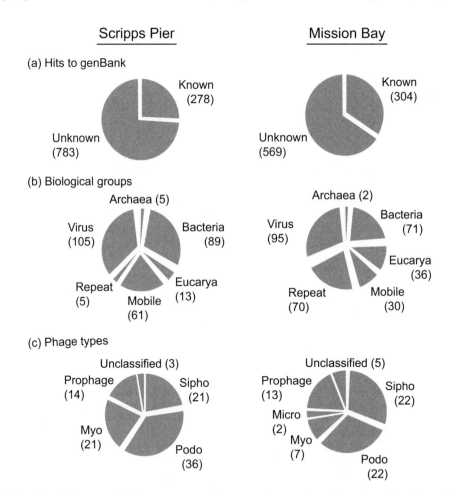

Figure 4.17   Overview of the content of viral genomes recovered from two marine coastal sampling stations, Scripps Pier and Mission Bay in California, USA. (a) A total of 1934 sequences were BLASTed to GenBank and 582 sequences produced a significant hit. (b) These sequences were classified according to sequence annotation, and 200 sequences were truly viral. (c) Among the 200 viral sequences 166 were from bacteriophages and these were classified according to the main phage families: Sipho, Siphoviridae (λ-like); Podo, Podoviridae (T7-like); Myo, Myoviridae (T4-like); Micro, Microviridae (ϕX174-like). From Breitbart *et al.* (2002), by permission of the National Academy of Sciences of the United States of America.

4.4.2 The soil metagenome

Soil organisms have been most valuable sources of all kinds of natural products ever since the Scottish bacteriologist Alexander Fleming discovered in 1928 that the soil fungus *Penicillium* produced a substance that killed *Staphylococcus* bacteria. Many other products derived from microbial secondary metabolites have been used to develop antibiotics, anticancer drugs, fungicides, immunosuppressive agents, enzyme inhibitors, antiparasitic agents, herbicides, insecticides, and growth promoters. Over the years, most of the microorganisms that can be cultured in the laboratory have been examined thoroughly for the production of compounds with biological activity, and biotechnological investigators have gained the impression that the limits of what these organisms can yield in terms of valuable products have been reached. However, as we have seen above, any environment, and certainly the soil, holds a great diversity of uncultured microorganisms that remain to be investigated. With the advent of metagenomic recombinant DNA technology (Handelsman *et al.* 2002) it became technically feasible to screen soil microorganisms for new functionalities without culturing them. This opportunity has raised great expectations and renewed interest in gene mining. The soil has been likened to Lady Bountiful (Rondon *et al.* 1999) and is considered a rich source for the discovery of novel natural products (Lorenz and Schleper 2002; Cowan *et al.* 2004; Daniel 2004, 2005).

How high is the probability of finding a novel product by functional screening of a metagenomic library? Gabor *et al.* (2004) explored this question in a theoretical way by analysing existing genome sequences of microorganisms. Assuming a random approach to expression cloning, it was argued that the probability of isolating an expressed gene in a metagenomic library depends on the mechanism by which that gene is expressed. The minimal requirements for gene expression in a host include the presence of a promoter for transcription and a ribosome-binding site for initiation of translation. Both of these sites must be recognized by the expression machinery of the host. If expression involves *trans*-acting factors from the host—for example special transcription factors, inducers, etc.—or if modifying enzymes are necessary for the gene product to become functional, the situation becomes much more complicated. Calculations were made for three modes of expression to estimate the number of clones that would have to be screened before a target gene was recovered with a probability of greater than 90%. For the most simple case, independent expression, the expected number of clones was found to depend on the size of the insert and decreased to around 3000 with an insert size increasing to 100 kbp. It was also estimated that 40% of the genes can be found in this way. So, if metagenomic screening effort covers a library of several thousands clones, each with an insert of 100 kbp, there is a fair chance that a designated gene will be found.

A pioneering study in soil metagenomics was the work of Rondon *et al.* (2000). These authors developed BAC libraries with DNA isolated from agricultural soil in Wisconsin, USA. The largest library had 24 576 clones with an average insert size of 44.5 kbp, whereas 10% of the clones had an insert size of between 70 and 80 kb. It was estimated that this library contained 1000 Mbp of DNA; given an average density in microbial genomes of one gene per 1000 bp, about 1 million genes were expected to be present in the library. Screening of the library for biological activities employed a variety of strategies. For example, to find clones expressing the enzyme cellulase, plates with the host cells were overlaid with agar containing brilliant red hydroxyethyl cellulose, and a yellow halo around the colony was taken as an indicator of cellulase activity. In this way, a great variety of enzymatic activities were screened, including β lactamase, keratinase, chitinase, and amylase.

A remarkable discovery coming from the metagenomic library screening conducted by Rondon *et al.* (2000) concerned a clone that had antibacterial activity against *Bacillus subtilis* and *Staphylococcus aureus*, but not against *E. coli*. The clone in which this activity was found was sequenced completely and it appeared to contain 29 ORFs, including a cluster of eight genes associated with phosphate transport. This showed that it is possible for BAC clones to contain complete, intact, operons. Fourteen of the 29 ORFs could not be assigned a function. Using transposon mutagenesis, the genes were mutated to see which one was responsible for the antibacterial properties of the clone. Finally a single candidate gene was identified, of which the predicted amino acid sequence had several repeat units (Fig. 4.18). The authors also considered the *hydrophobicity profile* of this molecule. This is a plot of scores along the sequence, in which each amino acid is given a number indicating the degree of hydrophobicity (see Lesk 2002). The profile of the unknown protein showed seven peaks, which is indicative of a membrane pump (amino acids anchored in the membrane have a high score for hydrophobicity if they are to be embedded stably in the lipophilic environment of the membrane). Recombinant expression of the protein confirmed

(a) MSFMKRFFCSCLTVAVILTACFSAAAQSE**GTLD**VSFNTTGVRYEDFGGAD

DKAMAVAV*QLDGKIVSVG*SSEVSGSGIDEAVVRYNSD**GTLD**SSFGTGGKV

TTAIGPGTSSDIAYSVVI*QSDGKIVVAG*SAAGISGTETDFAIVRYNAN**GT**

LDTSFGGTGKVTTPFGVATSADVANSVAL*QADGNIVPAG*YADDGSGADFA

LARYNTN**GSLD**ASFDTFGKTTTAIGAGTLGDFAQAVAI*QSDGKIVAAG*WT

EAASGLSIDFALARYNTN**GSLD**ATFDGDGKVITTVGSSTTFDLANAVLV*Q*

*ADGKIVAGG*FSDSLSSGADEALVRYNTN**GSLD**TSFDTDGIVITAIGPGTY

FDIAKAIVL*QPDGKIIAAG*YTDDLLVGFPSTDLALARYNVD**GSLD**TSFNA

DGKATIDLGGTEIINGAAIYAGNRIVVAGSSASNFLTARIWIATLVTAAP

VTVSGRITDERGRALKGVSVTLTDQDGVSSVASTNGFGYYRFTRVESGGT

HFLHATDRGYTFAPPVRIVDTKSDVSDADFVGTKQKGKPNSTR

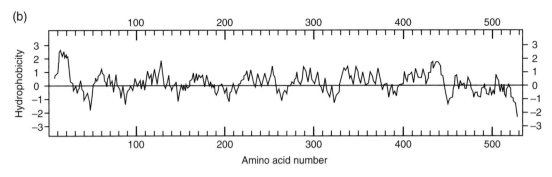

(b)

Figure 4.18 (a) Predicted amino acid sequence of a protein from an unknown soil organism, responsible for conferring antibacterial activity when expressed in *E. coli*. Amino acids are indicated by their standard single-letter codes. The ORF was discovered after screening a metagenomic BAC library made from DNA extracted from an agricultural soil. Repeated units are are shown underlined, in bold, or in italic. (b) Hydrophobicity profile of the ORF, showing seven clusters of hydrophobic amino acids (peaks above the line), which is indicative of a membrane protein. The horizontal axis shows amino acids numbered from N- to C-terminus and the vertical axis is a dimensionless score for the hydrophobicity of each amino acid. For details see the text. After Rondon *et al.* (2000), by permission of the American Society for Microbiology.

that it conferred antibacterial activity to the host; however, the protein itself was not active. The sequence shows significant homology with a gene in the genome of *Bdellovibrio bacteriovorus*, a delta-proteobacterium whose genomic sequence was established recently (Rendulic *et al.* 2004). *Bdellovibrio* is a predator of other bacteria and its genome is considered a valuable reservoir of antimicrobial substances. The nature of the possibly novel mechanism of antibacterial activity in the soil metagenomic library remains unknown to date (M. Rondon, personal communication).

Functional metagenomic screening exercises similar to Rondon *et al.* (2000) have been applied by a several other authors (Henne *et al.* 2000; Gillespie *et al.* 2002; Knietsch *et al.* 2003; Voget *et al.* 2003; Piel *et al.* 2004; Riesenfeld *et al.* 2004b). These studies were all motivated by the search for products with a medical or biotechnological relevance. Metagenomic profiling for ecological functions (e.g. crucial links in biogeochemical cycles, organic matter decomposition, and allelochemical production) has not yet been attempted.

Important insights into soil microbial communities have also come from sequence-driven screening of soil metagenomic libraries. Quaiser *et al.* (2002) were interested in the Crenarchaeota, a group of mesophilic Archaea that are frequently detected in soil communities; some of them are especially common in the rhizosphere. The Crenarchaeota are only known from their 16S rRNA sequences and have never been brought into pure culture. Quaiser *et al.* (2002) screened a fosmid metagenomic library developed from DNA

extracted from a calcareous grassland near Darmstadt, Germany, using a 16S rRNA probe. A clone of 33 925 bp which contained a crenarchaeotic 16S rRNA gene was sequenced entirely. It contained 17 ORFs, of which five could not be assigned a function. An overview of the organization of the clone is given in Fig. 4.19.

Among several others, an interesting cluster of genes was encoded in the crenarchaeote clone, which showed high similarity to the *fixABCX* gene cluster of nitrogen-fixing bacteria. In a phylogenetic analysis, the *fixA* gene of this operon clustered with *fixA* genes from two other Archaea, both hyperthermophilic crenarchaeotes. Together they seem to form a separate group which goes back to the ancestor of the Crenarchaeota, rather than to other Archaea or Bacteria via lateral gene transfer. As mentioned in Section 4.3, *fix* genes encode electron-transport flavoproteins and are co-expressed with the nitrogenase (*nif*) gene cluster in nitrogen-fixing bacteria. Presumably electron-transport flavoproteins provide the electron transport to the nitrogenase complex. In the symbiotic plasmid of *Rhizobium etli* the *fix* genes are tightly associated with the *nif* cluster (González *et al.* 2003). The *fixABCX* operon is highly conserved in diazotrophs as well as in a wide variety of other bacterial and archaeal species; however, its function remains unknown in non-nitrogen fixing species. In *E. coli fix* genes are related to the carnitine pathway, a transport system of long-chain fatty acids into the mitochondrion, prior to fatty acid oxidation. None of the obligately aerobic Archaea that contain the *fix* genes are capable of

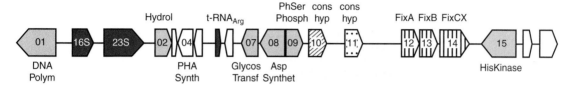

Figure 4.19 Schematic representation of a 34 kbp archaeal fosmid clone recovered from a calcareous grassland near Darmstadt, Germany. Different shadings indicate the most significant phylogenetic affinity of putative protein-coding genes to the Archaea (diagonal stripes), Bacteria (dots), Archaea and Bacteria (vertical stripes), or Archaea, Bacteria, and Eucarya (light grey). Genes without homology assignment are white. 01, Family B DNA polymerase; 02, α/β hydrolase; 04, polyhydroxyalkanoate synthase; 07, glycosyl transferase group 1; 08, asparagine synthetase; 9, phosphoserine phosphatase; 10, conserved hypothetical protein; 11, transmembrane protein; 12–14, *fixABCX*; 15, sensory transduction histidine kinase. From Quaiser *et al.* (2002), by permission of Blackwell Science.

nitrogen fixation; nitrogen fixation in Archaea is limited to strictly anaerobic euryarchaeotes such as *Methanosarcina* and *Methanococcus*. The intriguing presence of the *fix* gene cluster in Crenarchaeota cannot yet be evaluated from a functional point of view. Several other genes have been recovered from the same metagenomic libraries, some of which are unique to mesophilic Crenarchaeota; however, specific physiological traits have not yet been identified (Treusch *et al.* 2004).

Another group of microorganisms that has enjoyed interest from soil metagenomics research are the Acidobacteria, a relatively unknown phylum of the bacterial kingdom. Molecular ecological investigations had frequently detected 16S rRNA

sequences from Acidobacteria in soil; sometimes more than 30% of the sequences in a clone library belong to this group. Until recently only three species of Acidobacteria were known, but the work of Quaiser *et al.* (2003) has shown that 15 clones in a soil metagenomic library contained sequences from this group. Six of these clones were sequenced and several genes were identified. Phylogenies on the basis of a phylogenetic marker for Archaea, the purine-biosynthesis gene *purF*, were consistent with 16S rRNA-based phylogenies and showed that the Acidobacteria should be considered a separate phylum of the Bacteria, sharing a common ancestor with the Proteobacteria (Fig. 4.20). The analysis of the metagenomic clones

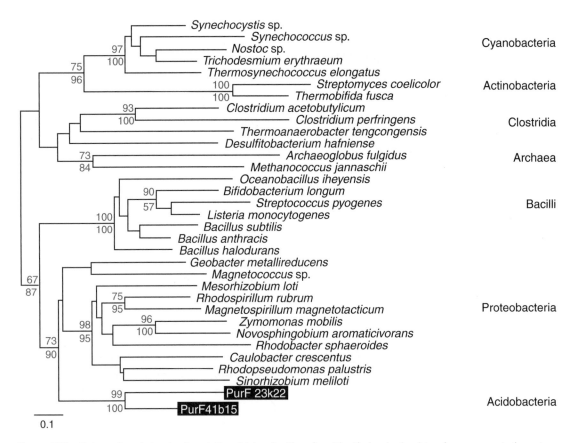

Figure 4.20 Phylogenetic analysis using the neighbour-joining algorithm of *purF* (amidophosphoribosyl transferase, a gene in the purine biosynthesis) sequences of several prokaryotes. Numbers on the nodes are bootstrap values. The position of two sequences from Acidobacteria, found in a metagenomics library of grassland soil DNA, demonstrates that this group should be considered a separate lineage within the bacterial kingdom. The scale bar indicates the proportion of amino acid difference. From Quaiser *et al.* (2003), by permission of Blackwell Science.

also revealed many deduced protein sequences, but only few gave hints of specific metabolic traits in Acidobacteria. A similar high abundance of Acidobacteria was found in a census of rRNA genes in another soil metagenomic library by Liles *et al.* (2003). As in the case of the Crenarchaeota, the ecological status of this abundant but uncultured group of soil bacteria remains largely unknown.

4.4.3 Genomes in extreme environments

Extreme environments usually harbour only a few species that have developed special characteristics to survive conditions that are lethal to most other organisms. The small and specialized communities found in extreme environments are interesting objects for ecological research. Since such communities involve only a few species, interactions between them and with the environment remain tractable, and each species fills a complete niche. Some extreme environments are considered models for extraterrestrial conditions and the study of such environments is motivated by the fact that microorganisms may leave biosignatures in rocks when they become fossilized (Walker *et al.* 2005). Extreme environments are also a rewarding object of study in genomics. Among the habitats investigated by ecological genomicists are polar deserts, deep-sea hydrothermal vents, and acid-mine drainage.

One of the most exciting studies in community genomics is due to Tyson *et al.* (2004), who investigated the metabolism of a small community of prokaryotes in an acid-mine drainage biofilm by means of the near-complete reconstruction of their genomes. *Acid-mine drainage* is a common environmental problem occurring in many coal and iron mines when pyrite (FeS_2) comes into contact with the air. This initially leads to the slow chemical oxidation of sulphide to sulphate and lowering of the pH. As we have seen above (Section 4.3) under low pH conditions it becomes profitable for iron-oxidizing microorganisms such as *Ac. ferrooxidans* to gain energy from the oxidation of Fe^{2+} to Fe^{3+}. Since this reaction allows only little energy gain large amounts of Fe^{2+} have to be oxidized. The resulting Fe^{3+} ions may react back with pyrite

to oxidize the metal sulphide bond, leading to more sulphate and a further lowering of the pH. Once the cycle of pyrite oxidation has started and acidophilic microorganisms have established themselves, the reaction tends to propagate: Fe^{3+} is regenerated and large quantities of acid are produced. The overall reaction is:

$$FeS_2 + 14Fe^{3+} + 8H_2O \rightarrow 15Fe^{2+} + 2SO_4^{2-} + 16H^+$$

Acid water and reduced iron will leach from the system and when this comes into contact with neutral surface water Fe^{2+} oxidizes spontaneously to Fe^{3+} and a precipitate of $Fe(OH)_3$ is formed. Microorganisms play a central role in the generation of acid-mine drainage and they may develop a pink biofim growing under very acid conditions (down to pH 0.5).

Tyson *et al.* (2004) developed a small-insert plasmid library from an acid-mine drainage biofilm in the Richmond Mine, California, USA, for WGS sequencing. With a sequence coverage of up to 10 times, the shotgun data-set could be assembled into near-complete genomes of two major members of the community, a bacterium from the *Leptospirillum* group II, related to *L. ferrooxidans* (Nitrospirae), and a previously unknown, uncultured archaeon designated *Ferroplasma* type II (Euryarchaeota, Thermoplasmatales). Partial genome sequences were obtained for another *Leptospirillum* species (*Leptospirillum* group III), a species of *Sulfobacillus*, and a *Ferroplasma* type I. The sequences of *Leptospirillum* type II, although arguably from different species, were nearly all the same; the average rate of nucleotide polymorphism was no greater than 0.08%. In the *Ferroplasma* type II genome, there was however a fair degree of polymorphism (2.2%), and these polymorphisms had a peculiar distribution in which 'hot spots' were interspersed with longer homogeneous regions, suggesting that the *Ferroplasma* type II genome should be considered a mosaic of at least three different strains, combined with each other through recombination. Although this detailed reconstruction of population-genetic structure from an environmental genome was already unique, Tyson *et al.* (2004) were also able to identify a significant number of functional genes in the two dominant species and link these to the geochemical processes in the biofilm (Fig. 4.21).

In the genome of *Leptospirillum* group II 63% of the ORFs could be assigned a function, while in *Ferroplasma* type II the corresponding figure was 58%. Both species are iron oxidizers but seem to use different systems to gain energy from this reaction. *Leptospirillum* has a system comparable to the one discussed in Section 4.3 for *Ac. ferrooxidans*, in which a periplasmic cytochrome oxidizes Fe^{2+} and transfers the electron to a membrane-bound electron-transport chain (Fig. 4.11). However, the genome of *Leptospirillum* does not encode rusticyanin, but rather a 'red cytochrome' with the same function. In *Ferroplasma* (which lacks a cell wall), a putative membrane-bound 'blue copper protein' conducts the initial oxidation of Fe^{2+}. This copper-containing protein shared sequence characteristics with both rusticyanin of *Ac. ferrooxidans* and sulphocyanin of the sulphur-oxidizing archaeon *Su. solfataricus* (see Fig. 4.10). The precise relationships between these electron-transport proteins remain to be elucidated, but from the data at hand it can already be concluded that different species may use different proteins for similar functions, and that there are homologies between the proteins used for iron and sulphur oxidation. The situation seems to be comparable to the electron-shuttling proteins used in iron reduction discussed above: a comparison between *Geobacter*, *Shewanella*, and *Desulfovibrio* showed that an evolutionarily diverse set of proteins may be used for similar functions in different species (see Section 4.3).

The genome of *Leptospirillum* encoded all the genes of the Calvin cycle and so *Leptospirillum* is probably a chemoautotroph. The lack of genes from the Calvin cycle in *Ferroplasma* and the presence of a large number of ATP-binding cassette (ABC)-type sugar transporters suggests that this species is heterotrophic. Interestingly, neither of the two species had genes for nitrogen fixation, but these were found in a third, incompletely reconstructed species, *Leptospirillum* group III. It is remarkable that the latter species, which has a much lower abundance than *Leptospirillum* group II, plays such a key role in the community.

Eukaryotic heterotrophs have also been identified in the acid-mine drainage communities, including two distinct groups of ascomycete fungi (Baker *et al.* 2004). Thus the community as a whole is a self-sustaining unit, independent of organic input from the surface; it can fix nitrogen and carbon from the air and uses iron oxidation as a prime mechanism for energy generation. The composition of the community with a few dominant genome types is consistent with the extreme character of the environment, providing only a few niches.

Obviously, microorganisms living in the extremely acid conditions of acid-mine drainage must be very tolerant to low pH. One aspect of tolerance is a high expression of proteins involved in combating oxidative stress and protein denaturation. We will see in Chapter 6 how such defence mechanisms provide protection against cellular stress. A proteomics study of the Richmond Mine acid-mine drainage biofilm confirmed that stress-defence proteins were a very dominant feature of the community proteome (Ram *et al.* 2005). Another key feature of acid tolerance is the nature of the cell membrane. A cell membrane with very low proton permeability is found in the acido-thermophilic archaeon *Picrophilus torridus*, which has been sequenced recently (Fütterer *et al.* 2004). The genome of *P. torridus* encodes an extraordinary large number of secondary transport systems and this suggests that the steep proton gradient across the cell membrane is used extensively for transport of metabolites. Acidophilic microorganisms must also be tolerant to extremely high levels of dissolved metal and metalloid ions, which reach concentrations in the upper millimolar range. Dopson *et al.* (2003) reviewed the various mechanisms by which microorganisms can achieve tolerance to As, Cu, Zn, Cd, Ni, Hg, and other metals. Most of the tolerance mechanisms in acidophilic microorganisms involve P-type ATPase efflux pumps, sometimes combined with reductases (in the case of Hg and As). The locus conferring As resistance has been particularly well studied. The *ars* operon encodes a reductase, an efflux pump, and a regulatory protein. A genome-wide inventory of heavy metal-tolerance genes

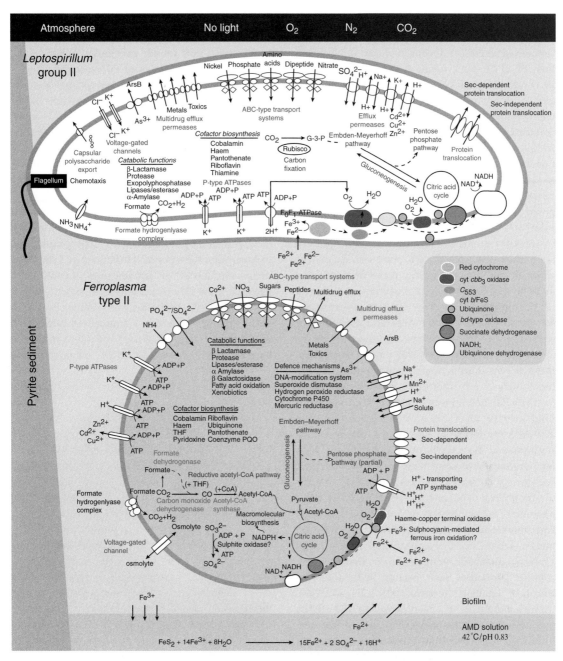

Figure 4.21 Cell metabolism of the genomes of *Leptospirillum* group II and *Ferroplasma* type II in an acid-mine drainage (AMD) biofilm with pyrite sediment and an acid-mine drainage solution of pH 0.83. Note the inner and outer membrane of the Gram-negative bacterium *Leptospirillum* and the presence of iron-oxidizing red cytochrome in the periplasm, coupled to a short electron-transport chain donating electrons to O_2. In the archaeon *Ferroplasma* type II, which lacks a cell wall, a membrane-bound sulphocyanin-mediated iron-oxidation system is assumed to be present. Both species have a complete citric acid cycle and several catabolic functions. *Leptospirillum* also has a complete Calvin cycle for CO_2 fixation, while *Ferroplasma* has plenty ABC-type transporters for sugar uptake. The overall reaction combining microbial Fe^{2+} oxidation with reduction of Fe^{3+} by reaction with pyrite leading to acid generation is shown below. THF, tetrahydrofolate. After Tyson *et al.* (2004) by permission of Nature Publishing Group.

was recently made for the non-acidophilic species *Pseudomonas putida*, the genome of which was sequenced in 2002 (Cánovas *et al.* 2003). No fewer than 61 ORFs were found that were likely to be involved in metal(loid) tolerance or homeostasis and some systems appeared to be duplicated. Loci were found for metal-specific uptake pumps, metal-reduction enzymes, chelating proteins, and metal-efflux pumps as well as several transcriptional regulators. Extrapolation of this machinery to acidophilic species may not be warranted, however, because the extreme acid environment poses additional requirements beyond 'normal' metal tolerance.

Another extreme habitat investigated by genome researchers are *hydrothermal vent systems* on the ocean floors. The sequencing of the first archaeon in 1996, *Me. jannaschii*, has illustrated the interest in this environment. The interest is especially strong because of the antiquity and evolutionary significance of these systems, allowing inferences about the very first microbial pathways. Hydrothermal vents on the ocean floor are places where the Earth's crust drifts apart, allowing seawater to mix with hot minerals, leading to a spurts of fluid at rates of 1–2 m/s with temperatures of 270–380°C. A range of specialized, thermophilic, Archaea and Bacteria have adapted to the high temperatures in these systems with growth optima between of 60 (thermophiles) and 105°C (hyperthermophiles). The energy generated by these microbial communities is derived from inorganic and geothermal sources (sulphide, H_2, reduced metals, CO_2, CH_4; Reysenbach and Shock 2002). The main metabolic strategies are methanogenesis, sulphur reduction (see Section 4.3), and the so-called knallgas reaction ($O_2 + 2H_2 \rightarrow 2H_2O$). Culture-independent screening using 16 S rRNA probes has shown the presence of several Archaea, including a new phylum, the Korarchaeota, only discovered in 1999 in a hot spring at Yellowstone National Park, USA, the Obsidian Pool. Another new archeal phylum, the Nanoarchaeota, was reported in 2002 (Huber *et al.* 2002). This phylum is represented by tiny cells that live attached to a sulphur-reducing archaeon of the genus *Ignicoccus*. They appeared to possess a 16 S rRNA gene that

has many base pair substitutions in regions of the gene considered to be universal; for this reason the group was not found previously when using regular 16 S PCR primers or fluorescent probes.

Surprisingly, thermal-vent ecosystems are also colonized by animals. A peculiar example is the polychaete worm, *Alvinella pompejana* (Pompeii worm), that is extremely tolerant to high temperature (Fig. 4.22). With a body temperature that can reach 80°C, this species is the most thermotolerant animal known. Population-genetic studies have shown that despite the island-like structure of their habitat, hydrothermal polychaete worms are able to disperse via larval stages, and genetic differentiation is determined by hydrography and topography of the ocean (Hurtado *et al.* 2004). The dorsal surface of *Al. pompejana* is covered with bacteria of the group Epsilonproteobacteria, which are considered symbionts, providing the worm with organic material. Campbell *et al.* (2003) developed a fosmid library of the symbiontic bacteria and showed that it contained two

Figure 4.22 Picture of *Alvinella pompejana*, a polychaete worm occurring near hydrothermal vent ecosystems on the ocean floor, carrying a dense layer of symbiontic bacteria on its dorsal surface. The Pompeii worm is the most thermotolerant animal known. Courtesy of the University of Delaware.

enzymes from the reversed citric acid cycle, a system used for CO_2 fixation in green sulphur bacteria in lieu of the Calvin cycle (see Table 4.6). Therefore it is likely that these symbionts are chemoautotrophs, fixing CO_2 in the same way as green sulphur bacteria.

Similar symbiotic relationships occur in the Vestimentifera, a peculiar group of tube-living marine animals, formerly classified as a separate animal phylum (Pogonophora), but now considered an order of the Annelida on the basis of molecular data. Vestimentiferans lack a normal intestinal tract but have developed a specialized tissue, the *trophosome*, in which symbiontic chemolithotroph (sulphur-oxidizing) bacteria live, nourishing the animal. Molecular investigations of these symbionts living in different hosts and different geographical sites have shown that the transmission from one generation to another is mainly horizontal; that is, the larvae acquire the symbionts from their environment, prior to attachment (Di Meo *et al.* 2000).

Community genomic studies in extreme environments are likely to deliver more surprises, when complete metagenomic libraries have been sequenced. Several of such projects are underway. Hydrothermal vent systems are more complicated than other extreme environments (deserts, acid-mine drainage) because of their relatively rich biodiversity; however, at the same time they offer exciting opportunities for investigating community syntrophy and peculiar symbiotic relationships.

4.5 Genomic approaches to biodiversity and ecosystem function: an appraisal

Microbial ecology has developed a wide variety of methods that can be used to study functions in ecosystems and it can rely on a rich background of biochemistry and genetics of the organisms involved. Genes associated with crucial links in all major biogeochemical cycles have been identified. The presence and diversity of these genes can be studied without consideration of the species in which they are present, as indicators of an ecosystem's capacity to perform specific functions. Most of these genes are known from microbial

biochemistry and microbial genetics but the possibilities offered by this strong background have not yet been exploited fully in an ecological context. Ecological studies have focused on genes from the nitrogen cycle, in particular denitrification genes. The few genomics studies that have been conducted with the aim of answering questions about biodiversity and ecological function seem to indicate a positive correlation between overall function and species biodiversity, supporting the rivet hypothesis (see Section 4.1). Community genomics studies also confirm that species carrying out indispensable key functions are not always among the dominant members of the community.

Ideally, assessment of functions in the environment would use transcription profiling, starting with extraction of functional mRNAs, as is done in model eukaryotes. This is more difficult in prokaryotes, however, and in microbial studies it is mostly limited to profiling pure cultures, rather than communities in the field. Instead, the potential for microbial functions is assessed from the presence, not the expression, of functional genes. With this limitation, the various microarray approaches hold good promise for large-scale profiling and monitoring. Attention needs to be paid to improving sensitivity and specificity before microarrays can be routinely applied for detecting functional genes. Oligonucleotide microarrays with 16S rRNA probes seem to be best for taxonomic surveys, especially since a large database on aligned 16S sequences is now available. However, technological improvement is also necessary in this area, to allow these phylochips to be used without prior PCR amplification of environmental DNA.

Microbial ecology faces a tough problem, in that many organisms in the environment cannot be cultured in the laboratory and so remain undescribed as proper species. With every survey of environmental samples, new sequences are detected and a levelling-off of the collector's curve is not yet in sight, not even for common habitats such as lake sediments and forest soils. The most common system to classify microorganisms is the SSU rRNA gene, although some groups remain unnoticed in this way, as demonstrated by the recent discovery of a completely new archaeal phylum, the Nanoarchaeota. Theoretical arguments, based on lognormal distributions of species over abundance classes, predict that prokaryote biodiversity may be an order of magnitude greater than even the maximum indicated by current genomics surveys.

Recent studies have also shown that spatial aspects and geographic barriers are more important in microbial community composition than thought before (Whitaker *et al.* 2003). We have seen evidence for this in the aeroplankton surveys discussed in Section 4.2, where two different sampling locations revealed differing community compositions, suggesting that there are local sources of airborne microbes. Studies of the relationship between structure and function in microbial communities should include an assessment of the sources of biodiversity, as in community studies of island biogeography (Curtis and Sloan 2004).

Community genomics is a recent field where exciting new discoveries may be expected. It has been made possible by new techniques of library construction and massive sequencing capacity. The two studies published in 2004 on Sargasso Sea microplankton and on acid-mine drainage biofilms are particularly impressive and seem to indicate the dawn of a major new direction of research. The question can be asked, are there limits to what brute-force sequencing of the environment may reveal? For the moment a focus on simple communities in extreme environments seems to be most profitable in the light of ecological theory. Given the relatively small number of species in such systems, a full reconstruction of their genomes is now within reach. It is also interesting to investigate the genomic background of the mutual interdependencies of species in such communities, varying from syntrophy to symbiosis.

The community genomics or metagenomics approach has also brought renewed interest in bioprospecting; that is, exploring novel functions by screening metagenomic libraries in which unknown genes from the environment are expressed in a heterologous host. This strategy has until now mostly been focused on functions of biotechnological relevance (new enzymes, antibiotics pathways); however, we see no reason why the same approach could not be applied to solving ecological questions.

Has ecological genomics solved the basic ecological question about biodiversity and ecological functions? Not yet. Only preliminary answers can be given at the moment, but all indicators suggest a major breakthrough soon.

CHAPTER 5

Life-history patterns

The great diversity of life cycles in nature, from short-lived annual plants to long-lived trees, from prolific reproducers such as oysters to economical types such as the albatros, has fascinated many biologists. An important scientific question has arisen from this fascination: can the life cycle of a species be understood in the context of the environment in which it lives? Attempts to answer this question have given rise to a large body of ecological literature, including observational data (population census, age distributions), experimental data (life-history manipulation by food or temperature), genetic data (heritable variation in life-history traits), and theoretical models (demography and quantitative genetics). Life histories of model organisms have also been analysed using molecular and genomic approaches. In this chapter we will visit this exciting blend of life-history ecology and genomics.

5.1 The core of life-history theory

When Charles Darwin travelled with the H.M.S. Beagle to Tierra del Fuego and the Falkland Islands in 1834, he was surprised to find, on counting the eggs of a large white *Doris* (a sea slug), how extraordinarily numerous they were. The slugs produce their eggs as a long ribbon, rolled up in a cone, and adhered to the rock. The inquisitive naturalist, extrapolating from a part of the structure, estimated the total number of eggs in one spire to be at least 600 000. Yet the animal itself was certainly not very common. Although he often searched under the stones, he could find only seven individuals. He then added in a footnote (Darwin 1845):

No fallacy is more common with naturalists than that the numbers of an individual species depend on its powers of propagation.

If the abundance of a species does not depend on the number of offspring, why are there such large differences between species in reproductive styles? That such differences exist is obvious. The mathematical biologist A.J. Lotka distinguished between the 'lavish type' and the 'economical type' (Lotka 1924). Many marine animals, including Darwin's slug, obviously belong to the first type, whereas humans, with their low birth rate and long lifespan, are an example of the second type. Also within relatively homogeneous groups of organisms large differences between species may exist. For example, in birds clutch size varies from one egg in petrels and condors up to 20 eggs in pheasants and partridges. Likewise, among lizards the number of eggs laid in a season varies per species from 2 to 20. An even more extreme variation may be observed among plant species. This diversity in reproductive output across species is usually positively correlated with juvenile mortality rate and negatively with longevity, so each species with stationary population size strikes a balance between mortality and natality; however, the weights on each side of the balance vary enormously between species.

An answer to the question of why the power of propagation differs so much between species comes from life-history theory. Characteristic for this area of population ecology is the recognition that the various *life-history traits*, also called *vital rates*, such as juvenile mortality, age at maturity, adult body size, clutch size, and longevity, cannot be isolated from each other. The theory considers all these traits jointly, including the interrelationships among them. An important concept is the presence of *trade-offs*. A trade-off is a negative correlation between two life-history traits in such a way that an increase of one trait (e.g. clutch size) imposes a cost to another (e.g. chick mortality). However, the term trade-off is also used in a broader sense to indicate any negative correlation between two traits, whether they are causally related or not. Given the trade-off structure among life-history traits, the theory attempts to explain patterns across species by assuming that the life history as a whole is subjected to natural selection and is optimized with respect to the environmental conditions to which the organism is exposed, subject to lineage-specific constraints. Life-history theory has developed into a major field of population ecology and the subject is summarized in several textbooks, such as those by Stearns (1992) and Roff (2002).

Life histories can be described using the formalism of *demography*, the science of age-structured populations and their dynamics. The aim of demographic analysis is to develop models in which life-history traits, such as mortality and fertility, are linked to population size and age structure. In human demography such models are used to forecast future population growth and composition, given fertility and mortality schedules. In ecology, the same models are used to estimate the optimal life history, given a trade-off structure among the vital rates and a criterion for optimality, which is usually taken as the population growth rate. In other words, it is assumed that survival and fertility of a species are optimized in such a way that, accounting for constraints from trade-offs, population growth rate is at a maximum. In addition to trade-offs, vital rates are also constrained by *lineage-specific effects*, which arise from the body plan and the physiological limits posed by the phylogenetic history of the group to which the species belongs, which, for example, prevents a bird from producing as many eggs as a mussel.

As an example of how arguments are developed in life-history theory, we consider a relatively simple model outlined by Roff (2002), who considers a hypothetical animal in a constant environment with a simplified life history about which the following two assumptions are made. First, mortality rate, indicated by θ, is constant. This implies that throughout life a constant fraction of a birth cohort is removed and the survival of the cohort is described by a negative exponential. Such a survival curve is often seen in species for which mortality mainly comes from external sources, such as from predators. Second, fertility is constant within one life cycle, but there are different options, depending on the age at which reproduction starts. If reproduction starts later, fertility is higher; a linear relationship is assumed between fertility and the age at maturity. Such a relationship is often seen in organisms in which reproductive capacity increases with body size. By postponing reproduction, the animal can reach a larger body size and consequently acheive greater fertility.

These assumptions are displayed graphically in Fig. 5.1. Three possible fertility scenarios are plotted: one that starts at age 5 and maintains a reproductive output of 15 per time unit (the total number of eggs laid by a female surviving to age 20 would be 225), one that starts at age 10 and produces 40 eggs per time unit (total number of eggs produced from age 10 to 20 would be 400), and one that starts at age 15 and produces 65 eggs per time unit (total output 325 eggs). If all animals survive to age 20, the best of these three scenarios would be the second; however, a still-better scenario is to start at age 11, since this would provide a total reproductive output of 405 eggs ($= (20 - 11) \times 45$). However, reproductive output must be corrected for the fewer and fewer animals remaining to produce eggs and so the actual optimum will be lower than 11. How much? A quantitative valuation can be given by considering a common criterion for optimality, net reproductive rate, R_0, which is given by

$$R_0 = \int_{\alpha}^{\infty} l(x)m(x)\mathrm{d}x$$

where $m(x)$ is fertility (number of offspring produced per time unit by individuals aged x), $l(x)$ is survival from birth to age x (as a fraction),

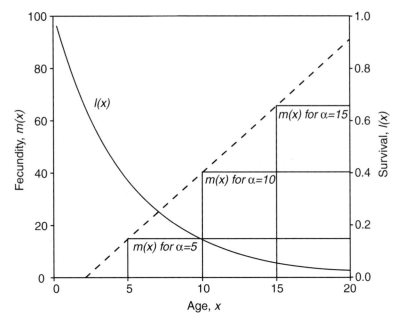

Figure 5.1 A hypothetical life history in which survival is a single exponential given by $l(x) = e^{-\theta x}$, where θ is the rate of mortality, and fertility, $m(x)$, is constant with age, but dependent on age at maturity, α, as $m = a + b\alpha$. Parameters are chosen arbitrarily as follows: $\theta = 0.4$, $a = -10$, and $b = 5$. Three different fertility scenarios are shown, for $\alpha = 5$, 10, and 15. Using the net reproductive rate R_0 as a criterion (see text) it can be shown that the optimal value for α is 7. After Roff (2002), reproduced with permission from Sinauer Associates.

and α is the age at maturity (i.e. first reproduction). Net reproductive rate is the average number of individuals in the next generation by which an individual of the present generation is replaced. Although Roff (2002) uses R_0 as a criterion for optimality, most authors prefer a parameter called intrinsic population growth rate, which can be calculated from $l(x)$ and $m(x)$ in a similar but more complicated way. R_0 ignores variation in generation time but is preferred in field studies with more or less stationary populations (Kozlowski 1993). The use of both net reproductive rate and intrinsic growth rate is supported by the *central theorem of demography*, which states that any population with time-invariant fertility and mortality schedules will, after some generations, attain a stable age distribution and then grow exponentially. A strong implication of the central theorem is that the growth factor per generation, R_0, can be estimated beforehand, directly from the life history; we need not wait until exponential growth is actually realized. In this way, R_0 can be

considered an indicator measuring the fitness of a life history.

The equation above shows that to estimate R_0 one needs information about fertility and mortality rates for each age class; in the case considered by Roff (2002), $l(x) = e^{-\theta x}$ and $m(\alpha) = a + b\alpha$, where θ is the rate of mortality (set at 0.2 in the example), and a and b are constants (set at -10 and 5, respectively). This parameterization leads to

$$R_0(\alpha) = \frac{1}{\theta}(a + b\alpha)e^{-\theta\alpha}$$

The optimal age at maturity, $\hat{\alpha}$, is that which maximizes R_0, and is found by evaluating

$$\frac{\partial R_0}{\partial \alpha} = \left(\frac{b}{\theta} - a - \alpha b\right)e^{-\theta\alpha} = 0$$

which produces

$$\hat{\alpha} = \frac{1}{\theta} - \frac{a}{b}$$

So conditions promoting postponement of reproduction (large $\hat{\alpha}$) include a low rate of mortality (small θ) and a slow increase of fertility with age (small b). If two related species live under conditions of unequal mortality, everything else being equal, the one living under the highest mortality regime should have the shortest juvenile period. In the numerical example of Fig. 5.1, $\hat{\alpha} = 7$ and the net reproductive rate of the optimal life history is 30.8.

This simple example shows how mathematical models can help to draw inferences on the optimality of life-history traits if some basic attributes of the species and its environment are given. Whereas age at maturity, body size, and clutch size are important traits for animals, for plants traits such as germination fraction, shoot–root allocation, flowering time, and number of seeds are considered. Much more complicated elaborations of the basic demographic principles are discussed in Roff (2002), including fluctuating and predictably changing environments. A crucial role in many models is played by trade-offs; in the present case it was assumed that the organism could increase its fertility by postponing maturity and growing larger first, and that there was a linear relationship between the two. Such a mechanism is often considered a consequence of *energy allocation*: what is spent on one side cannot be spent on the other side. The idea of trade-offs due to energy allocation is very old and can be traced back to the 'loi de balancement' proposed by Geoffroy Saint Hilaire in 1818 (Leroi 2001). Darwin (1859) noted that artificial selection of domestic animals and plants showed many examples of correlated reponses or 'compensation of growth'. He referred to both Geoffroy and Goethe in stating 'if nourishment flows to one part or organ in excess, it rarely flows, at least in excess, to another part; thus it is difficult to get a cow to give much milk and to fatten readily'.

The principle of energy acquisition and allocation was developed by Kooijman (2000) into a systematic physiology-based framework of growth, reproduction, and aging, the *dynamic energy budget model* (DEB model). In this model, food uptake is assumed to be proportional to body surface and assimilated energy is converted into reserves. The reserve pool is in dynamic equilibrium with a mobile pool available to all organs of the body. A fixed proportion (κ) of the circulating pool is spent on growth plus maintenance, and the remaining portion, 1–κ on development plus reproduction. This aspect of the model is designated as the κ *rule* for allocation. Energy taken up by somatic tissues is first used for maintenance, and the remainder is used for growth. In this way growth competes directly with maintenance, but reproduction competes with growth only through the κ rule. This model can explain why many animals continue to grow after the onset of reproduction. Their growth slows down due to the increasing maintenance costs of a larger body, not directly through competition with reproduction. In the model, the onset of reproduction, which is due to the 1–κ flux being redirected from development to reproduction, does not create the discontinuities and inconsistencies that are present in other allocation models. The great variety of examples discussed in Kooijman's (2000) book illustrates that the DEB model is a powerful instrument for analysing energetic relationships since it argues from first principles rooted in thermodynamics and emphasizes the similarities across organisms as different as yeast, waterfleas, parasites, fish, and birds.

Despite the importance of allocations and trade-offs in life-history theory, reliable empirical measurements are difficult. This is especially annoying because often the outcome of an optimization procedure depends critically on the shape of a trade-off function, for example a convex relationship between reproduction and survival predicts *iteroparity* (repeated ongoing reproduction), whereas a concave relationship predicts *semelparity* (a single, large reproductive output followed by death). Empirical studies are hardly able to distinguish between these two forms of trade-off. In addition, trade-offs may be masked by fluctuation in the resource that is the subject of allocation. For example, if the total energy available for growth and reproduction increases due to increasing food intake, the allocation between them becomes invisible, and the correlation between growth and reproduction at the phenotypic level may turn from negative to positive (Van Noordwijk and De Jong 1986). There has been a tendency to measure trade-offs in terms of negative genetic correlations between life-history traits, either by pedigree analysis or by selection, using the formalism of quantitative genetics; however, such estimates have not produced satisfactory results because very large sample sizes are needed to resolve the presence of genetic correlations (Roff 2002).

In addition to energy allocation, two other classes of mechanism can cause negative associations between life-history traits: *negative (antagonistic) pleiotropy* and *hormonal signals* (Zera and Harshman 2001; Leroi *et al.* 2005). Geneticists define pleiotropy as the phenomenon that one gene affects two or more phenotypic traits. Negative pleiotropy arises when expression of a single gene affects one trait in a positive way and another trait in a negative way. This can also be true for hormonal control: the same hormone may affect one process in a positive direction, and another in a negative direction. Negative pleiotropy may be a more common mechanism for negative association between life-history traits than energy allocation. Leroi *et al.* (2005) examined several classes of genes involved with the regulation of longevity and concluded that many of them have antagonistic effects on life-history traits. Additional evidence for the importance of negative pleiotropy comes from the literature on tolerance to pesticides (Van Straalen and Hoffmann 2000). Many pesticide tolerances are associated with apparent 'costs' to vitality or reproductive capacity, but these costs are more often due to metabolic side effects of a gene mutation that confers tolerance, than to an energy drain towards detoxification of the pesticide.

A new framework for addressing questions about life-history patterns may come from genome-wide gene-expression studies. Stearns and Magwene (2003) suggested that trade-offs can be seen as *conflicts over gene expression*. This argument is illustrated in Fig. 5.2. If two functions in an organism—for example, reproduction and longevity—are regulated by two sets of genes, a trade-off between the functions may arise if the two sets overlap and if, for example, one function calls for upregulation and the other for downregulation. Such a trade-off can also arise from

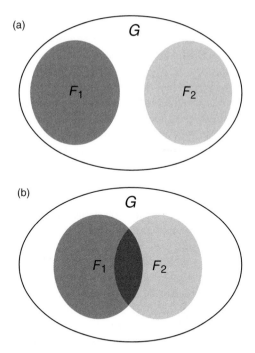

Figure 5.2 Theoretical illustration of how trade-offs between life-history traits might be explained by conflicts over gene expression. G is the set of all expressed genes in an organism, F_1 is the set whose expression changes in response to function 1, and F_2 to function 2. If there is overlap between F_1 and F_2, as in (b), a trade-off will arise if genes need to be upregulated in response to 1 and downregulated in response to 2. After Stearns and Magwene (2003) by permission of the University of Chicago Press.

conflicting signals downstream of gene expression, as we will see below.

Finally, we want to point out that life-history traits always result from interaction between genotypic determinants and the environment. Even those traits that are under strong genetic control, for example clutch size in birds, may vary depending on environmental conditions. The way in which a life-history trait is shaped by an environmental factor can itself be considered an aspect of the life-history pattern. For example, some birds will tune their clutch size to environmental food supply, taking advantage of years with abundant food availability and minimizing reproduction in bad years; other birds just lay the same number of eggs irrespective of the environment. This aspect of a life history is denoted as *phenotypic plasticity*, and the function describing a life-history trait as a function of an environmental factor is called a *norm of reaction*. A reaction norm is a property of a genotype; different genotypes may have different slopes of the reaction norm, allowing natural selection to act on plasticity. Until now plasticity as an adaptive trait has been analysed mainly by statistical methods (multivariate quantitative genetics), but genomic technology is opening up new perspectives for providing a genetic basis to phenotypic plasticity (Pigliucci 1996).

In this chapter we will explore the genomic basis of life-history patterns, emphasizing proximate causation in the life history. There is a lot of literature on the evolution of longevity and aging that is left undiscussed here; the reader is referred to Kirkwood and Austad (2000) and Partridge (2001) for reviews of theories and experimental data. The molecular work on aging, which is mainly on *C. elegans*, *Drosophila*, and mouse, is often inspired by its possible extrapolation to human gerontology and is not very well integrated in ecological studies. We nevertheless believe that a molecular connection can provide a mechanistic basis for life-history theory and so increase its power to explain the variety of life-history patterns in nature.

5.2 Longevity and aging

The nematode *C. elegans* has developed into a classical model for the study of aging. Most of this research is motivated by a possible connection with aging in humans, but the new insights accumulated over the last few years are of equal relevance to ecology and life-history theory. It also appears that the properties of the molecular machinery regulating aging are shared with *Drosophila* and are even conserved across the whole animal kingdom. The main reason why *C. elegans* became a model for aging was due to the discovery of mutants that showed extended longevity. It turned out that the genes mutated to extend longevity were the same as those associated with the formation of dauer larvae. The dauer larva is a developmentally arrested stage, the entry of which is triggered by adverse conditions such as

food shortage or crowding (see Section 3.3). Here we will consider the signalling pathways associated with dauer formation as well as extended longevity.

5.2.1 The insulin signalling pathway

The first report of single-gene mutations that affected lifespan and reproduction in *C. elegans* dates back to 1988. The gene involved was called *age-1*, and mutations in this locus increased longevity and decreased hermaphrodite fertility (Friedman and Johnson 1988). Later, Kenyon *et al.* (1993) discovered that two other single-gene mutations could affect lifespan. One of these longevity-regulating genes was known as *daf-2* (daf from dauer formation); animals mutated at the *daf-2* locus lived twice as long as the wild-type worms (Fig. 5.3). It was also shown that a second gene, *daf-16*, is required for the lifespan-extending effect of *daf-2*, because the double mutant, *daf-2; daf-16*, had a normal lifespan (Fig. 5.3). Interestingly, the reproductive output of the *daf-2* mutant was hardly decreased (the brood size was 212 ± 36 eggs in the long-lived mutant versus 278 ± 35 in the wild type) and it was also shown that ablation of the germ-line precursor cells, effectively sterilizing the adult, did not increase lifespan. Thus the increased

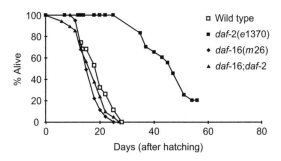

Figure 5.3 Survival curves for *C. elegans* mutated in the *daf-2* gene (*daf-2(e1370)*), compared with a control group (wild type). Survival curves are also shown for nematodes mutated in the *daf-16* gene (*daf-16(m26)*) and in both genes (*daf-16;daf-2*). Median survival times are 17 days for *m26*, 17 days for *m26;e1370*, 19 days for the wild type, and 46 days for *e1370*. The comparison of mutants demonstrates that *daf-2* dysfunction increases longevity by more than a factor of 2, and this effect requires a functional copy of *daf-16*. From Kenyon *et al.* (1993) by permission of Nature Publishing Group.

lifespan of the *daf-2* mutant was not a consequence of lower reproduction. Later several other genes were identified that can extend longevity when altered in *C. elegans*, and in total approximately 70 are known today. These genes include regulators of metabolism, genes involved in sensory perception, and reproduction genes. In addition, several genes regulating longevity were found to be associated with defence against oxidative stress. One specific group of genes receiving much attention appeared to encode proteins of the *insulin signalling pathway*.

Insulin is a peptide hormone that in vertebrates is secreted from groups of cells associated with diverticula of the gut, in mammals taking the form of the islets of Langerhans in the pancreas. Insulin-like peptides and insulin-like growth factors (IGFs) are also present in invertebrates, usually in their highest concentrations in the gut. Insulin in mammals has a central role in carbohydrate and lipid metabolism, the best-known effect being increased uptake of glucose from the blood by muscles and adipose tissue and the formation of glycogen by the liver. Insulin and IGF do not enter their target cells but instead react with a receptor protein in the cell membrane, the *insulin/IGF receptor*. This protein has an extracellular domain, binding insulin or IGF, and a cytosolic domain, which acts as a tyrosine kinase: when activated it catalyses the phosphorylation of tyrosine residues in cytosolic proteins. These proteins in turn activate others in a complicated cascade, finally leading to phosphorylation of a cytosolic protein known as DAF-16. DAF-16 is a transcription factor of the so-called forkhead family. When active, this transcription factor switches on a series of genes that form a programme of dauer-larva formation. As long as DAF-16 is phosphorylated by DAF-2 signalling, it cannot enter the nucleus and so is effectively inactivated as a transcription factor (Lin *et al.* 2001). The secretion of insulin-like peptides is under control of the nervous system, which receives sensory input from the mouth region. So the whole system seems to be targeted towards translating information about food availability in the environment into either normal development (DAF-16 inhibited by activated DAF-2) or a dauer

programme (DAF-16 activated by relaxation of DAF-2; Braeckman *et al.* 2001; Olsen *et al.* 2003; Fig. 5.4). The pathway is referred to as insulin/IGF-1 signalling.

The fact that mutations in the DAF-2/DAF-16 pathway regulate longevity as well as dauer formation strongly suggest that the effect of *daf-2* knockout on longevity is basically a dauer programme 'mis-expressed' in the adult. Indeed, Dillin *et al.* (2002), using RNAi to suppress *daf-2* and *daf-16* at different times in the life cycle, found that the same pathway regulates aging in the adult and dauer formation in the larvae. Jones *et al.* (2001), using SAGE (see Section 2.3) applied to dauer and non-dauer populations of *C. elegans*, showed that the dauer-specific transcriptome was greatly enriched in genes that previously were known to regulate longevity. The study also showed that dauer formation is not to be considered a true resting stage from a genetic point of view, because no fewer than 18% of the genes detected were specifically upregulated in the dauer larva, compared to a mixed-stage population. Similarly, Wang and Kim (2003) in a microarray study classified 1984 genes as dauer-regulated, which is 11% of the *C. elegans* genome. The dauer-enriched genes include several enzymes characteristic for anaerobic metabolism (Holt and Riddle 2003).

The long-lived non-dauer *daf-2* mutant illustrates that it is possible to uncouple part of the dauer programme (the part that confers extended longevity) from the main programme leading to quiescence. The mutations in *daf-2* and other genes of the insulin/IGF-1 signalling pathway that affect longevity are weak mutations. Strong mutations in the same genes cause the *C. elegans* larvae to go into dauer dormancy regardless of environmental cues. This indicates that there may be thresholds in the levels of endocrine signalling such that a mild decrease of signalling is already sufficient to start the anti-aging programme, whereas a further decrease is necessary to enter the dauer stage. These thresholds could also be dependent on temperature; some *age-1* mutants that develop into long-lived adults at normal culture temperature will go into dauer diapause at high temperature (27°C).

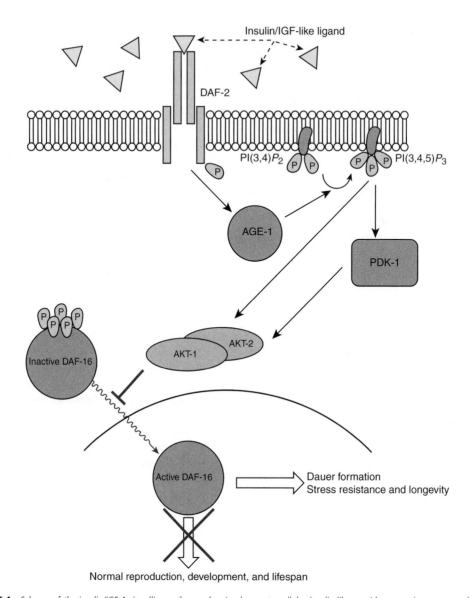

Figure 5.4 Scheme of the insulin/IGF-1 signalling pathway, showing how extracellular insulin-like peptides may trigger a cascade of events, starting with binding to DAF-2, a receptor protein in the cell membrane and eventually leading to phosphorylation of DAF-16, a transcription factor of the forkhead family. By decreased insulin/IGF-1 signalling DAF-16 is activated, allowed to enter the nucleus, trigger a programme of dauer formation, stress resistance, and longevity, and suppress normal reproduction. P, phosphate group; PDK-1, phosphoinositide-dependent kinase 1; PI(3,4)P_2, phosphatidylinositol 3,4-bisphosphate; PI(3,4,5)P_3, phosphatidylinositol 3,4,5-trisphosphate. After Olsen *et al.* (2003), with permission from Springer.

Increased longevity similar to decreased DAF-2 signalling is also seen when the animals are cultured under *dietary restriction,* also called *caloric restriction.* This refers to a diet in which food intake is limited to 30–40% of the intake shown by animals fed *ad libitum.* In *C. elegans* this can be achieved by diluting the bacteria in the medium or by applying an axenic medium with heat-killed *E. coli* cells. Because the insulin signalling pathway is associated with carbohydrate metabolism and the apparent cues for dauer induction come from food availability, it seems logical to conclude that dietary restriction promotes longevity through the same DAF-2/DAF-16 pathway. However, this appears not to be the case. Houthoofd *et al.* (2003) did experiments with *daf-2* and *daf-16* mutants both exposed to dietary restriction (Table 5.1). Their data show clearly that dietary restriction causes an increase in the median survival time by a factor of two and a half, both in the wild type and in the *daf-16* mutant. In addition, a restricted diet can boost the median lifespan of the *daf-2* mutant (which is already increased by a factor of 1.7 compared with the wild type), to a record value of 90.9 days (maximum, 136 days). In human terms these animals would correspond to healthy 500 years old! Interestingly, extension of lifespan is this way is not accompanied by decreased metabolic activity; actually, respiration was elevated substantially by dietary restriction as were activities of antioxidant enzymes, such as superoxide dismutase (Houthoofd *et al.* 2002).

Another mutation that extends lifespan in *C. elegans* seems to act mostly independent of the insulin/IGF-1 signalling pathway. Mutations in the so-called *clk-1* gene (the name derives from clock biological timing abnormality) increase longevity in association with a slowing down of the rate of many processes, including cell division, rhythmic behaviour, rate of feeding, and mitochondrial respiration. The correlation between longevity and metabolic rate supports the *rate of living hypothesis* of aging, which states that aging is

a consequence of accumulating metabolic damage, in particular from endogenous oxygen radicals (Finkel and Holbrook 2000; Hekimi and Guarante 2003). *Reactive oxygen species* such as superoxide anion (O_2^-), hydroxyl radical ($OH^\bullet$), and hydrogen peroxide (H_2O_2) are amply generated by the electron-transport chain in the mitochondrion. The major source is complex III, in which ubiquinone, alias coenzyme Q, resides. This electron-transport molecule has an unstable intermediate that can easily donate electrons directly to molecular oxygen rather than to complex III. The mutated *clk-1* protein cannot perform the final step in the biosynthesis of ubiquinone and as a consequence the precursor, demethoxyubiquinone, is incorporated in the electron-transport chain. Paradoxically, this alternative component appears to be less prone to the production of reactive oxygen species. Another role for *clk-1* has been suggested by Branicky *et al.* (2000). The protein could have a regulatory role by reporting to the nucleus on the metabolic state of the mitochondria, in such a way that the rate of living may be adjusted to the generation of energy. If *clk* is mutated nuclear genes do not receive the right information on respiration and would set the rate of living at a default level lower than normal. This theory is interesting because it would imply that the effect of respiration on aging acts through a metabolic switch, rather than through a direct impact of damage accumulation (Guarante and Kenyon 2000).

In addition to mutations, surgical operations can be applied to *C. elegans* to increase longevity. An interesting effect is seen after removal of the germ line. As noted in Chapter 3, *C. elegans* has a completely determinate developmental pattern in which the destination of every cell is fixed as soon as it comes into existence; the adult has exactly 959 cells, not including eggs and sperm cells, which descend from the zygote in a fixed lineage. Two cells, Z2 and Z3, give rise to the germ line by continuous division during development. The germ cells differentiate into sperm during the L4 stage or oocytes during adulthood (see Fig. 3.18). Removing the germ-line precursor cells by means of a laser extends the lifespan of *C. elegans* by

Table 5.1 Survival times of different populations of *C. elegans* showing that effects of dietary restriction increase longevity independent of the insulin signalling pathway

Population	Medium lifespan (days; mean ± S.E.)	
	On normal medium	Under dietary restriction on axenic medium
N2 (wild type)	14.4 ± 0.1	36.4 ± 0.2
daf-16(*m26*) mutant	12.9 ± 0.1	32.6 ± 0.2
daf-2(*e1370*) mutant	24.3 ± 0.2	90.9 ± 0.4

Source: After Houthoofd *et al.* (2003).

about 60% and this effect requires the presence of DAF-16. Animals that lack germ cells due to a mutation are also long-lived. In a series of elegant experiments, measuring the survival curves of various mutants, Arantes-Oliveira *et al.* (2002) showed that the lifespan-suppressive effect of the germ line is not dependent on the sperm or oocytes themselves, but only on proliferating, active germ-line precursor cells. So, despite the obvious negative correlation between reproduction and longevity in this system, there does not seem to be a simple trade-off in the classical sense that energy allocated to reproduction would detract from maintenance and so increase the rate of aging. Rather, a signal from the germ-line stem cells directs both aging and reproduction, maybe by altering the production of a steroid hormone or by altering the response to such a hormone (Arantes-Oliveira *et al.* 2002).

An important issue in the metabolic network affecting aging in *C. elegans* is the question of which genes act upstream and which downstream in the insulin signalling pathway. Taking DAF-16 as a reference point, *upstream* genes are defined as the ones that affect expression or activity of DAF-16. To this category belong the neurosecretory signals leading to secretion of insulin-like factors, the genes encoding the insulin-like peptides themselves, the insulin/IGF receptor DAF-2, and the various genes in the signalling cascade leading to phosphorylation of DAF-16, such as *age-1* and *pdk-1*. The *downstream* genes include the ones that are regulated by the transcription factor DAF-16 and whose expression contributes to longevity and stress resistance. The difference between upstream and downstream genes is difficult to make when looking at gene expression as such, but can be unravelled by studying mutants that are knocked out at crucial positions in the pathway. For example, Murakami and Johnson (2001) studied a transmembrane tyrosine kinase gene called *old-1*, which if overexpressed in *C. elegans* increases longevity by a factor of 1.5; using transgenic nematodes in which the *old-1* gene was fused to green fluorescent protein the authors observed that the positive effect of *old-1* on longevity was absent in a mutant in which *daf-16* was knocked out. This

makes it likely that *old-1* is regulated by DAF-16; that is, it is downstream of DAF-16.

5.2.2 Genome-wide analysis of lifespan modulation

The various single-gene mutation studies and other manipulations have demonstrated that the effect of insulin/IGF signalling on longevity in *C. elegans* is linked to a large number of other processes, such as reproduction, lipid metabolism, diapause entry, and stress resistance. This has made the situation increasingly complex and it became difficult to see the bigger picture. However, recent genome-wide microarray studies (Murphy *et al.* 2003; Golden and Melov 2004; McElwee *et al.* 2004) have enforced an important breakthrough, presenting an interpretative framework for earlier data and even a new theory of aging.

Murphy *et al.* (2003) aimed to identify all the genes that act downstream of DAF-16 by means of a cDNA microarray survey. These authors focused on genes that showed opposite expression profiles in *daf-2-* versus *daf-16*-knockout mutants. They also treated animals with RNAi of selected genes to confirm that these genes had an effect on longevity. A division in two classes was made: class-1 (lifespan-extending) genes were upregulated in *daf-2* mutants and in *daf-2* (RNAi) animals, but downregulated in animals in which both *daf-16* and *daf-2* were inhibited by RNAi. The second class of genes (lifespan-shortening) was defined by genes that displayed the opposite profile. A relatively succinct list of genes could be identified as belonging to either class and the most prominent ones of each class are listed in Table 5.2.

Inspecting the genes in the two classes, it is obvious that the first class contains many genes of *stress-defence systems* (Table 5.2). Heat-shock proteins are molecular chaperones that support the folding of other proteins; many of them are highly inducible by several stress factors, including a heat shock, in which they were first described. Cytochrome P450 is an enzyme that catalyses the oxidation of aromatic lipophilic compounds, including many xenobiotics, but also endogenic substances such as steroids. Catalase is an enzyme

Table 5.2 Overview of genes in *C. elegans* acting downstream of DAF-16, identified by differential expression using a microarray and reduced expression of *daf-2* and *daf-16* by RNAi

Gene	Brief description
Class 1: upregulated by DAF-16 and positive regulators of longevity	
ctl-2	Peroxisomal catalase
dod-1	Member of cytochrome P450 family
hsp-16.1	Heat-shock protein
lys-7	Enzyme associated with response to pathogenic bacteria
dod-2	Thaumatin (sweet protein) associated with plant pathogenesis
hsp12.6	Heat-shock protein, α crystalline
mtl-1	Metallothionein-like, cadmium-binding protein
gei-7	Member of family of malate synthase/isocitrate lyase
dod-3	Protein of unknown function
dod-4	Aquaporin, member of family of transmembrane channels
Class 2: downregulated by DAF-16 and negative regulators of longevity	
dod-17	Protein of unknown function
nuc-1	Endonuclease associated with apoptosis
dod-18	Member of *Maf*-like transcription factors
dod-19	Protein of unknown function
gcy-6	Putative guanylate cyclase, catalysing cGMP second messenger
dod-20	Protein of unknown function
dod-21	Protein of unknown function
vit-5	Vitellogenin, 170 kDa yolk protein
mtl-2	Protein of unknown function
dod-22	Protein of unknown function

Notes: Only the first 10 most prominent genes are shown for each class.
Source: From Murphy *et al*. (2003).

that supports the catalysis of hydrogen peroxide, a very reactive oxygen species that may cause damage to membranes. It is also striking that class 1 includes several proteins that are known to be involved in antibacterial defence. Stress-defence systems will be discussed in more detail in Chapter 6. The class 2 genes listed in Table 5.2 are less well defined; in addition to a yolk protein, class 2 includes many proteins of unknown function. Several of the differential expressions reported in Table 5.2 were also found in a microarray study by Golden and Melov (2004), who compared a *daf-2* mutant with the wild type.

How DAF-16 upregulates or downregulates all these genes is not yet clear. Lee *et al*. (2003) identified 17 genes in the genome of *C. elegans* that were orthologous between *C. elegans* and *D. melanogaster* and had a putative binding site for DAF-16 in their promoter (between the start

site and 1 kbp upstream). Expression analysis confirmed that six of them were differentially expressed between a *daf-2* and a *daf-2*; *daf-16* mutant, as expected in the case of genes with a causal relationship to the insulin/IGF-1 signalling pathway. Interestingly, and in accordance with Murphy *et al*. (2003), both the upregulated and the downregulated genes had the consensus binding motif. However, none of the 17 genes identified by Lee *et al*. (2003) is shared with the list of Murphy *et al*. (2003), which indicates that other binding sites and *trans*-acting factors other than DAF-16 may be involved.

Murphy *et al*. (2003) also discovered that a gene encoding an insulin-like peptide, *ins-7*, was present among the class 2 members. This protein is not only downstream of the DAF-16 pathway, but, being insulin-like, also acts as an activator of DAF-2. So, any positive signal on DAF-2 will amplify the pathway, via inhibition of DAF-16, relieving the negative regulation of *ins-7* expression and INS-7 stimulation of DAF-2. This positive-feedback loop in the system was thought to contribute to synchrony across cells in the animal. When dauer larvae sense the presence of favourable food conditions and the DAF-2 pathway is activated in some cells by increasing insulin-like peptides, an amplified signal to other cells will prevent the animals emerging from the dauer stage with a mixture of dauer and non-dauer cells.

An overview of the aging-regulator system of *C. elegans* is given in Fig. 5.5 (Gems and McElwee 2003). The bigger picture integrates many aspects of previous studies. It explains why there are many genes with a small additive effect on aging: they are all regulated by the same transcription factor. The presence of yolk proteins in class 2 is consistent with the link between aging and reproduction, while the presence of antioxidant

enzymes in class 1 is in accordance with the well-known relationship between oxygen radicals, cell damage, and aging. Another confirmation of the picture is found in Hsu *et al.* (2003) who showed that the transcription factor HSF-1, which is a regulator of the heat-shock response, also influences aging. Overexpression of *hsf-1* extends lifespan, while reducing it shortens lifespan. So the two transcription factors DAF-16 and HSF-1 partly regulate the same genes with similar effects on aging.

Another genome-wide survey of insulin/IGF-1 signalling-mediated longevity in *C. elegans* is found in the work of McElwee *et al.* (2003, 2004). These authors compared the expression profiles of dauer larvae with those of *daf-2* mutants and argued that genes promoting longevity would have an expression signature similar to the dauer profile. It turned out that in fact around 21% of genes upregulated in dauer larvae are also

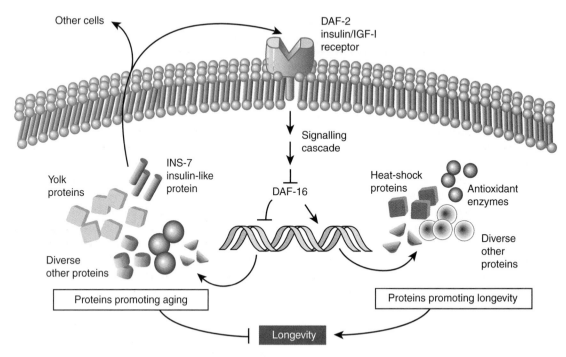

Figure 5.5 The insulin signalling pathway of *C. elegans* converges on repression of the transcription factor DAF-16. DAF-16 itself downregulates expression of proteins promoting aging and upregulates expression of proteins promoting longevity. Among the factors promoting aging is an insulin-like protein, INS-7, that acts upon DAF-2 as a positive-feedback link in the network and a signal to other cells. After Gems and McElwee (2003), by permission of Nature Publishing Group.

upregulated in the *daf-2* mutants, and a similar pattern holds for the downregulated genes. Like Murphy *et al.* (2003), McElwee *et al.* (2004) noted a prominent representation of genes associated with stress-defence systems among the positive regulators of longevity. Several of these genes are known to be associated with the so-called *drug metabolism* or *biotransformation system*. This system of interacting enzymes, which is studied extensively in toxicology, involves two phases in which lipophilic, often aromatic, compounds are first activated and then conjugated to form water-soluble complexes that can be excreted. Phase I of the biotransformation pathway is conducted by enzymes of the *cytochrome P450* family. These are haem proteins that can oxidize aromatic compounds to phenols, epoxides, and quinones, which can then be subjected to *conjugation* by enzymes of phase II, which attach endogenic compounds such as sulphate, glucose, glucuronic acid, or glutathione to the activated product of phase I. The system acts against an enormous variety of lipophilic compounds including many xenobiotics (drugs, environmental pollutants, and plant secondary metabolites) as well as endogenic aromatics such as steroid hormones. We will learn more about this defence system in Section 6.2.

The upregulation of many enzymes of the drug metabolism system in long-lived *daf-2* mutants suggested to McElwee *et al.* (2004) that this system plays a central role in protection from the metabolic damage that accumulates with age (Fig. 5.6). Various cellular processes as well as xenobiotics produce ample reactive molecules that can cause permanent damage to cellular constituents. The biotransformation system can prevent such damage and upregulation of its enzymes is assumed to promote longevity. This new theory of aging is attractive since it assigns a central biological role to the biotransformation system which did obviously not evolve only to metabolize man-made chemicals such as drugs. On the other hand, the new theory could turn out to be a bit too simplistic, since biotransformation does not only detoxify chemicals, it can also activate chemicals to intermediates that cause more harm than the substrate itself. The detoxifying role of the biotransformation

system depends on a delicate balance between phase I and phase II enzyme activity; a general upregulation of all enzymes would be very ineffective and could even be damaging. In addition, many biotransformation enzymes have a great number of isoforms, with different induction profiles, where each substrate needs another isoform to be metabolized effectively.

5.2.3 Longevity-regulating systems across species

One might think that the regulation of lifespan seen in *C. elegans* is tied so specifically to the dauer stage, which is absent in insects and vertebrates, that it does not have a validity outside nematodes. However, the converse is true! Fundamental aspects of the longevity programme elaborated in *C. elegans* appear to be conserved across organisms as widely different as yeast, *Drosophila*, and mouse; that is, the aging mechanisms are 'public' rather than 'private' (Gems and Partridge 2001; Partridge and Gems 2002a; Partridge and Pletcher 2003). This has stimulated hope for extrapolation to human longevity assurance, but it is also of relevance to ecology, since we may expect that several aspects of the genomic determination of longevity in model species are also valid for non-models in an ecological context.

In *D. melanogaster*, as in nematodes, the discovery of mutants with elongated lifespans has been the trigger for aging research. A first hint that regulation of aging was conserved between nematodes and fruit flies came from the observation that, as in *C. elegans*, *D. melanogaster* mutants disturbed in the insulin/IGF signalling pathway have a longer lifespan. The proteins involved in the insulin/IGF-1 signalling pathway of *Drosophila* are INR (insulin/IGF receptor; homologous with DAF-2 in *C. elegans*), an insulin receptor substrate designated CHICO, a phosphoinositide 3-kinase, and a protein kinase B. Most attention has been paid to mutations in the gene encoding CHICO, which confer a dwarf phenotype with reduced fecundity, named after the smallest of the Marx brothers. Clancy *et al.* (2001), studying heterozygous and homozygous *D. melanogaster chico*[1]

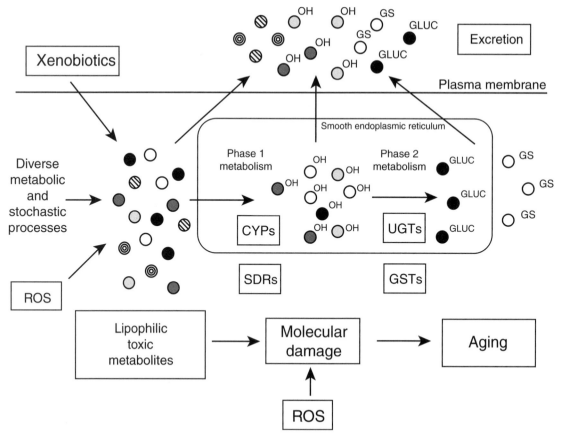

Figure 5.6 Schematic representation of the theory by McElwee *et al.* which holds that a major contribution to the cellular aging process comes from lipophilic toxins generated by cellular activity and xenobiotics, which are biotransformed by phase I and phase II metabolism. ROS, reactive oxygen species; CYPs, cytochrome P450s; SDRs, short-chain dehydrogenase/reductases; UGTs, uridinediphosphate glucosyltransferases and uridinediphosphate glucuronosyltransferases; GSTs, glutathione S-transferases; GLUC, glucosyl or glucuronosyl; GS, glutathionyl. From McElwee *et al.* (2004), by permission of the American Society for Biochemistry and Molecular Biology.

mutants, were able to show that female median lifespan was increased from 48 to 57 days in the heterozygote and further to 63 days in the homozygote. The longevity-extending effect could be 'rescued' by introducing a P-element containing a *chico*(+) construct in the heterozygote *chico¹*. This reduced the median survival time back to a normal value of 49 days. This type of genetic manipulation, by which not only mutants but also engineered *rescue transgenes* are studied, is necessary by present molecular genetic standards to prove that a gene is causally linked to an observed phenotype.

Interestingly, the effect of *chico¹* acted more strongly in females than in males. The male

heterozygous *chico¹* showed only a small increase in lifespan, whereas the homozygotes lived for even less time than the wild-type flies. This suggests that female reproductive tissues might have something to do with the regulation of longevity in *Drosophila*. Mutations in the other components of the insulin signalling pathway, e.g. in *InR*, the *Drosophila* homologue of *daf-2*, produced longer lifespans in mutants with a mild reduction of signalling (Tatar *et al.* 2001). In addition, Clancy *et al.* (2001) noted that the effect of *chico¹* was not due to reduced fecundity per se. This was obvious when *chico¹* was introduced in a strain carrying a mutation that blocks oogenesis and causes complete sterility. *chico¹* was still able to extend the lifespan of these sterile flies. Also, *chico¹* flies did not have a lower metabolism per unit of body mass (Hulbert *et al.* 2004).

Another commonality between the lifespan-regulatory mechanisms of *C. elegans* and *D. melanogaster* is the involvement of a forkhead transcription factor homologous to the mammalian *Foxo*, which is called *dFoxo* in *Drosophila* and is a homologue of the nematode *daf*-16. Overexpression of *dFoxo* in *Drosophila* fat body increased female lifespan by 20–50% and reduced fecundity by 50% (Giannakou *et al.* 2004). In accordance with Clancy *et al.* (2001) there was no effect in male flies.

Like *C. elegans*, *Drosophila* also shows the effects of dietary restriction. The response of lifespan to food concentration during larval culture shows an optimum curve, in which the optimum falls below normal. Low food density shortens lifespan due a starvation effect, and high density shortens lifespan due to accelerated aging (Fig. 5.7). Interestingly, in this work the response of *chico¹* mutants to food was very similar to the wild-type

flies, but was shifted to a higher food density (Clancy *et al.* 2002). The data can be explained by assuming that *chico¹* flies are less sensitive to food concentration, but respond similarly otherwise. This suggests that the metabolic changes induced by *chico¹* are in some way similar to those of dietary restriction and act along overlapping mechanisms. However, this conclusion is somewhat at variance with work on *C. elegans* (see above) and mice (Bartke *et al.* 2001), which has suggested that dietary restriction and reduced insulin signalling rely on two separate mechanisms. As in the case of reduced insulin/IGF-1 signalling, the lifespan-extending effect of dietary restriction is not a consequence of reduced fertility (Mair *et al.* 2004).

We should note that median survival time is only a summary statistic of the survival curve and it does not capture the difference between two treatments well when the shape of the curve is also affected. In principle, an increase in median lifespan can be brought about by two mechanisms: first, by a reduction in the rate of mortality at all ages, extending the survival curve by making it

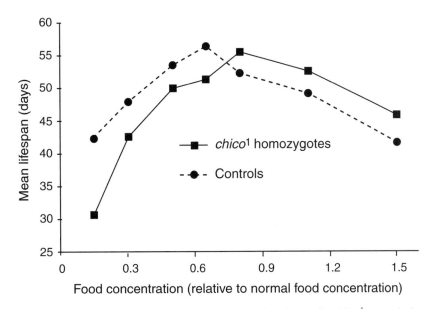

Figure 5.7 Interaction between dietary restriction (dilution of food) and reduced insulin signalling (*chico¹* mutant, backcrossed into the wild-type strain) on adult lifespan. Food concentration is given as a fraction of normal food concentration. Median lifespan was estimated from a survival curve recorded by monitoring deaths every 2–3 days. Reprinted with permission from Clancy *et al.* (2002). Copyright 2002 AAAS.

less steep, and second, by postponing the age at which mortality begins, causing a shift of the curve towards higher ages. The effect of dietary restriction on *Drosophila* is obviously of the first type. There does not seem to be a memory effect of previous food intake: flies that were subjected to dietary restriction after having been fed *ad libitum* for different times all showed the same immediate reduction of mortality rate (Mair *et al.* 2003). This would imply that dietary restriction does not exert its beneficial effect on lifespan by preventing accumulation of damage. This is in accordance with the situation in mice, where it appears that dietary restriction slows the rate of aging (changing the slope of the survival curve), while mutations in insulin/IGF-1 signalling delay the onset of aging (Bartke *et al.* 2001).

Summarizing, the insulin/IGF-1 signalling pathway is definitely involved in regulation of longevity in *Drosophila*, as it is in nematodes, but the effects in *Drosophila* are slightly more complicated; they seem to depend more on the rate of pathway activity than in *C. elegans*. Also, the effects of dietary restriction on lifespan seem to be less than in *C. elegans* and some aspects of the stress response, for example heat resistance, that are induced by *daf-2* mutation in *C. elegans* are not seen in *chico*[1] in *Drosophila*. Finally, reduced insulin/IGF-1 signalling activity has an effect on lifespan only in female *Drosophila*, but this aspect cannot be compared with *C. elegans*, because all experiments in that species have been done with hermaphrodites, not with males.

The evolutionary conservation of longevity regulation is not limited to *C. elegans* and *Drosophila*, but extends to other animals. Holzenberger *et al.* (2003) were inspired by the work on nematodes and fruit flies and investigated the longevity of mice with defective IGF signalling. Whereas *C. elegans* has only one insulin-like receptor, mammals have several; the one that is a homologue of DAF-2 and *InR* is insulin-like growth factor-type 1 receptor, IGF-1R. Holzenberger *et al.* (2003) studied a transgenic mouse in which the essential exon 3 of the *igf-1R* gene was deleted. Controls were included to show that the knockout allele was transcribed but not translated into a functional

product; in the absence of dosage compensation by the wild-type allele the heterozygote had about half the level of IGF-1R protein. This mutant mouse lived 26% longer than the wild type, but the effect was more pronounced in females (33%) compared with males (16%, not significantly different from the wild type). The mutant was also appreciably more resistant to oxidative stress, which was tested by injection of paraquat. Paraquat is a herbicide and a notorious redox cycler, producing large amounts of oxygen radicals, especially superoxide anion, when activated by biotransformation. In addition, Holzenberger *et al.* (2003) noted that the *igf-1R* mutation had hardly any effect on growth and no effect at all on physical activity, food intake, and fertility of the mice. In another study Blüher *et al.* (2003) studied mice with a fat-specific insulin receptor-knockout and these animals likewise showed an extended lifespan, which in contrast to Holzenberger *et al.* (2003) held for both females and males and was accompanied with reduced adiposity (less fat and a leaner body).

Some aspects of longevity regulation in animals may even be linked to yeast (Guarante and Kenyon 2000; Hekimi and Guarante 2003). A key regulator of aging in yeast is a gene involved in silencing chromatin, called *sir2* (silent information regulator 2). *Gene silencing* is a process by which segments of a chromosome are excluded from transcription. SIR-2 silences specific targets of chromatin by deacetylation of histones (proteins that are packed with the DNA in chromosomes), which introduces a local change in chromatin by which DNA is rendered inaccessible to transcription. In yeast, SIR-2 is involved with the regulation of the sexual cycle, and it mediates the formation of spores. Upregulation of *sir2* leads to a longer lifespan (greater number of cell divisons before the larger (mother) cell shows signs of senescence). A longer lifespan has also been reported for nematodes with a *sir2* duplication (Tissenbaum and Guarante 2001). In addition, it has been shown in yeast that caloric restriction, in the sense of limited glucose supply, increases longevity as well as upregulates *sir2*. Finally, *sir2* is also upregulated in long-lived fruit flies under dietary restriction (Rogina *et al.* 2002).

So, all in all, there is considerable agreement between the three genomic model species of nematode, fruit fly, and mouse, in at least the main aspects of longevity regulation. Some aspects of the system may even extend to yeast. The picture is most complete in *C. elegans*, but without doubt genome-wide surveys of aging-related gene expression will shed more light on similarities across species soon. The similarities were reviewed by Tatar *et al.* (2003) and are reproduced here in Fig. 5.8. Upstream of the network there is a sensory mechanism providing input to the central nervous system; this controls the secretion of insulin-like peptides, which suppress the insulin/IGF-1 signalling pathway, starting with *daf-2* in *C. elegans*, *InR* in *Drosophila*, and *igf-1R* in mouse. This then leads to the activation of a forkhead transcription factor, which triggers a great variety of gene expressions and leads to a complicated network, including steroid hormones and signals from the germ line. The final outcome is that growth and reproduction are decreased and anti-aging mechanisms promoted. The division of the body into gonad and soma in this network is crucial (Fig. 5.8).

5.2.4   Trade-off or independent control?

The results discussed above illustrate that the integration of genomics into life-history theory is now in full flight. At the same time, some of the results have come as a kind of shock, because some mutants seem to invalidate a fundamental theorem of life-history theory, the trade-off between life-history attributes. Experimental studies suggest that reproduction, metabolic rate, and longevity can sometimes be uncoupled from each other, which is in flagrant conflict with the idea that there should be a 'cost to reproduction' to prevent Darwin's demon—an hypothetical organism that produces an infinite number of offspring directly after birth and lives forever—from conquering the world (Barnes and Partridge 2003). Is there really such a conflict?

In many of the experimental studies reviewed above there is an obvious negative correlation between longevity and fertility. Fig. 5.9 summarizes the data on the different *daf-2* genotypes in

C. elegans (Leroi 2001). From the viewpoint of life-history theory, it is sufficient to know this relationship and use it to draw inferences about the ultimate effects for the optimal life history. From the viewpoint of ecological genomics, however, it is necessary to understand why this negative correlation is there. It then turns out that decreasing reproduction is not the *cause* of the increased longevity, but that the two processes are regulated by a common mechanism with two different consequences, one usually (but not always) suppressing reproduction, the other increasing stress resistance and (more in females than in males) longevity. Above all, the studies show that the negative relationship in Fig. 5.9 is not due to competition between soma and gonad for a limited resource.

Leroi (2001) argued that life-history ecologists too often cannot resist the temptation to postulate some kind of energy allocation wherever they find a negative correlation between two life-history traits. Geoffroy's 'loi de balancement' is assumed to reign whether it applies or not. This conclusion has been challenged, however, by Lessels and Colegrave (2001) and Barnes and Partridge (2003), who argued that the lack of longevity extension seen when the gonad of *C. elegans* is ablated does not negate the presence of an energy-allocation mechanism, if resources once mobilized but not exploited for reproduction are lost. In other words, 'removing the bucket will not stop the tap'. Barnes and Partridge (2003) argue that the idea of a cost of reproduction has survived the challenge in this case.

Our feeling is that the issue is mainly semantic. Life-history ecologists tend to use the term trade-off in a general sense, without specifying what kind of mechanism is behind a negative correlation. This is perfectly reasonable because life-history theory seeks explanations in evolutionary terms and looks for ultimate causation, rather than proximate mechanisms. Still, we feel that there should be some understanding between these two worlds and evolutionary explanations should be consistent with the underlying physiology and molecular genetics. In particular, if a trade-off is assumed to imply allocation, it should be made clear what kind of resource is being allocated.

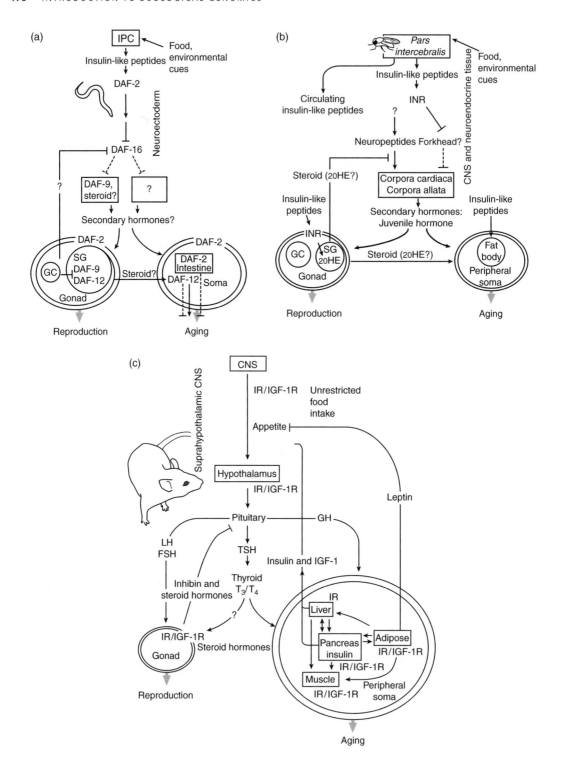

However intricate the genetic or hormonal regulatory pathways, no genomic model organism can escape the energy-conservation laws of thermodynamics. An issue which tends to be neglected in this respect is that in many experiments the actual intake of energy is not measured. This is often due to the difficulty of measuring ingestion and assimilation in animals that take up food from the very substrate in which they live (nematodes, earthworms, fly larvae, flour beetles, etc.). In many allocation arguments the intake of food per unit body mass or surface area is just considered constant, or is assumed to depend only on environmental availability (e.g. prey density). However, food intake is also under control of internal drives. Many animals will eat more when there is a higher internal demand, for example due to the onset of egg production. If this goes unnoticed to the experimenter any energy allocation will be masked.

The question may be asked, why are lifespan mutants not more common in nature, if they live longer and some even maintain normal reproduction? Partridge and Gems (2002b) argued that the life-shortening effect of insulin/IGF-1 signalling could be a negative pleiotropic effect of a fitness increase at earlier ages, the advantage being that efficient insulin/IGF-1 signalling promotes fertility in *Drosophila* and unarrested development in *C. elegans*. Another part of the answer could lie in the conditions acting upon life-history traits in the wild. Walker *et al.* (2000) demonstrated that long-lived nematodes carrying a mutation in *age-1* quickly disappeared from a mixed culture in which they had to compete with wild-type worms under periodic starvation periods. So there are certainly fitness costs to a long life, but they are not always visible under standard laboratory conditions. Unfortunately little is known about the field ecology of *C. elegans* and this seriously hampers a

better understanding of the ecological relevance of life-history mutations (see Section 3.3).

A final remark concerns the ecological relevance of a long life. In almost all organisms in the wild, the life-time of the average individual falls far below the lifespan seen under protected conditions in captivity or laboratory culture. Everyone who has observed great tit parents flying to and fro all day to feed their nestlings knows what fate awaits the great majority of caterpillars in the surroundings. In an earlier career, studying the demography of field populations of Collembola in forest soils, one of us estimated the fraction of hatchlings that would survive to lay their first clutch as 0.7–6.9% and the life expectancy at hatching as 2.0–3.8 weeks (Van Straalen 1985); the large majority of juveniles disappears in the guts of beetles, mites and spiders. Still it is not very difficult to keep individuals of the same species alive in the laboratory for more than 2 years. So do laboratory observations on longevity have any relevance for the field? Kirkwood and Austad (2000) argued that the principal determinant in the evolution of longevity is the level of extrinsic mortality. Animals that live a short life in the wild due to extrinsic sources, are expected to live also short (but longer than in the wild) if the external sources of mortality are taken away in cage or laboratory; the pattern of species differences remains the same. So there could indeed be a relationship between aging phenomena observed in the laboratory and the vicissitudes of outdoor life, but it remains quite a challenge for ecological genomics to demonstrate that relationship.

5.3 Gene-expression profiles in the life cycle

The analysis of the Roff (2002) model given at the beginning of this chapter has shown that age at maturation is a very important life-history trait.

Figure 5.8 Model for endocrine regulation of longevity in (a) *C. elegans*, (b) *D. melanogaster*, and (c) *Mus musculus*, showing how environmental cues are translated into neuroendocrine signals, acting upon an insulin/IGF signalling pathway and triggering a variety of hormonal responses, including steroid responses. The ultimate effect is that different priorities are given to gonad versus soma; that is, reproduction and growth versus aging. CNS, central nervous system; FSH, follicle-stimulating hormone; GC, germ-line cells; GH, growth factor; 20HE, 20-hydroxy-ecdysone; INR, insulin receptor; IPC, insulin-producing cells; IR/IGF-1R, insulin receptor/IGF-1 receptor; LH, luteinizing hormone; SG, somatic gonad tissue; TSH, thyroid-stimulating hormone; T_3, 3,3',5-tri-iodothyronine; T_4, thyroxine. Reprinted with permission from Tatar *et al.* (2003). Copyright 2003 AAAS.

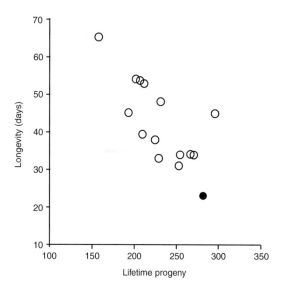

Figure 5.9 Negative correlation between reproduction and median longevity of different genotypes of *daf-2* mutants of *C. elegans*, which have different levels of insulin/IGF signalling. The wild type is indicated by a filled circle. The rather strong negative correlation ($r = -0.75$) between the two processes would be seen as evidence for a trade-off in life-history theory; however, molecular analysis of the mechanism provides no evidence for resource allocation, but rather suggests the action of a hormonal signal with two opposite effects. Reprinted from Leroi (2001), with permission of Elsevier.

It is a well-known fact in life-history theory that when it comes to maximizing intrinsic population growth rate, decreasing the age at first reproduction has the greatest effect, especially when reproductive effort is already high. So it is interesting to explore how expression of genes develops with age and how genomic networks underly the timing of important events in the life history, such as the onset of reproduction. The question is, are there specific expression programmes that mark the various life-history stages?

5.3.1 Developmental stage

Gene-expression profiles tend to change markedly in organisms that develop through distinct stages. This is obvious from work conducted on the model organisms *C. elegans* and *D. melanogaster* (Reinke and White 2002). An exemplary study with nematodes was conducted by Hill *et al.* (2000). These authors isolated oocytes, eggs, juvenile worms, egg-laying adults, and 2-week-old adults (see Fig. 3.19) and profiled the transcriptome of each stage using an oligonucleotide microarray. The statistical technique of 'self-organizing maps' (see Section 2.4) was used to cluster the genes into groups that had similar expression changes throughout the stages. In total, 4221 genes were classified and 36 different clusters were recognized, of which four, A–D, are shown in Fig. 5.10.

It is obvious that each life stage has its characteristic expression profile. In the four groups in Fig. 5.10 cluster A can be considered adult genes, cluster B are larval genes, cluster C are reproduction and egg development genes, and cluster D represents exclusive egg genes. However, none of the clusters, except possibly D, is expressed in one stage only. There are marker genes for each category; for example, *vit-6*, a gene encoding a vitellogenin protein, falls into class A, a cuticular collagen, *dpy-13*, is typical for class B, *mom-2*, required for polarization of the MDS cell (one of the blastomeres in the four-cell stage), is characteristic of class C. Class D appeared to contain several transcription factors and rare messages, but many of these genes had very low expression levels and the data were not considered reliable; nevertheless, class D seemed to contain mostly egg-specific genes (Fig. 5.10). In short, the genes falling into each category reflect the most prominent activity in a life stage.

A similar transcription-profiling study was done by Jiang *et al.* (2001), who used microarrays spotted with PCR fragments from a genomic library rather than oligonucleotide chips. These authors recognized 25 groups of genes with distinct expression patterns across the stages. Particular attention was paid to a group of genes called *cyclins*, which encode proteins involved with cell division. Cyclins are named after their periodic appearance in the cell cycle and they activate cyclin-dependent kinases, which trigger various events in the cell cycle. In *C. elegans*, almost all cell divisions occur in the egg stage, in which the number of cells increases from one to around 700, and in the L4 larva and the adult, in which there is extensive proliferation of cells in the gonads. The *C. elegans* genome encodes seven cyclin genes,

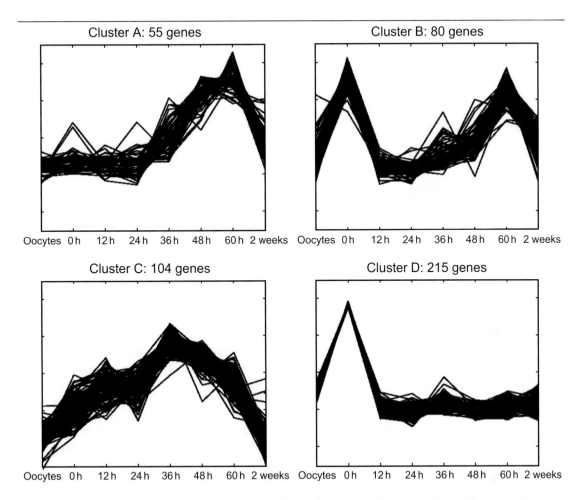

Cluster A: 55 genes

Oocytes 0h 12h 24h 36h 48h 60h 2 weeks

Cluster B: 80 genes

Oocytes 0h 12h 24h 36h 48h 60h 2 weeks

Cluster C: 104 genes

Oocytes 0h 12h 24h 36h 48h 60h 2 weeks

Cluster D: 215 genes

Oocytes 0h 12h 24h 36h 48h 60h 2 weeks

Figure 5.10 Showing changes of gene expression through the life cycle of *C. elegans* for four clusters of genes. The number of genes in each cluster is indicated above the panel. The profiles were all normalized to the same amplitude, so no units are indicated on the *y*-axis. Reprinted with permission from Hill *et al.* (2000). Copyright 2000 AAAS.

which have a similar expression pattern over the stages (Fig. 5.11). They show peak expressions in the embryo, the L4 stage, and the adult, as expected.

Another perspective on the *C. elegans* data may be obtained by clustering the genes according to their similarity with other genomic models. Hill *et al.* (2000) classified genes as 'core' genes (shared among yeast, nematode, and fruit fly), 'animal' genes (shared between nematode and fruit fly), and 'worm' genes (unique to the nematode). When comparing these gene categories over the life stages the proportion of core genes decreased from

egg to adult and the 'worm' genes increased (Fig. 5.12). So, one can say that in the course of development, *C. elegans* becomes more and more a nematode and loses some characteristics of an average animal. The genes expressed in the early stages are most similar to other animals. This pattern is reminiscent of Ernst Haeckel's *biogenetic law*, which states that 'ontogeny recapitulates phylogeny'; reformulated in genomic wording the law may read that genes expressed in early development tend to be more evolutionarily conserved than genes expressed in adult life. It would be interesting to explore whether similar tendencies

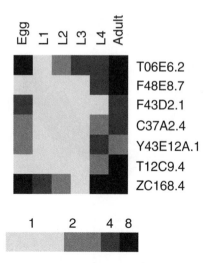

Figure 5.11   Expression of seven cyclin genes through the life stages of *C. elegans*. Expressions were assessed with a microarray spotted with PCR products from a genomic library, and hybridization was relative to cDNA from a mixed-stage population. After Jiang *et al.* (2001), by permission of the National Academy of Sciences of the United States of America.

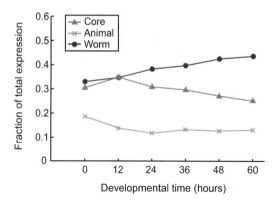

Figure 5.12   Fraction of genes assigned to three categories ('core', shared among yeast, nematode, and fruit fly; 'animal', shared between nematode and fruit fly; 'worm', unique to nematode) among 4221 genes showing differential expression during the life cyle of *C. elegans*, as detected by oligonucleotide microarray transcription profiling. Reprinted with permission from Hill *et al.* (2000). Copyright 2000 AAAS.

are valid for other species; with the availability of more and more genome-wide stage-specific expression profiles, a thorough test of the pattern discovered by Hill *et al.* (2000) would be feasible.

Also in *Drosophila*, a considerable part of the gene complement shows stage-specific gene

expression. Here comparisons were made between eggs, larvae, pupae, and adults (Arbeitman *et al.* 2002). Interestingly, many genes in *Drosophila* appear to be expressed in two waves during development, with embryonic expressions being recapitulated in pupae, and larval expressions recapitulated in adults. Judging from their expression profiles, pupae are like eggs and adults are like larvae. This is understandable from the processes going on in the various stages. The pupa is a stage of intense reorganization of the body, in which most of the larval tissues degenerate and many adult tissues are newly developed from the imaginal discs, processes which are comparable with the extensive cell differentiation in the egg, and apparently involving partly the same genes. The larval stage is characterized by extensive energy acquisition and growth, which is parallelled in the adult by energy acquisition directed to reproduction.

Arbeitman *et al.* (2002) also showed that in the early egg stage a very large proportion of the transcripts are maternal; that is, they are deposited by the mother during oogenesis. Most of these maternal transcripts degrade during the first hours of embryogenesis, but some persist on to the first larva. Maternally derived messengers indicate a mechanism by which the mother can direct the development of her offspring in a non-genetic way. Such *maternal effects* are very common in insects. Well-known examples can be found in the mechanisms of developmental plasticity in response to seasonal change in temperate climates. The photoperiod experienced by an ovipositing female insect often determines the likelihood that the offspring will go into diapause. This is achieved by adding maternal factors to the egg that will direct its development into a diapausing stage. Such phenomena are a well-known source of annoyance to entomologists doing quantitative genetics experiments, because maternal effects confound any estimation of heritability from parent–offspring comparisons. However, maternal effects are now increasingly recognized as being shaped by natural selection to act as a mechanism by which maternal experience is translated into increased fitness of the offspring (Mousseau and Fox 1998). The study of

Arbeitman *et al.* (2002) shows how powerful the mother can be in dominating the transcriptome of the early egg.

Recently the diversity of life-cycle transcript changes has been enriched with a new phenomenon, *exon-specific expression*. Stolc *et al.* (2004) profiled the transcriptome of *D. melanogaster* eggs, larvae, pupae, and adults using an oligonucleotide microarray with no less than 179 972 36-mer probes. In this microarray, exons of the same gene were targeted by different probes, allowing the researchers to profile not only the fluctuations of gene expression of individual genes, but also the changes in expression of exons from the same gene. This is especially relevant if genes undergo *alternative splicing*; that is, the mature mRNA is composed of a subset of exons from the primary transcript, depending on the conditions. Using exon-specific probes on the microarray, Stolc *et al.* (2004) could indicate several genes for which the abundance of exons was not the same in the different stages. For example, for one gene designated as CG8946, exon 1 showed a peak in larva, whereas exon 4 peaked in pupa, and four other exons were expressed synchronously (Fig. 5.13b). For some genes the expression of different exons was even anticorrelated (Fig. 5.13c). It is likely that such stage-dependent exon-specific expressions are crucially important in the developmental process of each stage. These results imply, quite disquietingly, that a vast amount of variation in gene expression may be missed in microarray studies that use cDNAs or that assess only a subset of exons from each gene.

5.3.2 Diapause

Many organisms have a dormant stage in which they shut down activities such as locomotion, and develop tolerance against adverse conditions. The best-known example of life-cycle dormancy is the phenomenon of diapause in insects with seasonal phenologies. Depending on the species, diapause may apply to the egg, larva, pupa, or adult. Are there specific gene-expression patterns characteristic of diapause stages? Surprisingly, this question can only be answered in a rudimentary sense.

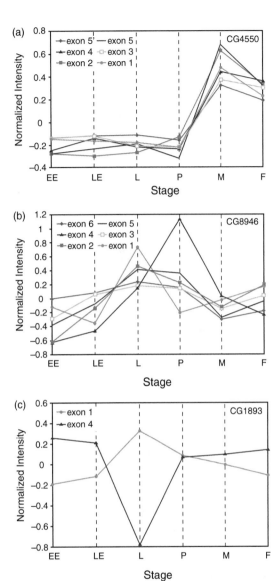

Figure 5.13 Exon-specific expression profiles for three genes over the life stages of *D. melanogaster* (EE, early egg; LE, late egg; L, larva; P, pupa; M, male; F, female), assessed using a microarray in which different oligonucleotide probes targeted different exons of the same gene. Expression was normalized by the mean and the standard deviation. (a) In gene CG4550 (*ninA*, a G-protein-coupled photoreceptor expressed in the eye) all exons are expressed synchronously. (b) In gene CG8946 (*Sly*, a sphingosine-1-phosphate lyase involved with phospholipid metabolism) exon 1 peaks in larvae and exon 4 peaks in pupae. (c) The two exons of gene CG1893 (encoding a product with phospholipid scramblase activity) show anticorrelated expression over the stages. Reprinted with permission from Stolc *et al.* (2004). Copyright 2004 AAAS.

Although there is a tremendous amount of literature on environmental regulation of diapause (especially photoperiod), the circadian clock, and the hormonal changes associated with diapause entry and termination, the molecular genetics of diapause have received little attention (Denlinger 2002). This may be due to the fact that *D. melanogaster* is not the best model for studying diapause, since in this species diapause consists of only a weak reproductive arrest. Better models are the silkworm, *Bombyx mori*, which has a facultative egg diapause, and the flesh fly, *Sarcophaga crassipalpis*, which has pupal diapause. Many agricultural pest species have been used for diapause research, for obvious reasons.

Photoperiod is an important driver of diapause entry in many insects, since it is an accurate predictor of seasonal change. It is widely assumed that the circadian clock provides the mechanism by which insects can perceive photoperiod. Photoperiodic clock genes are known from the *Drosophila* genome and several of them have been sequenced in other insects such as *Bo. mori*, the tobacco hornworm *Manduca sexta*, and *S. crassipalpis* (Goto and Denlinger 2002). The gene *period* (*per*) has been examined in relation to diapause, but how the circadian clock triggers the entry into diapause is not clear at all. It turns out that *Drosophila* carrying a null mutant of *per* can enter diapause just as well as the wild type, although they do not have a circadian rhythm.

Denlinger (2002), drawing together a large number of mostly pre-genomic studies, provides an overview of genes that have their expression regulated by diapause. Table 5.3 gives a summary of this information. Genes are subdivided into six categories, depending on whether they are up- or downregulated early or late in diapause, or not modulated at all. The great majority of genes is downregulated, although some are specifically upregulated. This demonstrates that, like the dauer stage of *C. elegans*, insect diapause does not imply a simple shutting down of the genome but involves an active and specific programme of gene regulation. This is also evident from the fact that in individuals destined for diapause the stage preceding the diapause is longer than normal.

This stage includes physiological preparations and sometimes specific behaviours aimed at locating suitable microhabitats for hibernation or aestivation.

Some of the differential expressions in Table 5.3 could be due to master genes, regulating the various processes during diapause; other genes may act downstream of regulatory cascades, being up- or downregulated as a consequence of master genes. Still other genes have specific functions in the diapause, such as those related to cold-hardiness, defence against microbial infection, etc. Some genes are expressed intermittently during diapause, possibly in response to the periodic boosts of respiration triggered by pulses of juvenile hormone.

Table 5.3 shows that the picture of diapause-specific gene expression in insects is still incomplete. Some gene expressions are reminiscent of dormancies in other species. This is especially applicable to the upregulation of heat-shock protein 70, which is consistently reported in insect diapause and is similar to the pattern seen in the *C. elegans* dauer larva (see above) and several other organisms, even fungi. The universal involvement of heat-shock proteins in dormancy and life extension is striking and must relate to a very fundamental role of these proteins in the cell (see also Section 6.2). In insect diapause, not all heat-shock protein are upregulated, however: some are not regulated at all, and *Hsp90* is actually downregulated (Table 5.3).

5.3.3 Adult life and sex

After the attainment of adulthood, gene-expression profiles tend to change only little. This is evident both in *C. elegans* (Lund *et al.* 2002) and in *D. melanogaster* (Zou *et al.* 2000; Jin *et al.* 2001). Applying rigid statistical criteria, Lund *et al.* (2002) found only 164 genes, representing about 1% of the *C. elegans* genome, to change from the first day of adult life to senescence (19 days). Many of these genes were related to stress resistance. The 27 heat-shock genes on the microarray all showed a common expression profile, rising over the first part of adult life and decreasing later in life.

Table 5.3   Overview of gene products regulated by diapause in insects

Gene or gene product	Function	Species
Not influenced by diapause		
Heat-shock cognate protein 70 (*hsc70*)	Cellular stress response	*S. crassipalpis, C. fumifera*
28 S ribosomal protein	Protein synthesis	*S. crassipalpis*
Cyclins E, p21, and p53	Cell cycle	*S. crassipalpis*
Glutathione S-transferase	Detoxification	*C. fumiferana*
Ubiquitin	Cellular stress response	*C. fumiferana*
Downregulated throughout diapause		
Mitochondrial phosphate transport protein	Respiration	*C. fumiferana*
Midgut enzymes	Food assimilation	*L. dispar*
Heat-shock protein 90 (*hsp90*)	Cellular stress response	*S. crassipalpis*
Proliferating cell nuclear antigen (*pcna*)	Cell cycle	*S. crassipalpis*
cdc2-related Ser/Thr kinase	Cell cycle	*Bo. mori*
Upregulated throughout diapause		
Heat-shock protein 23 (*hsp23*)	Cellular stress response	*S. crassipalpis*
Heat-shock protein 70 (*hsp70*)	Cellular stress response	*S. crassipalpis,*
		R. pomonella, O. nubilalis
Heat-shock protein 70A (*hsp70A*)	Cellular stress response	*L. decemlineata*
E26 transforming sequence (ETS) protein	Transcription factor	*Bo. mori*
7.9 kDa peptide	Function unknown	*G. atrocyanea*
Alkaline phosphatase	Various functions	*L. dispar*
Upregulated in early diapause, downregulated in late diapause		
pScD41, clone from brain-enrichment library	Function unknown, possibly a retrotransposon	*S. crassipalpis*
Samui	Cold-induced activator of sorbitol dehydrogenase	*Bo. mori*
55 kDa protein in gut	Function unknown	*L. dispar*
Downregulated in early diapause, upregulated in late diapause		
Ultraspiracle (*usp*)	Ecdyson receptor	*S. crassipalpis*
Sorbitol dehydrogenase (*sdh*)	Conversion of sorbitol into glycogen	*Bo. mori*
defensin	Microbial defence	*C. fumiferana*
45 kDa actin-like protein in brain	Function unknown	*L. dispar*
Genes expressed intermittently during diapause		
60 S ribosomal protein PO	Protein synthesis	*S. crassipalpis*

Notes: Species investigated were *S. crassipalpis* (flesh fly; Diptera, Sarcophagidae), *Choristoneura fumiferana* (spruce budworm; Lepidoptera, Tortricidae), *Lymantria dispar* (gypsy moth; Lepidoptera, Lymantriidae), *Bo. mori* (silkworm; Lepidoptera, Bombycidae), *Rhagoletis pomonella* (apple maggot fly; Diptera, Tephritidae), *Ostrinia nubilalis* (European corn borer; Lepidoptera, Pyralidae), *Leptinotarsa decemlineata* (Colorado potato beetle; Coleoptera, Chrysomelidae), *Gastrophysa atrocyanea* (leaf beetle; Coleoptera, Chrysomelidae).
Source: After Denlinger (2002).

Another category of differential expressions consisted of insulin-like proteins. Three of the insulin genes increased in expression, but one, *Ins-7*, showed a decreased expression during life. *Ins-7* was identified in the study of Murphy *et al*. (2003) as belonging to a group downregulated by DAF-16 and contributing to a positive-feedback loop in the insulin/IGF-1 signalling signalling pathway (see Section 5.2). Decreased *Ins-7* expression during life, as found by Lund *et al*. (2002), is consistent with upregulation of stress-responsive genes by increased DAF-16 activity. So it seems that normal aging in *C. elegans* is not accompanied by a specific genetic programme. The changing expression profiles seem to indicate a response to accumulating cellular damage from reactive oxygen species and unfolded proteins, rather than being part of the developmental repertoire.

Also in *Drosophila* the effects of age on gene expression in normal (wild-type) flies tend to be weak. Zou *et al.* (2000) measured genome-wide changes in expression of flies between 3 and 50 days of age, using an EST microarray. A total of 127 genes was found to be regulated by age, and these were classified according to three categories: genes associated with reproduction, metabolic genes, and stress-responsive genes. In all categories there was a trend towards decreasing expression with age, but among the stress-responsive genes some were upregulated. In a similar study, Jin *et al.* (2001) compared 1 and 6 week-old flies of different sexes and from different strains using replicated microarray hybridizations. Expression profiles were mostly affected by sex, less so by strain, and only weakly by age. Analysis of variance showed that sex, in combination with the sex × strain interaction component, accounted for between 60 and 90% of the variation in gene expression. Some of the genes that were age-related in the study of Zou *et al.* (2000) were confirmed as age-related by statistical significance in the study of Jin *et al.* (2001), but several others were not. The large influence of sex in this study is a notable result. Much of the sex bias can probably be attributed to reproductive activities that are naturally different between male and female.

In *C. elegans* most studies are done with hermaphrodites, but Jiang *et al.* (2001) compared expression profiles of hermaphrodites and males. No fewer than 2171 genes (12% of all the genes in the genome) were found to be sex-regulated. About half of the male-specific genes were expressed in the gonad and the sperm, and the other half were expressed in the soma. Male *C. elegans* have an elaborate mating behaviour to find the hermaphrodite's vulva with their tails and to ejaculate sperm into it. In conjunction with this behaviour, males have 115 neurons and neuronal support cells not found in the hermaphrodite, whereas hermaphrodites have only eight sex-specific neurons. Sex-specific expression patterns

of the whole animal may partly be due to genes expressed specifically in these cells. One of the groups of sperm-enriched genes is found in the L4 larva, which is the major sperm-producing stage of the hermaphrodite.

The sex bias in gene expression in *Drosophila* and *C. elegans* is corroborated by the fact that on the phenotypic level strains of the same species tend to be more different in one of the sexes than in the other. On a larger scale, the same phenomenon is seen in the differences between closely related species of fruit fly: such differences tend to be larger in males compared to females. A comparison of gene-expression profiles of the sibling species *Drosophila simulans* and *D. melanogaster* showed that 50% of the genes that differed in expression between the species did so in a sex-dependent manner. Almost all the genes that differed by more than a factor of four were male-specific genes (Ranz *et al.* 2003). These data underline the importance of sex-dependent selection in the differentiation of the *Drosophila* species cluster.

The relatively minor age-specific expression changes observed in the studies reviewed above are more or less at variance with a detailed study by Pletcher *et al.* (2002) on transcript profiles in aging *D. melanogaster*. Although these authors analysed only females and so the effects of age cannot be weighed against the effects of sex, the differential expressions they observed indicated that nearly 23% of fruit fly genes were modulated by age, a figure much higher than reported by Jin *et al.* (2001) and Zou *et al.* (2000), and by other studies on rodents and rhesus monkeys. There is, however, some degree of similarity across studies in the types of genes regulated by age. Pletcher *et al.* (2002) found decreasing expression for genes involved in chorion formation, which is obviously due to reproductive senescence, and upregulation of stress-response genes such as cytochrome P450, as in the other studies. Interestingly, aging was also associated with increased expression of antibacterial peptides. This suggests that microbial infection is an important factor in the life of *Drosophila* in laboratory culture.

Whereas some of the age-specific expression profiles are only valid for certain model organisms, and maybe only under specific culturing conditions, ecological genomics of life histories should aim at finding regulatory signatures of life-history events that are of more general validity. Such signatures can be detected when expression profiles are compared across species. One of the first studies to use this approach was done by McCarroll *et al.* (2004). These authors analysed microarray data from *C. elegans* and *D. melanogaster*, in which expressions were compared between two stages, young adults (0 days for *C. elegans*, 3 days for *D. melanogaster*) and mature adults (6 days for *C. elegans*, 23 days for *D. melanogaster*). The measurements were then paired systematically between orthologous genes from the two organisms. The authors were interested in genes with the same expression ratio in the two species; that is, genes which if downregulated with aging in *Drosophila* are also downregulated with aging in *C. elegans*. The methodology is illustrated in Fig. 5.14.

Classifying the genes by Gene Ontology categories, McCarroll *et al.* (2004) identified three categories of shared transcriptional profiles with decreasing expression during adult life:

• genes involved in oxidative metabolism (respiratory chain, citric acid cycle),
• genes involved in catabolism (peptidases) and DNA repair, and
• genes involved in molecular transport functions, such as ion transporters and ABC transporters.

So, the shared transcriptional profile of aging seems to involve a decreased commitment to energy generation and active movement of ions, transmitters, and nutrients. Interestingly, these changes are implemented abruptly early in adult life, and only little further change occurred later in life. So it seems that, if there is a developmentally timed transcriptional regulation of aging, it is a programme that is associated with the onset of adulthood. What we see later in life, for example upregulation of stress-response genes in the studies reviewed above, is more a consequence than a cause of aging.

The type of *comparative functional genomics* applied by McCarroll *et al.* (2004) is a promising

new strategy for identifying the general genomic basis of life-history events and to remove some of the noise and contingency that (not unlike ecological studies) seems to be inherent to many microarray-based transcription profiles.

5.3.4 Flowering time in *Arabidopsis*

The developmental patterns of plants show considerably more flexibility than those of animals. Most of the development process in plants occurs post-embryonically through the action of shoot and root apical meristems. Consequently, there is ample opportunity for plants to adapt their morphology to specific environmental conditions, whereas the body plan of animals is more or less fixed and only certain aspects of it may depend on the environment. One of the major developmental transitions in the life of a plant is the switch from vegetative growth to reproductive development. The timing of this switch in relation to environmental conditions is of crucial adaptive value; fitness will be lost if reproduction and seed set fall in an unfavourable season or are not synchronized with other members of a population in out-crossing species. This implies that flowering time, also called *bolting time*, and especially the way in which it is modulated by environmental cues, is an important life-history trait optimized by natural selection.

The major model for the genomics of flowering time is *A. thaliana*. There is some work on other Brassicaceae and on crops such as corn and rice, but this is relatively little in comparison with *Arabidopsis* and therefore we will focus initially on this model. Despite its present widespread distribution (Fig. 3.23), *Arabidopsis* has its origin in a northern, seasonal climate and so the timing of the reproductive switch with respect to daylength and temperature is a crucial aspect of its life history. *A. thaliana* is a facultative long-day plant, which implies that bolting is stimulated by the long days of early spring. Bolting also requires *vernalization*: a prolonged period of cold (3–8 weeks at 4°C or lower), necessary to allow flowering in spring.

Through the analysis of mutants a large number of genes has been identified that in some way or

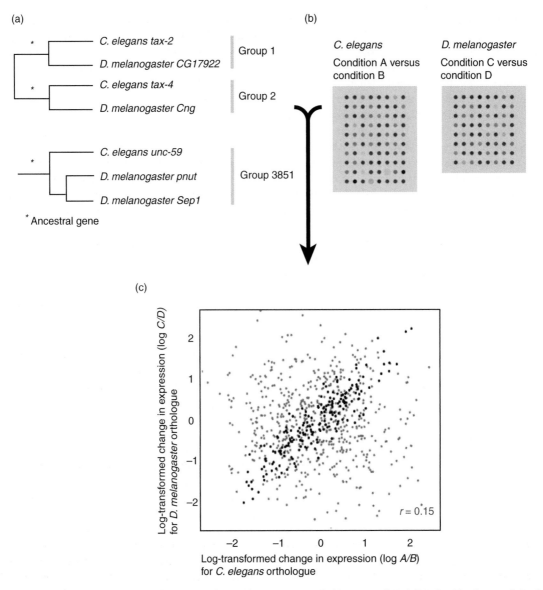

Figure 5.14 Illustrating the principle of comparative functional genomics, as applied by McCarroll *et al.* (2004) to identify transcriptional profiles of aging, shared across *C. elegans* and *D. melanogaster*. (a) Phylogenetic analysis identifies orthologous pairs between the species. If more than one paralogue is present within a species, the most conserved orthologous gene pair is identified. (b) Transcription profiling with microarrays is applied to reveal the relative change in expression when comparing two conditions (e.g. young and old age). (c) Groups of genes are selected according to the Gene Ontology annotation system and the expression ratio for the two species is plotted in a correlation diagram. If there is a significant correlation, the group is said to have a conserved regulation and may be considered an expression signature of the treatment applied in the two species. From McCarroll *et al.* (2004), by permission of Nature Publishing Group.

another influence flowering time. Putterill *et al.* (2004) provide an annotated list of 51 genes. Interestingly, most of their products act as regulatory proteins, for example transcription factors, RNA binding proteins, signal transducers, and kinases. The number of genes involved illustrates that flowering time is an extreme example of polygenic control. Genetic research has shown that the flowering-time genes fit into four regulatory networks (Koornneef *et al.* 1998; Mouradov *et al.* 2002; Ratcliffe and Riechmann 2002; Simpson and Dean 2002; Putterill *et al.* 2004), which will be discussed in detail below.

The 'default' life cycle of *Arabidopsis* seems to be a direct floral transition early in life, and this can be achieved by mutations in some of the flowering-time genes. Such 'rapid-cycling' varieties are useful under laboratory conditions, but in the wild type the floral transition is actively repressed. The gene *FLOWERING LOCUS C* (*FLC*) is a crucial element of this repression; it encodes a MADS-box transcription factor, which regulates several floral-identity genes in the apical meristem. Consequently, conditions that stimulate bolting and shorten flowering time must remove the repressive action exerted by the *FLC* locus. This way of regulation, in which an environmental signal removes a repressor, is thought to be more stable and more specific than a system in which the environmental signal acts as a direct positive regulator (Casal *et al.* 2004). We saw above that a similar system of double-negative regulation governed the dauer transition in *C. elegans*.

The MADS-box genes, to which *FLC* belongs, are a large family of very old transcription factors, which have diversified in plants more than in animals (Becker and Theißen 2003). They derive their name from the *MADS box*, a highly conserved 180 bp consensus sequence, which like the homeobox in the *Hox* genes encodes a DNA-binding domain. There are two types of MADS-box gene, type I and II, which are both represented in plants, fungi, and animals, and which are assumed to derive from a very old duplication in the stem of the eukaryotes, more than 1000 million years ago. Little is known about the type I genes in plants; type II genes have diversified considerably and have given rise to genes that control all the developmental processes in plants, such as the ontogeny of roots, flowers, seeds, and fruits. Loss-of-function mutations in MADS-box genes cause homeotic transformations (replacing, for example, sepals with carpels or petals with stamens), indicating that these genes act to determine the identity of a meristem or a primordium.

The type II MADS-box genes in plants belong to two families, MIKC*-type genes and MIKC^c-type genes, where MIKC^c is subdivided into four subfamilies (Becker and Theißen 2003). *FLC* belongs to one of the MIKC^c subfamilies.

As indicated above, four interacting pathways control flowering time in *A. thaliana*: the photoperiod and light-quality pathway, the vernalization-response pathway, the autonomous pathway, and the gibberellin signalling pathway. These pathways are interconnected and converge on *FLC* and other flowering regulators such as *FT* and *SOC1* to activate the floral-identity genes *AP1* and *LFY* (Fig. 5.15).

The *photoperiod-response pathway* integrates information from the daylength to promote flowering in response to long days. A crucial gene in this pathway is *CO* (*CONSTANS*), named after the mutant that confers insensitivity to daylength. *CO* induces early flowering by activating *FT*, which regulates the floral-identity gene *AP1* (Fig. 5.15). The expression of *CO* is influenced by the circadian clock, a system of autoregulated genes with feedback loops such that clock proteins regulate their own expression and appear in an oscillating fashion. The clock is entrained by photoreceptors such as phytochromes (*PHYA* and *PHYB*) and cryptochromes (*CRY1* and *CRY2*). Entrainment ensures that the period of the circadian rhythm is synchronized with the daily cycle of light and dark. The circadian feedback loop generates a series of oscillating outputs including rhythmic expression of *CO*. The peak of expression falls in the second half of the day; however, to exert its action, the CO protein must be stabilized and this occurs only in the light. CO is stabilized by PHYA in far-red light and by CRY1/CRY2 in blue light (Valverde *et al.* 2004). This recent insight in the post-translational regulation of CO provides an explanation for the *external coincidence model*, which holds that photoperiodic responses are regulated by a signal with circadian expression that must fall in the light to trigger the response. In the case of *CO*, the peak of expression at the end of the day coincides with the light only on long days; CO is then stabilized and may act upon *FT*. In short days, the peak of *CO* expression falls in the

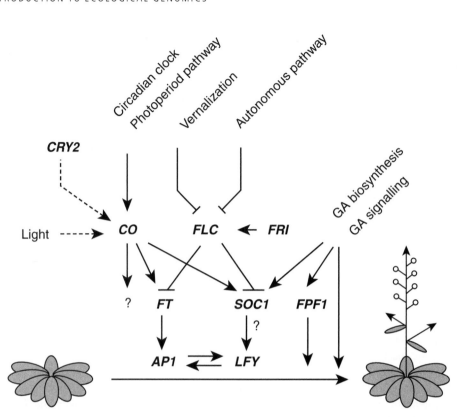

Figure 5.15 Overview of pathways in control of floral transition in *A. thaliana*. Four inputs are indicated, with a selection of the key genes involved. *CRY2, CRYPTOCHROME 2; CO, CONSTANS; FLC, FLOWERING LOCUS C; FRI, FRIGIDA; FT, FLOWERING LOCUS T; SOC1, SUPPRESSOR OF OVEREXPRESSION OF CONSTANS 1; FPF1, FLORAL PROMOTING FACTOR 1; AP1, APETALA 1; LFY, LEAFY*; GA, gibberellic acid. Arrows indicate promotion; bars indicate repression. After Mouradov *et al.* (2002). Copyright American Society of Plant Biologists.

dark, the CO protein is then unstable, *FT* expression is not upregulated, and flowering is delayed (Yanovsky and Kay 2003; Hayama and Coupland 2004; Putterill *et al.* 2004; Schepens *et al.* 2004).

A surprisingly large part of the genome of *A. thaliana* undergoes day–night oscillations. A genome-wide transcription-profiling study by Harmer *et al.* (2000) classified 6% of the probes on an oligonucleotide gene chip as cycling with a period of between 20 and 28 h. Four categories were recognized: (i) light-harvesting centres, photosynthesis genes, phytochromes, and cryptochromes, (ii) genes involved with photoprotective pigment pathways, such as flavonoids and anthocyanins, (iii) enzymes involved in resistance to chill and drought, for example catalysing lipid modification, and (iv) enzymes of the carbon, nitrogen, and

sulphur pathways. The authors were also able to identify a conserved 9 bp motif in the promoters of 35 cycling genes, which suggests that these genes are regulated by the same transcription factor. A similar study was reported by Schaffer *et al.* (2001), who used a microarray with 11 521 *Arabidopsis* ESTs showing that 11% of the genes were expressed diurnally. Obviously, such widespread rhythmicity in the genome reflects the pervasive influence of light on the metabolism of phototrophic organisms; at the same time, it reminds experimenters how important it is to standardize the time of the day when harvesting RNA in studies looking at gene expression in plants.

The *vernalization pathway* for the control of flowering time integrates information from the past temperature regime. A crucial gene in this pathway is *FRI* (*FRIGIDA*). This gene promotes the transcription of *FLC* and so represses the floral transition. Mutations in *FRI* remove this positive regulation and cause loss of vernalization requirement. The *FRI* locus is particularly relevant in the ecology of *Arabidopsis*, because it shows natural variation associated with flowering time. When a comparison is made of flowering times of *Arabidopsis* accessions of different geographic origin, plants from low latitude tend to flower earlier in a common garden than plants from high latitude. However, this latitudinal cline is only found in ecotypes that have a functional *FRI* gene (Stinchcombe *et al.* 2004), suggesting that *FRI* mediates the transduction of latitudinal information into regulation of flowering time.

An important issue in understanding the response to vernalization has been the question of how vernalized plants can remember a cold treatment and flower several weeks later, even when temperatures are higher for some time after the cold signal. It turns out that cold treatment induces an altered state in the shoot apex which can be passed on through mitotic cell divisions, even in the absence of cold. Recent research has shown that gene silencing due to changes in chromatin structure is the basis of this cellular memory (Bastow *et al.* 2004; Sung and Amasino 2004).

We know from basic biochemistry that in eukaryotic chromosomes the DNA double helix is wound around groups of small globular proteins, histones, forming nucleosomes. These histones have tails protruding outward from a nucleosome, in which the lysine residues are normally acetylated. Because histones are highly positively charged proteins, acetylation of the tails is nessessary to prevent the formation of aggregates of nucleosomes and to maintain a loose chromatin structure in which DNA is accessible to transcription.

Sung and Amasino (2004) identified a regulatory gene in *Arabidopsis* called *VERNALIZATION INSENSITIVE 3* (*VIN3*), which, in conjunction with two other vernalization genes, *VNR1* and *VNR2*, inactivates *FLC* by local deacetylation of histones. *Histone deacetylation* causes condensation of chromatin and shuts off DNA from transcription. This process is catalysed by histone deacetylase complex (HDAC), a cluster of molecules involving a DNA-binding protein and an acetyltransferase enzyme. The condensed state of chromatin is transferred to the daughter cells when cells divide and so provides a mechanism of inheritance that does not involve alterations in DNA. This type of inheritance is called *epigenetic*.

VIN3 is the most upstream component of the vernalization pathway identified so far, but it is still not known how the protein senses cold. Sung and Amasino (2004) suggest that VIN3 might be a receptor for phosphoinositides (a group of phospholipids) in the nucleus, and could perceive changes in the composition of these compounds that occur during cold exposure.

The third pathway for regulating flowering time in *Arabidopsis*, the *autonomous pathway*, integrates information from the developmental stage of the plant. In the default state it represses *FLC* like the vernalization pathway (Fig. 5.15). *Arabidopsis* mutants of the autonomous pathway are early-flowering but retain the photoperiodic response, and so the flowering signal is independent of environmental cues (hence the use of the term autonomous). However, work by Blázquez *et al.* (2003) has shown that expression of genes in the autonomous pathway depends on temperature; this would represent a system of thermosensory control of flowering time that acts in parallel to vernalization. The autonomous pathway involves a group of six different genes, which act upon *FLC* in two different ways. One of the mechanisms involves inactivation of *FLC* by histone deacetylation, like in the vernalization pathway (He *et al.* 2003). This is mediated by a locus *FLD*, which encodes a protein that forms part of an HDAC. Another gene product of the autonomous pathway, FVE, is probably part of the same HDAC (Amasino 2004; Ausín *et al.* 2004; Kim *et al.* 2004). In a series of elegant experiments He *et al.* (2003) were able to show that a specific region in the first intron of *FLC* acted as a binding site for the HDAC (Fig. 5.16). Mutations in *FLD*, as well as deletion of a 294 bp region from intron 1, prevented binding of HDAC to the *FLC* gene, causing continued transcription of *FLC* and late flowering.

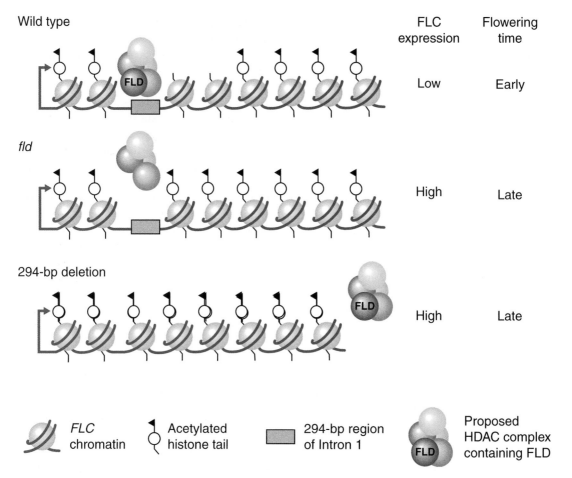

Wild type

FLD

FLC expression: Low

Flowering time: Early

fld

FLC expression: High

Flowering time: Late

294-bp deletion

FLD

FLC expression: High

Flowering time: Late

FLC chromatin	Acetylated histone tail	294-bp region of Intron 1	Proposed HDAC complex containing FLD

Figure 5.16 Regulation of *FLOWERING LOCUS C* (*FLC*), a floral repressor of *Arabidopsis*, by *FLD* (*FLOWERING LOCUS D*). In the wild type, *FLC* expression is suppressed by deacetylation of histones in the vicinity of FLC, conducted by an HDAC, of which FLD is a part. If *FLD* is mutated or if a 294 bp region in the first intron of *FLC* is deleted, HDAC is no longer able to deacetylate the *FLC* histones, alleviating *FLC* from transcription suppression and causing late flowering. Reprinted with permission from Bastow and Dean (2003). Copyright 2003 AAAS.

The second regulatory input on *FLC* from the autonomous pathway comes from the floral-promotion genes *FCA* and *FY*. The way in which these genes regulate *FLC* expression represents another complicated but beautiful example of gene regulation underlying life-history traits (Eckardt 2002; Macknight *et al.* 2002; Amasino 2003; MacDonald and McMahon 2003; Quesada *et al.* 2003; Simpson *et al.* 2003). The *FCA* gene includes 20 introns, which are spliced out during mRNA assembly; however, introns 3 and 13 are spliced alternatively, leading to four different transcripts, designated as α, β, γ, and δ. Only the γ transcript

functions in the control of flowering time; it encodes a protein that, together with the FY protein, suppresses *FLC* mRNA and thus stimulates flowering. In addition, FY and FCA proteins feed back upon the activity of the *FCA* gene because they promote the formation of the inactive β transcript, by polyadenylation in intron 3. This negative-feedback loop is essential to maintain a low concentration of active transcripts of *FCA*. When the autoregulation breaks down, active FCA is formed, causing inhibition of FLC mRNA and induction of flowering (Fig. 5.17). The *alternative splicing* of *FCA* transcripts provides a mechanism

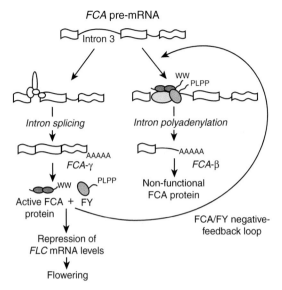

FCA pre-mRNA

Intron 3

Intron splicing

Intron polyadenylation

AAAAA
FCA-γ

AAAAA
FCA-β

Active FCA + FY
protein

Non-functional
FCA protein

Repression of
FLC mRNA levels

FCA/FY negative-
feedback loop

Flowering

Figure 5.17   Scheme of the input from *FCA* and *FY* in the regulation of flowering time in *Arabidopsis*. Intron 3 of the *FCA* pre-mRNA transcript can be spliced out in two different ways, one of which is promoted by FY and FCA proteins, leading to intron polyadenylation and a non-functional *β* transcript. The other pathway leads to a fully functional *γ* transcript, which in conjunction with FY represses *FLC* mRNA. Since FLC is a flowering repressor, this stimulates flowering. WW, a 35–40-amino acid protein domain engaging in protein–protein interactions; PLPP, amino acid motif of the FY protein, interacting with the WW domain of FCA. After Putterill *et al.* (2004), by permission of John Wiley & Sons.

by which *FLC* expression can be fine-tuned during development and limited to certain tissues.

It is not known what endogenous signal is directing the autonomous pathway or triggers expression of *FLD*, *FVE*, *FCA*, and *FY*. Plants must pass through a juvenile phase before they can flower, but it is unclear how this is sensed internally. It might be that the pathway is not regulated dynamically, but functions constitutively to maintain high levels of *FLC* expression throughout early development (Simpson and Dean 2002). In addition, parts of the pathway may act to transfer information about environmental temperatures, as indicated above. The inhibitory effect of *FLC* on flowering is due its repressive effect on *FT* (*FLOWERING LOCUS T*; Fig. 5.15). Recently another gene suppressing *FT* has been identified, designated *EBS* (*EARLY BOLTING IN SHORT*

DAYS; Gómez-Mena *et al.* 2001; Piñeiro *et al.* 2003). The sequence of this gene suggests that it is part of a chromatin-remodelling complex, and this suggests another case of epigenesis in flowering-time regulation.

The fourth pathway acting upon flowering time is the *gibberellin signalling pathway* (Fig. 5.15). Gibberellic acid (GA) is a plant hormone with important effects on growth regulation, promoting germination, stem elongation, and fruit growth. Mutations in GA biosynthesis and GA receptors disrupt the floral transition, and have many other effects on plant development. One way in which gibberellins promote flowering is by increasing transcriptional activity of the floral meristem-identity gene *LFY* (*LEAFY*), through upregulation of a MADS transcription factor gene, *SOC1* (*SUPPRESSOR OF OVEREXPRESSION OF CON-STANS 1*). It is not known which endogeneous or exogeneous signals are integrated by the GA input in the flowering-time network.

Most of the insights discussed above have been obtained through analysis of mutants and naturally occurring varieties. Genome-wide expression-profiling studies have confirmed the regulatory networks derived from single-gene mutants, but at the same time have suggested the involvement of even more genes. Schmid *et al.* (2003) used microarray-based transcription to obtain a genome-wide picture of floral transition in *Arabidopsis*. Plants mutated in different flowering-time genes were grown initially under short days and then transferred to long days, and gene-expression profiles determined over 7 days. Schmid *et al.* (2003) noted that many genes were repressed, rather than upregulated, by transition to long days. This is at variance with the scheme outlined above, because until then repressors had been identified only in the vernalization and autonomous pathways, not in the photoperiodic pathway (Fig. 5.15). Two of the repressor genes were identified and named *SCHLAFMÜTZE* (*SMZ*; meaning night cap) and *SCHNARCHZAPFEN* (*SNZ*; meaning snoring uvula). Knockout mutants of *SNZ* and *SMZ* flowered normally, showing that the two genes have a redundant function, and could not have been detected by genetic screens using mutants. They both encode a so-called AP2 domain, which makes them potential binding proteins for *micro-RNAs*, which are short, single-stranded, RNA molecules that can base-pair with complementary sequences in mRNA, blocking their translation. Regulation of mRNA by microRNA has been implicated as an important mechanism in plant morphogenesis. The study of Schmid *et al.* (2003) suggests that microRNAs can also play a role in the regulation of flowering responses to daylength.

5.3.5 Regulation of flowering time in other plants

Is all the detailed knowledge about regulation of flowering time in *Arabidopsis* applicable to other plant species? Similarities are found in the Brassicaceae, the plant family to which *Arabidopsis* belongs (Kole *et al.* 2001). In *Brassica rapa* several flowering-time QTLs were identified and high-resolution mapping showed that one of them, *VFR2*, was homologous to *Arabidopsis FLC*. It seemed to have the same phenotypic effect: late flowering, downregulated by vernalization in the biennual genotype of *B. rapa*. Kim *et al.* (2003) identified genes sharing strong homology with *AGAMOUS-LIKE 20*, a MADS-box transcription factor downstream of *FLC* in *B. rapa* as well as in two other Brassicaceae, *Cardamine flexuosa* and *Draba nemorosa*. The expression pattern of *AGL20* in *C. flexuosa* during the floral transition was similar to *Arabidopsis*, which together with the data from Kole *et al.* (2001) suggests that at least some aspects of the regulation of flowering time are conserved between *Arabidopsis* and other Brassicaeae.

It is likely that most of the flowering-time genes identified in *Arabidopsis* are also present in plants outside the Brassicaceae, but they do not necessarily have the same function. For example, Petersen *et al.* (2004) recovered MADS-box genes of perennial ryegrass, *Lolium perenne*, which were differentially expressed during transition from vegetative to reproductive growth induced by vernalization. They used a differential-display technique (see Section 2.1, Fig. 2.3) with one PCR primer targetting a monocotyledon-specific

conservative region in the MADS-box gene. After sequencing nine ryegrass MADS-box genes and studying their expression with quantitative real-time PCR the authors noted both similarities and differences with the *Arabidopsis* system. Genes with sequence homology to the *Arabidopsis AP1* subfamily appeared to have a different expression pattern and possibly a different function in vernalization, compared to *Arabidopsis*.

A more detailed comparison of flowering-time regulation is possible between *Arabidopsis* and rice (Hayama *et al.* 2003; Griffiths *et al.* 2003; Hayama and Coupland 2004; Putterill *et al.* 2004). Like many economically important crops *Oryza sativa* is a short-day plant; it promotes flowering when daylength falls below a critical threshold. Would this suggest that the mechanisms for regulation of flowering are entirely different from the long-day plant *Arabidopsis*? Recent genetic studies have elucidated part of the photoperiodic control of flowering time in rice. Several rice genes have been shown to be orthologues of *Arabidopsis* genes: a gene named *Hd1* is homologous to *CO*, and *Hd3a* is an orthologue of *FT*. However, *CO* promotes expression of *FT* in *Arabidopsis* whereas *Hd1* suppresses expression of *Hd3a* in rice (Hayama *et al.* 2003). So the photoperiodic response is reversed by using the same set of regulatory genes, regulated differently. Fig. 5.18 provides an overview of the similarities in the regulation of flowering time between the two plant species.

Obviously, the usefulness of *A. thaliana* as a model for life-history investigation is limited by the fact that it represents only one type—a winter annual—out of the large variety of plant life histories. Biennuals, perennials, shrubs, and trees may have completely different ways of dealing with the problem of optimal timing of reproduction. Examination of the gene content of the recently completed *Populus trichocarpa* genome shows that most of the *Arabidopsis* flowering-time genes have counterparts in the poplar genome; however, a significant exception is the central floral repressor *FLC* which seems to lack an orthologue in poplar (Brunner and Nilsson 2004). The *FLC* subgroup of MADS-box genes seems to be specific to the Brassicaceae lineage. This raises

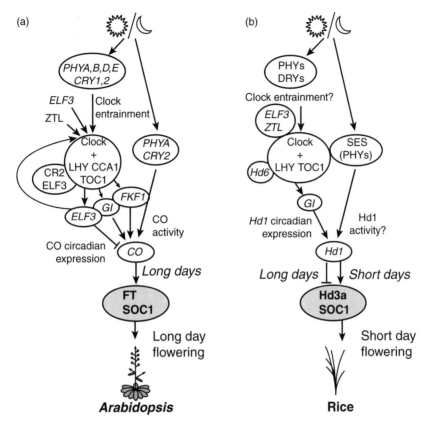

Figure 5.18 Similarities and differences between the photoperiodic control of flowering in (a) *Arabidopsis* and (b) rice. *Arabidopsis* flowering is promoted by long days while rice flowering is promoted by short days but repressed by long days. Despite the different phenotypic responses many elements of the regulatory network are the same. After Putterill *et al.* (2004) by permission of John Wiley & Sons.

questions as to the function of the other flowering-time genes in poplar. Despite the fact that a large part of the flowering time network may be conserved across species the functions of the genes are not necessarily the same in species with different life histories.

5.4 Phenotypic plasticity of life-history traits

Many life-history attributes do not attain fixed values but respond to environmental conditions in an adaptive fashion. Part of the variation in life histories over species stems from the fact that different species have adopted different ways of responding to environmental factors. The functional relationship between the phenotype and the

environment is called a *norm of reaction* (see Section 5.1). Life-history theory assumes that such a norm itself can be subject to natural selection and is optimized with respect to certain environmental conditions. We have discussed plant flowering time in the previous section as a life-history trait; however, its response to photoperiod and temperature can be seen as a reaction norm. In this section we will discuss more examples of phenotypic plasticity and its genomic underpinning.

There is considerable confusion as to how the primarily ecological concept of a reaction norm should be interpreted on the genetic level. One view holds that reaction norms are no more than a set of life-history traits measured in different environments, and that one has to analyse each trait in each environment as a separate variable, taking into account cross-environment correlations; this approach is also known as the *character-state view of phenotypic plasticity* (Roff 2002). Another approach emphasizes the continuity of the reaction norm across environments. This is especially appropriate if environments form a natural gradient, for example in the case of temperature. Plasticity is then defined by the response in the average environment, plus the slope of the reaction norm if it is linear. For non-linear reaction norms there is no obvious indicator of plasticity, although the first derivative of the reaction-norm value in the average environment may be used as a local measure of plasticity when that environment changes (De Jong 1995). This approach is known as the *reaction-norm view of phenotypic plasticity* (Roff 2002). It has been suggested that in addition to genes determining the average response, there should exist 'genes for plasticity'; that is, genes which determine the slope of the reaction norm. Natural selection emanating from variable environments could promote such plasticity genes. In particular, plasticity is favoured (Schlichting and Smith 2002):

- if environmental change is frequent,
- if environmental cues are reliable,
- if environmental variation is fine-grained in space or time,
- in the case of coarse-grained temporal variation, if change is predictable, or
- in the case of fine-grained temporal variation, if there is a predictable sequence.

The reaction-norm perspective is very much dominated by the terminology of quantitative genetics and by statistical analysis of phenotypic data across environments. Molecular data are changing this perspective (Pigliucci 1996; Schlichting and Smith 2002). The example of flowering time discussed above, as well as the cases discussed below, illustrate that it may be difficult to reconcile the quantitative-genetics views of plasticity with modern genomic insights. The concept of a plasticity gene does not have a clear interpretation on the molecular level. Rather, plastic life-history traits are determined by networks of gene expressions, which integrate multiple cues from the environment with signals from the internal metabolism in such a way that the difference between genes for plasticity and genes for an average response cannot be seen. In this section we discuss three cases—polyphenism and body size in insects, and shade avoidance in plants—to illustrate this argument. As we will see, the genomic underpinning of life-history plasticity is less well developed than some of the other issues discussed in this chapter, so the section is phrased in terms of examples rather than a general theory.

5.4.1 Polyphenetic development

Many insects will develop not only one phenotype but several at the same time or different phenotypes in a temporal sequence, phenomena known as *polyphenisms*. Obviously, when the same genotype develops different phenotypes, there must be developmental switches that are sensitive to environmental conditions. Therefore, polyphenetic development can be considered a prime case of phenotypic plasticity, also called discrete phenotypic plasticity to differentiate it from the continuous phenotypic plasticity described by reaction norms. Functional genomic analysis should be able to reveal the expression programmes underlying development of alternative phenotypes.

Spectacular examples of polyphenisms are found in insects that can develop different morphotypes in different seasons or in response to changes in the diet. Several examples are discussed by Nijhout (2003a), including spring and summer forms of nymphalid butterflies, gregarious and non-gregarious phases of locusts, soldier and worker castes in ants, and hornless and horned mating types in dung beetles. In all cases investigated, polyphenetic switching of the developmental pathway is triggered by a hormonal signal. Development is canalized into two or more alternative pathways, through an altered hormone titre, an altered threshold sensitivity to the hormone, an altered timing of hormone secretion, or an altered timing of the hormone-sensitive period. Often the environmental signal triggering an alternative condition falls considerably before the actual developmental switch itself; when a decision on a specific pathway has been made, development usually becomes irreversible after some time and the animal is destined to become a specific morph.

Seasonal polyphenisms in *Bicyclus* butterflies (family Satyridae) have been investigated in detail. There are about 70 species in the genus *Bicyclus*, which inhabit a variety of habitats in Central Africa. Most species have a wet-season morph with brown-coloured wings, and conspicuous eye-spots and bands, as well as a dry-season morph with dull colours and cryptic wing patterns. This seasonal diphenism can be understood as an adaptation to seasonal changes in the habitat, expressing camouflage against a background of dead vegetation in the dry season, when the but-terflies are less active, and predator deflection using eyespots in the wet season, when butterflies are more active. The two morphs can be produced in laboratory cultures maintained at different temperatures (Fig. 5.19). However, it is not certain whether temperature alone is the most important environmental cue determing the morphs, since

Figure 5.19 Seasonal polyphenism in *Bicyclus anyana* (Lepidoptera, Satyridae) includes a wet-season morph (top) and a dry-season morph (bottom). Courtesy of P.M. Brakefield, University of Leiden.

temperature can be negatively correlated with humidity in one place and positively in another (Roskam and Brakefield 1999).

As Fig. 5.19 illustrates, a major difference between seasonal morphs lies in the eyespots on the wings. Butterfly eyespots are pattern elements composed of concentric rings of coloured scales. These rings develop around an organizing centre, a so-called *developmental focus*, which during metamorphosis induces the surrounding scale-building cells to produce a designated pigment (Brunetti *et al.* 2001; Beldade and Brakefield 2002; Carroll *et al.* 2005). Four stages can be recognized in the development of eyespots on butterfly wings. The first step already takes place in the imaginal disc of the last larval instar, when subdivisions (fields) of the wing are defined. In the second stage, foci are established within specific fields. The establishment of foci takes place in each field separately, which explains why mutants of but-terflies with multiple eyespots on the wing can have one of the spots missing without any con-sequence to other spots (Monteiro *et al.* 2003). Specification of the eyespot focus is indicated by expression of a homeobox transcription factor, *Distal-less* (*Dll*), a *Hox* gene that has been cloned in *Drosophila* and many other species. Due to con-servation of *Hox* genes throughout the animal kingdom, antibodies against the *Drosophila* Hox proteins also provide reactivity to the orthologous Hox proteins of other insects, which allows the expression of these developmental proteins to be localized in developing butterfly wings (Brunetti *et al.* 2001; Beldade *et al.* 2002). In the third stage, which takes place in the early pupa, a signal from the focus induces the surrounding cells to adopt a certain colour. The type of pigment that these cells adopt seems to depend on the sensitivity thresh-olds of the responding cells, whereas the strength of the signal determines the size of the eyespot. The fate determination around the focus uses the *hedgehog* signalling pathway, one of the major so-called toolkit genes of developmental genetics (Carroll *et al.* 2005). This pathway, like the insulin signalling pathway discussed in Section 5.2, translates the binding of an extracellular ligand into regulation of a transcription factor, which in this case is *cubitus interruptus*. Finally, the fourth phase of eyespot formation is the pigmentation itself, which takes place in the late pupal stage.

Which developmental switches are made to produce the alternative seasonal morphs of butterflies like *Bicyclus* is not known precisely. Given the knowledge about eyespot formation summarized above, it is likely that the switch will involve regulation of developmental genes such as *Distal-less*, or one of the genes in the *hedgehog* signalling pathway. The involvement of ecdysteroid hormones in seasonal polyphenisms has been demonstrated for several butterfly species, so it seems reasonable to assume that the upstream part of the regulation consists of an endocrine signal, which in some way or another is sensitive to an environmental cue. Seasonal polyphenism represents a fascinating area of research where genomics, through developmental genetics, can meet ecology and evolution in a fruitful manner.

Another well-known case of polyphenism is represented by the castes of many social insects (termites, ants, wasps, and bees). The best-investigated caste system is that of the worker/queen polyphenism in the honey bee, *Apis mellifera*. Experiments have shown convincingly that the development of a larva into a worker or a queen is completely under environmental control, rather than reflecting a genetic predisposition of some larvae to follow a distinct developmental pathway. Larvae are induced to develop into queens when fed a rich mixture of food throughout development, including royal jelly, a secretion from the mandibular gland of nursing workers which contains a higher concentration of sugar than worker jelly. As a consequence, queen larvae develop faster, grow larger, and have larger corpora allata, the source of juvenile hormone, which controls the differentiation of oocyctes and the production of vitellogenic proteins by the fat body. Levels of juvenile hormone are considerably higher in queen larvae than in worker larvae, especially in the fifth instar, the stage just before pupation. Up to the fourth instar (2.5–3.5 days old) the queen developmental pathway is reversible, but after entry into the fifth stage a change of feeding regime has no further influence on the phenotype appearing after pupation. Obviously, the alternative developmental pathways are regulated by larval nutrition, mediated by endocrine signals, which in some way are translated into gene expressions.

Are the castes of honey bees characterized by specific expression signatures? In an early study, Evans and Wheeler (1999) looked at differential gene expression between queen and worker larvae of *Apis mellifera* using an SSH protocol (see Section 2.1.2, Fig. 2.4); this was followed later by macroarray expression profiling using 144 cDNAs from enriched libraries (Evans and Wheeler 2000, 2001). Gene expressions were compared between early fourth instar (bipotential) larvae and fifth instar larvae destined to workers or queens. Several loci were confirmed to be differentially expressed and, interestingly, many of these were downregulated in queen larvae and the data showed that workers resemble the bipotential young larvae more than queens. So the suggestion is that the worker programme is the default pathway, which must be altered actively to produce a queen programme, and this latter programme includes switching off many genes and turning on a limited set of queen genes.

Fig. 5.20 provides a summary of differential gene expression in honey bee castes. Young larvae (Y in Fig 5.20) overexpressed two heat-shock proteins and several proteins related to RNA processing. Among the genes specifically upregulated in worker larvae (W) is hexamerin 2, a member of a group of hexameric storage proteins, which serve as a source of energy during metamorphosis and adult life. Such hexameres also accumulate in the haemolymph of insects preparing for diapause (Denlinger 2002). A cytochrome P450 was also upregulated consistently in workers and this was also found in another bee species, *Melipona quadrifasciata* (Judice *et al.* 2004). The expression profile of queen larvae (Q) shows overexpression of ATP synthase and cytochrome oxidase I. This is consistent with earlier work by Corona *et al.* (1999) who, using differential display, identified three mitochondrial proteins, mitochondrial translation-iniation factor, cytochrome oxidase subunit 1, and cytochrome *c*, that were upregulated consistently in queen larvae compared with worker larvae.

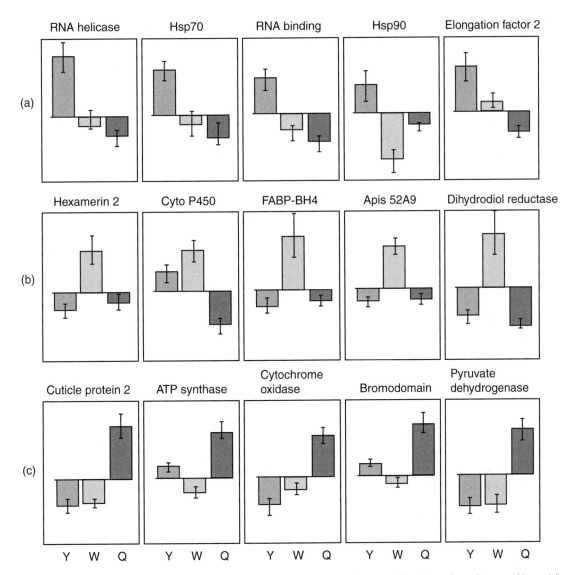

Figure 5.20 Overview of 15 genes of honey bee, *A. mellifera*, differentially expressed between fourth instar larvae (Y; young, bipotential), and fifth instar larvae destined to become workers (W) or queens (Q). Normalized expression levels are shown, with standard errors, estimated from non-competitive hybridization of cDNA samples with a macroarray of 144 probes, derived from SSH libraries. Shown are genes with high expression in (a) young larvae, (b) worker larvae, and (c) queen larvae. After Evans and Wheeler (2000), by permission of BioMed Central.

So, these limited data suggest that the worker larvae expression profile is characterized by allocation to storage and the queen larvae profile by genes reflecting higher mitochondrial respiration.

Developmental genomics of caste determination in honey bees is in its infancy. The studies conducted so far on honey bees caste determination are limited by the relatively small number of genes analysed. The recently completed genome sequence of *A. mellifera* will allow a significant acceleration of expression profiling. Several EST libraries have been developed for a more detailed analysis of the honey bee transcriptome (Whitfield *et al.* 2002; Nunes *et al.* 2004). Genome-wide analysis of honey bee castes will not only shed more light on the developmental and metabolic aspects of caste differentiation, but also on the biological basis of differential behaviours.

The approaches taken in the honey bee would also be applicable to other eusocial insects, including ants, a favourite object of study in ecology. Many ant species have a more complicated system of caste differentiation and task division than honey bees; castes may include workers as well as soldiers, males, and alate (winged) females. The intriguing variety of ant social behaviours would be a very rich source of genomic discovery; however, genome-wide studies on ants have not yet been attempted.

5.4.2 Body size

Body size is one of the most important life-history traits. Even though body size does not appear explicitly in demographic tables of fertility and mortality from which fitness is estimated (see Section 5.1), it is implicated in changes of these vital rates with age, for example because increasing size allows higher fertility and often entails a lower mortality (see Fig. 5.1). The adult body size of a species is also one of the few variables that predicts successfully a large number of ecological attributes, such as metabolic rate, and energy intake and assimilation. We saw in Chapter 3 that body size was also associated with population size, which is one the main determinants of genome size according to the neutral theory of population genetics (see Fig. 3.4). Body size is also tightly linked with questions of shape and form, because physical forces have essentially different impacts on small objects compared to large objects, including cells and organisms. In his marvelous book, *On Growth and Form*, D'Arcy Thompson (1917) expressed it succinctly: 'Size of body is no mere accident.'

Despite the almost universal importance of body size, its determination by physiology and genes continues to be a formidable vexing problem (Nijhout 2003b). The mechanisms that determine body size are obviously contingent on the underlying framework for regulating cell size, cell

number, and the size of organs; however, if an organism is to attain a characteristic body size (which many animals do) there must be additional mechanisms, responding to information generated all over the body. One way of looking at body size is to see it as a consequence of halting growth. The question then becomes not so much how to attain a fixed size but when to stop growing. This point of view is particularly applicable to holometabolous insects, which only grow in the larval stage and whose body size depends entirely on the mass reached just before pupation. Entomological research has shown that one of the major factors determining adult body size of insects is a threshold reached in the larval stage, the so-called *critical weight*. This is defined as the minimal weight above which further feeding and growth are not strictly required for successful pupation (Davidowitz *et al.* 2003). The strength of this concept is that it is causally linked to a series of endocrine events inducing the onset of pupation in the last larval instar. At the critical weight, the corpora allata stop producing juvenile hormone, which removes this hormone's suppression of prothoracicotropic hormone secretion, allowing the prothoracic glands to secrete ecdysone, which then sets into motion a cascade of events leading to metamorphosis. Larval growth stops when this sequence of endocrine events culminates in a peak of ecdysteroids, at which point the larva has grown to a size above the critical threshold, depending on the time delay for induction of ecdysteroid secretion and the photoperiod.

Detailed research on the tobacco hornworm, *Manduca sexta* (Lepidoptera, Sphingidae), has shown that additive genetic variation exists for critical weight, and this has allowed selection for adult body size in laboratory cultures. In addition, two other determinants of body size contribute to genetic variation of body size in *M. sexta*, prothoracicotropic hormone delay time and growth rate (D'Amico *et al.* 2001). From research on *Drosophila* we know that growth itself is influenced by signals from the insulin/IGF-1 signalling pathway (Brogiolo *et al.* 2001; see Section 5.2). Inspired by mammalian research, another signalling pathway has shown to be involved in the regulation of growth rate of *Drosophila*, which centers around a protein called *target of rapamycin* (TOR). This protein was first discovered in yeast through mutants resistant to the cytotoxic effects of the fungicide, rapamycin. TOR is part of a signalling cascade that interacts partly with insulin/IGF-1 signalling, is sensitive to amino acids in the haemolymph, and influences metabolism and growth (Britton *et al.* 2002; Oldham and Hafen 2003). Fig. 5.21 provides a schematic overview of the various determinants of body size in insects.

The physiological evidence reviewed makes it understandable that body size is both genetically determined and plastic. Indeed, numerous experiments with insects have shown that adult body size is affected by environmental conditions, the best-studied response being the effect of temperature. Rearing temperature has a consistent effect on body size: insects grow larger at low temperature and smaller at high temperature. This trend is also present in the field: populations at higher latitude attain a larger body size than tropical populations of the same species. In *D. melanogaster*, the latitudinal cline of body size

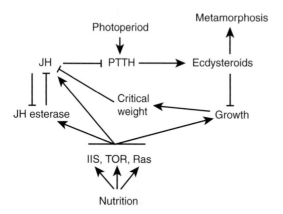

Figure 5.21 Tentative schematic overview of relationships involved in the determination of adult body size in insects, inspired by research on tobacco hornworm, *M. sexta*. Adult size is determined by cessation of larval growth before pupation, which is triggered by ecdysteroids that appear when the titre of juvenile hormone falls upon reaching the critical weight. JH, juvenile hormone; PTTH, prothoracicotropic hormone; IIS, insulin/IGF signalling pathway; TOR, target of rapamycin signalling pathway; Ras, Ras signalling pathway. Based on Nijhout (2003) and other sources.

has a high heritable component and is accompanied by clines of allele frequencies, starvation resistance, and chromosome inversions (De Jong and Bochdanovits 2003).

One of the few genomic approaches to the problem of body-size determination is found in the work by Bochdanovits *et al.* (2003) and Bochdanovits and De Jong (2004). Temperate and tropical fruit fly populations, each cultured at two temperatures, were examined in the third larval stage, just before pupation, by means of transcription profiling using an oligonucleotide microarray. Genes with differential expression were identified by analysis of variance and classified as differing between populations (45 genes), differing between rearing temperatures (200 genes), and showing a population–temperature interaction (31 genes). The data showed not only that a considerable part of the genome is modulated by temperature (18% of the genes detected in all treatments), but also that a fair number of them (16%) respond to temperature in a population-dependent manner. Expression of 19 genes showed a correlation with the adult body size of the isofemale line from which the larvae were collected; 16 of these showed a positive and three a negative correlation (Fig. 5.22). Assigning functions to these genes revealed that most of them fell into the category of cell growth and maintenance. Thus the gene-expression signature of body size in *Drosophila* is dominated by energy metabolism and morphogenesis.

Another aspect of body size in *Drosophila* is that it is negatively correlated with larval survivorship; genotypes with high larval plus pupal survival have a small adult size. Bochdanovits and De Jong (2004) examined whether this trade-off could be due to antagonistic pleiotropy, thereby testing the model proposed by Stearns and Magwene (2003), illustrated in Fig. 5.2. This was done by analysing correlations between gene expression and body size and sifting genes whose expression was positively correlated with body size but negatively with larval survival, or the other way around. The authors found 34 of such genes, of which 15 were low weight/high survival genes and 19 high weight/low survival genes (Table 5.4). The joint

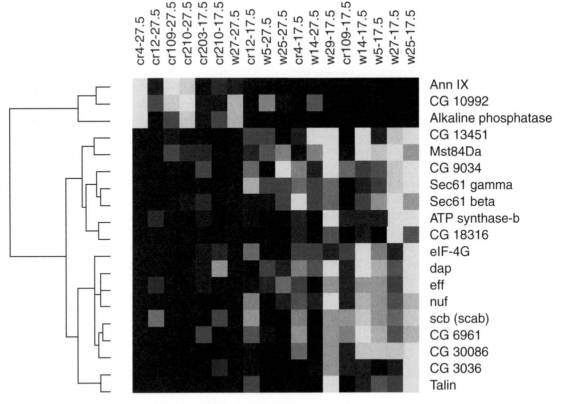

Figure 5.22 Heat diagram of gene expression and cluster analysis of 19 genes in larval *D. melanogaster*, assayed using oligonucleotide microarray hybridization. The experiments were done with two populations, each reared at two temperatures. A total of 1134 genes were detected in all treatment groups; 19 of these showed a correlation with adult body size of the isofemale line and are displayed here. The individuals listed along the top are coded according to population (cr, tropical; w, temperate) and rearing temperature (17.5 and 27.5 °C). The list on the right provides the identifiers of the genes according to FlyBase. Strength of expression is indicated by the darkness of the square; the darker the square, the stronger the expression. From Bochdanovits *et al.* (2003) by permission of Oxford University Press.

expression of these genes explained 86% of the trade-off.

The analysis of Bochdanovits and De Jong (2004) illustrates that trade-offs between life-history traits may indeed be due to genes with pleiotropic effects, but how these genes may exert such effects on the phenotype is far from clear. Among the 'trade-off genes' listed in Table 5.4, particular interest is raised by the Ras protein CG9611. Ras proteins are part of the Ras/mitogen-activated protein kinase (MAPK) signalling pathway, which like the insulin/IGF-1 signalling and TOR pathways transduces signals from extracellular growth factors (in this case the so-called epidermal growth factor, EGF) into metabolic action in the cell (see

Fig. 6.3). The Ras/MAPK pathway is known for its role in regulating cell growth, cell proliferation, and survival. The repeated occurrence of genes from signalling pathways involved in transducing nutritional and growth factor information into cellular metabolism is probably indicative of a crucial role of these pathways in the regulation of life-history traits and possible trade-offs among them.

5.4.3 Shade avoidance

The last plastic life-history trait to be discussed concerns the remarkable flexibility seen in plant growth when exposed to different light intensities

Table 5.4 Genes of *D. melanogaster* whose expression showed opposite signs of correlation with male body size and pre-adult survival

Name or identifier (FlyBase)	Molecular function	Biological function
Low-weight/high-survival genes		
Lep1	Larval cuticle protein	Structural integrity of larval cuticle
CAH1	Carbonic anhydrase I	Energy metabolism
CG10622	Succinate-CoA ligase	Carbohydrate metabolism
CG7834	Electron-transport flavoprotein	Mitochondrial respiration
CG8145	Nucleic acid binding	Cell proliferation
High-weight/low-survival genes		
Smt3	Ubiquitin-like protein	Cell growth and maintenance
CG14957	Protein involved in mRNA splicing	Cell growth and maintenance
Arf51F	ADP ribosylation factor 51F	Cell growth and maintenance
CG3164	ABC transporter	Nucleotide binding and transport
CG9611	Protein of Ras signalling pathway	Signal transduction
Est-P	Esterase P	Signal transduction
CG3843	Ribosomal protein L10A	Protein synthesis
ImpE2	Ecdysone-inducible gene E2	Imaginal disc morphogenesis
CG5171	Trehalose phosphatase	Carbohydrate metabolism
CG33317	Similar to RNA-binding protein	Unclear; incomplete sequence

Notes: In total 34 genes were found whose expression had a correlation falling into opposite tails (3.5 percentiles) of the correlation coefficient distributions; however, only 14 could be assigned a function and are listed here.
Source: From Bochdanovits and De Jong (2004), by permission of the Royal Society.

or light spectra. Part of this flexibility can be understood as a mechanism to avoid shade and therefore this type of phenotypic plasticity is known as the shade-avoidance syndrome. Shade avoidance is observed when plants are exposed to low ratios of red to far-red light, such as would prevail during growth under a closed canopy. The response involves suppression of branching, emphasis on vertical elongation, and early flowering. This is assumed to contribute to plant fitness, since it allows completion of the life cycle and seed set before being overgrown by competitors.

Plants have three groups of light-sensitive molecules that translate aspects of the light regime into metabolic action: phototropins, cryptochromes, and phytochromes (Chen *et al.* 2004). Phototropins are sensitive to UV-A and blue light; they are involved in the well-known phototropic responses of plants (roots grow away from light, stems bend towards light), and also in regulating chloroplast movements and stomatal opening. Cryptochromes are also sensitive to UV-A and blue light and are

involved in de-etiolation (the transition of dark-grown seedlings to phototrophically competent plants), photoperiod-dependent induction of flowering, and entrainment of the circadian oscillator. Phytochromes are sensitive to red and far-red light and are involved in seed germination and shade avoidance. Many of the light-dependent responses of plants are modulated by networks involving both cryptochromes and phytochromes (see the discussion of flowering time in Section 5.3), but the shade-avoidance response is regulated only by phytochromes and this is the reason for discussing them in a little more detail here.

There are five different phytochrome genes in *Arabidopsis*, designated *PHYA* to *PHYE*. Similar families of phytochromes are known in species throughout the plant kingdom, from algae to angiosperms. The proteins encoded by these genes differ in their spectral properties and the rates of conversion between active and inactive states; consequently they have different physiological roles. Mutants of *Arabidopsis* defective in one of the phytochrome genes have confirmed this differentiation of function. PHYB plays a predominant role in the shade-avoidance response, with redundant action of PHYD and PHYE (Schlichting and Smith 2002).

How can phytochromes translate light signals into gene expression? A large amount of detailed knowledge is available on the first steps in this process (Chen *et al.* 2004; Schepens *et al.* 2004). The light-interception component of the system is not the phytochrome itself, but a *chromophore* called phytochromobilin. The role of this chromophore is similar to the role of retinal bound to proteorhodopsin, as discussed in Section 4.4: it triggers a state transition of the main molecule. The phytochrome protein may undergo a conversion between two relatively stable states: a red-light-absorbing Pr form and a far-red-absorbing Pfr form. The Pfr form is assumed to be the active configuration.

Phytochromes are present as dimeric molecules in the cytoplasm. When at least one of the units is activated to the Pfr form, the molecule as a whole can be imported into the nucleus (Fig. 5.23). In the nucleus, phytochromes are localized in so-called *nuclear bodies*. In general, nuclear bodies, also called speckles, are considered subcompartments of the nucleus that carry out specific functions, such as mRNA splicing. The nature of the nuclear bodies recruiting activated phytochromes is unknown. They seem to represent structures by which phytochromes can transfer a signal to transcription activators. Both subunits of the phytochrome must be in their active form before localization in these bodies can occur.

Expression of a large number of genes is triggered by phytochromes. Transcription profiling using microarrays applied to *Arabidopsis* mutants defective in one of the phytochrome genes have revealed the genome-wide nature of the phytochrome signalling pathway (Ma *et al.* 2001; Quail 2002; Wang *et al.* 2002; Devlin *et al.* 2003). The first targets seem to be transcription factors, since these genes respond within 1 h of the light signal.

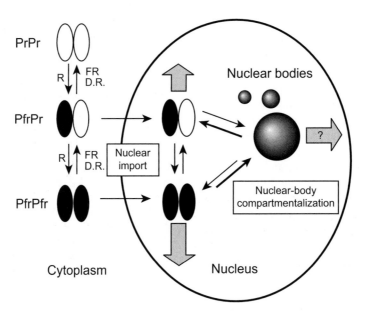

Figure 5.23 Relocalization of phytochromes in the cell, using PHYB as a model. Interception of red light by a chromophore attached to PHYB leads to activation of one or two of the subunits of the dimeric molecule (transition from Pr to Pfr). The activated molecule may be imported into the nucleus where it is compartmentalized into nuclear bodies. Localization into nuclear bodies is necessary for triggering gene expression. The PfrPfr form is incorporated preferentially into nuclear bodies over the PfrPr form. D.R., dark reversion (slow spontaneous decay of Pfr to Pr in the dark); R, red light; FR, far-red light. After Chen *et al.* (2004), reproduced with permission from Annual Reviews.

Presumably, many of the other expression changes are downstream targets of transcription factors. No less than 26 different cellular pathways were found to be regulated by phytochrome signalling—some suppressed, others promoted—in a co-ordinated fashion (Ma *et al.* 2001).

Devlin *et al.* (2003) proposed seven functional categories for the genes differentially regulated by shade in *Arabidopsis*. Genes related to photosynthesis, fatty acid metabolism, and redox metabolism were mostly downregulated. Genes acting upon the cell wall, including pectinesterases and pectate lyases, were mostly upregulated; these genes support loosening of the cell wall. Genes regulated by the plant growth hormone auxin made up a large proportion of the upregulated genes. These switches in the transcriptome are all understandable in light of the pronounced elongated growth and accelerated flowering shown by the shade-avoidance phenotype.

Several studies have demonstrated the importance of plant hormones (auxins and gibberellins) in regulating the shade-avoidance growth response. A role of the volatile plant hormone ethylene is also suggested. In a study on tobacco, Pierik *et al.* (2004) showed that ethylene-insensitive plants had a reduced response to shade. The effect of ethylene requires GA because plants with inhibited GA production showed hardly any ethylene-dependent response to shading.

The study of phytochrome-mediated shade avoidance, although far from complete, demonstrates that it is possible, in principle, to explain the process of phenotypic plasticity in terms of gene expression regulated by environmental signals (Schlichting and Smith 2002). With sufficient sophisticated knowledge of the signalling pathways and the expression programmes they trigger, it would be possible to account for the widely different plant phenotypes that may develop from the same genotype. On the other hand, in considering the large number of genes involved and the way their expression is organized into networks, rather than in linear causal chains, this could turn out to be an extremely difficult task. With this open-ended challenge we close our discussion of phenotypic plasticity.

5.5 Genomic approaches to life-history patterns: an appraisal

Genomic analyses of life histories, as reviewed in this chapter, have revealed a number of molecular principles underlying regulation and determination of key life-history events. These can be summarized as follows.

Environmental information (food, crowding, light, daylength, temperature) often modulates life-history traits and this is the basis of phenotypic plasticity. In animals, such information is typically processed by the nervous system, then translated into a hormonal signal, which acts upon signalling pathways to steer gene expressions in target cells. Signalling pathways consist of a membrane-bound extracellular receptor, connected to an intracellular system of kinases, able to trigger a cascade of biochemical events, finally leading to activation of a transcription factor, which then triggers gene expression. The involvement of the insulin signalling pathway in regulating growth, reproduction, and longevity is a prime example of this principle, but also the induction of insect diapause and the developmental switches in polyphenisms act similarly. In plants, phytochromes play a major role in the translation of light signals.

Second, we may note that if an organism has the capacity to follow more than one developmental option, for example a dormancy stage in addition to an active stage, or flowering in addition to vegetative growth, the alternative pathway is often constitutively present throughout life but repressed until an environmental cue triggers its appearance. The execution of the alternative programme then becomes a question of simply removing a repression, rather than staging the whole programme anew. So-called resting stages (e.g. diapause) are not quiescent at all from a molecular point of view. Despite the fact that many genes are downregulated during diapause, the entry, maintenance, and exit of diapause involves active upregulation of several other genes. We have seen indications of such 'hidden' developmental programmes in nematodes (the dauer larva programme) and in *Arabidopsis* (the floral transition). Interestingly, mutations in

C. elegans show that it is possible to uncouple part of this programme (life extension) from the main pathway leading to dauer formation, and thus to confer increased longevity to the non-dauer adult.

Third, we have seen that expression of life-history traits often comes with intricate mechanisms of regulation that go further than simple transcriptional regulation. We have seen four examples of this: alternative splicing, gene silencing, RNAi, and post-translational regulation of protein stability. Alternative splicing was seen in a *Drosophila* study showing that different splice variants of the same gene are expressed in different life-history stages, and in *Arabidopsis*, where one of the loci for flowering induction is suppressed by a feedback loop promoting an inactive splice variant of the gene. Alternative splicing seems to be a mechanism by which gene expression can be fine-tuned in time (developmental stage) or space (specific tissues). The second type of non-transcriptional regulation, gene silencing, was observed in the vernalization response of *Arabidopsis*, where histone deacetylation is employed to suppress floral suppression and allow bolting. Gene silencing was also involved in mutations conferring lifespan extension in yeast, nematodes, and fruit flies. Such epigenetic regulation seems to act as a memory of environmental events, which can trigger an adequate response in cells that do not witness the event themselves but are imprinted by their mother cell. The third type of non-transcriptional regulation is RNAi. Both microRNAs (implicated in the photoperiodic pathway of the *Arabidopsis* floral transition) and RNA-binding proteins (in the autonomous pathway) have been discussed above as means to selectively inhibit mRNAs. Finally, post-translational regulation was noted in the light-dependent stabilization of CONSTANS protein by phytochromes and crytochromes in the photoperiodic pathway regulating flowering time in *Arabidopsis*.

The fourth molecular principle illustrated in this chapter is that life-history traits are underpinned by a complex network of gene expression and feedback loops. Networks are characterized by upstream and downstream functional units. In the upstream part, a simple trigger may act upon a signalling pathway. If such upstream genes are mutated, this often has very large effects on the phenotype and may introduce a syndrome of correlated altered life-history traits (see the *daf-2* mutants of *C. elegans*). Downstream of the network we often see a host of gene expressions, which are triggered by a limited number of transcription factors, acting upon all genes sharing a certain motif in their promoter. An obvious example of a downstream gene-expression cascade was the microarray analysis of longevity in *C. elegans*, in which a large number of genes upregulated as well as downregulated by the transcription factor DAF-16 were identified. Another characteristic of a network is the principle of convergence, which holds that different pathways are integrated by acting (some positively, some negatively) on a single integrator gene. The floral repressor *FLC*, which integrates the autonomous and the vernalization pathway in *A. thaliana*, is a clear example of this. A third property of networks is redundancy or parallelism. This is especially obvious in organisms with duplicated genomes, where more than one copy of the same gene exists and mutants do not have recognizable phenotypes due to another copy of the gene performing a similar function. The recent discovery of *SCHLAFMÜTZE* and *SCHNARCHZAPFEN* in the photoperiodic pathway of *Arabidopsis* floral transition is a clear example of this.

A final molecular lesson from the examples given in this chapter concerns the extrapolation of genomic information across species. As discussed in Chapter 3, comparative genomics has developed a whole gamut of instruments by which fine- and coarse-scaled comparisons of genomic sequences are made. In this chapter we have seen one example, the photoperiodic response of *Arabidopsis* and rice, where the genes themselves were conserved between species, but regulated in a different way, with the consequence that *Arabidopsis* flowering is stimulated by long days and rice flowering is stimulated by short days. If this principle of similar structure but different function in related species is common, homology of coding sequences is insufficient for extrapolating across species; rather, the basis for extrapolation should lie in functional comparative genomics. A similar disparity between the transcriptome and the genome was presented as an example in Chapter 1, when gene expression in different organs of *H. sapiens* and *Pan troglodytes* was compared (Fig. 1.7). In the present chapter we have seen that the first methodologies to analyse transcriptional profiles shared across species are now being developed.

Is ecological genomics about to make a major contribution to life-history theory? Obviously the genomics revolution has already intruded into the analysis of life histories, but we see four major limitations. In the first place, the outcomes of microarray studies published so far sometimes differ widely between similar studies. This is especially obvious from the appallingly small overlap of genes reported in different studies to be regulated by age in *Drosophila*. Maybe what is lacking here is is a good definition of the conditions under which test animals are cultured or exposed. Animal physiologists know that physiological responses of animals are influenced by many environmental factors, some of them difficult to standardize, such as food quality, air humidity, microbial infection, pheromones from conspecifics, etc. Consequently, a genome-wide gene-expression profile may partly reflect such non-standardized aspects of the test organism. Only replication across studies and robust statistical analysis can reveal the universal responses among the ones that are just contingent on specific experimental conditions.

A second limitation is that most of the studies conducted to date have focused on a few genetic model species. Although some generalities have been indicated in the sections above, the regulation of aging though the insulin/IGF-1 signalling pathway being a case in point, most of the work is concentrated on *C. elegans*, *Drosophila*, and *Arabidopsis*, species that represent only a tiny fraction of the array of life histories in nature.

Broadening the comparative basis will surely benefit a further integration of life-history theory and ecological genomics.

A third limitation is the focus on laboratory observation. Because any life history, especially mortality, has an important component external to the organism (predation, microbial infection, competition, spatial heterogeneity), which is difficult to mimic in the laboratory, the relevance of gene-expression profiles and mutants only observed in the laboratory can be doubted. An example illustrating this argument is a study by Weinig *et al.* (2002), who demonstrated several QTLs for flowering time in *Arabidopsis* that are only expressed under certain field conditions and thus not found in laboratory mutants. A good strategy for ecological genomics may be to exploit the natural variation of genomic programmes in wild organisms and then to work out mutations and expression profiles relating to these 'eco-variable' loci.

Finally, we note that many gene-expression studies have not yet left the descriptive stage. It is one step to draw up an inventory of transcriptional profiles associated with some life-history event, but quite another to explain the causal relationship between these profiles and the life history. Only in the cases of longevity in *C. elegans* and flowering time in *Arabidopsis* is such an understanding coming within reach.

Despite these limitations we see a glorious future for a further merger between ecological genomics and life-history theory. This will involve establishing a link between crucial phenotypic phenomena, such as trade-off and plasticity on the one hand and gene expression, pleiotropy, and molecular signalling on the other. We expect that such bifaceted discussion about fundamental concepts of life-history theory will contribute to a smooth transition between evolutionary explanation, the underlying physiology, and molecular genetics.

CHAPTER 6

Stress responses

Stress is a fundamental aspect of life and a major aspect of natural selection in the wild. Ecologists have studied the responses of plants and animals to environmental stress factors since the 1960s. Previously known as physiological ecology or ecophysiology these studies are now often called stress ecology. The study of stress responses on the genomic level has produced new insights into the mechanisms that enable plants and animals to survive in harsh environments and that limit the distribution of species. Biochemical studies have shown that, on the cellular level, there is surprising degree of uniformity in the stress responses of different species, even to widely different environmental stress factors. Genomic studies have reinforced this idea while at the same time providing new insights into the coherence of the cellular stress response. Stress is evoked in an organism at the edges of its ecological niche. The extent to which the organism is able to deal with such stresses determines the limits of its ecological amplitude. There is therefore a logical link between genomic analysis of the stress response and the ecology of the species. In this chapter we aim to introduce the reader to the ecological genomics of stress analysis.

6.1 Stress and the ecological niche

Even the most casual observer of natural systems will note that many species tend to occupy a characteristic place in nature. The idea is demonstrated most vividly by gradient studies, for example the distribution of plants on a salt marsh, where each species tends to be limited to a certain zone by a combination of soil texture, redox potential, salt, and lime. Ecologists have invented the concept of an ecological niche to organize their thoughts about the ways in which organisms fit into their environment. The inception of this concept in the ecological literature is attributed to the American ornithologist Joseph Grinnell with his now classical paper on the California thrasher published in 1917, but the most widely used and influential elaboration of the niche concept is due to Hutchinson (1957). Hutchinson defined the niche in terms of any number of conditions and resources that limit the distribution of a species. The niche was pictured as an n-dimensional hypervolume that envelops those values of continuously varying environmental factors that allow long-term survival of the species. As an illustration of the Hutchinsonian niche concept we reproduce a two-dimensional picture of fitness in the collembolan *Folsomia candida* as a function of zinc exposure and food density (Fig. 6.1; Noël *et al.* in press).

Hutchinson (1957) realized that a distinction should be made between the *fundamental niche*, which comprises all the conditions under which a species potentially may occur, and the *realized niche*, which is usually more narrow than the fundamental niche due to competition in the field. The fundamental niche is observed in laboratory experiments and in the field when competitors are absent. At the edges of its fundamental niche a species is less well equipped to face competition with others and so when competition is important, it will give away at the edges and occupy a smaller section of the gradient, the realized niche. The formal definitions of the ecological niche by Hutchinson (1957) have spurred a great variety of

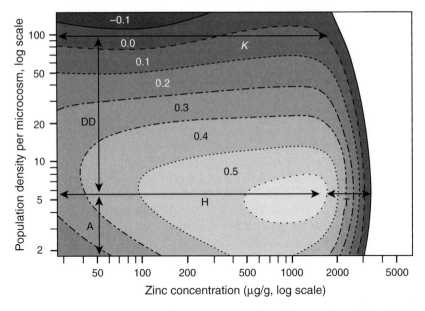

Figure 6.1 Graph illustrating the niche concept in two dimensions. Contours of population growth rate of the soil-dwelling collembolan *Folsomia candida* are plotted as a function of dietary zinc concentration and initial population density per microcosm (reflecting food density). Four different ranges of effect are indicated. At high exposure levels zinc becomes toxic and suppresses fitness (T); with increasing population size fitness decreases due to crowding (DD, inverse density dependence; *K*, carrying capacity); at low exposure levels zinc supports fitness by a stimulatory effect on growth, a phenomenon known as hormesis (H); at low food density fitness decreases due to absence of inter-individual communication stimulating consumption (Allee effect; A). Courtesy of H. Noël and R.M. Sibly, University of Reading.

studies aiming to explain community structure in the wild from the properties of individual species and their responses to environmental factors.

The ecological niche concept touches the very heart of ecology—the relationship between species and their environment—but this has not prevented the proliferation of a great deal of confusion in the literature. Several reviews have pointed out the historical context of this confusion, which is partly due to independent elaboration by Joseph Grinnell and Charles Elton (Chase and Leibold 2003). On the one hand the concept emphasizes the requirements of species, but on the other hand it includes the species' role or impact on its environment. This Janus-faced property of the niche has caused such confusion that some ecologists in the 1970s have suggested to avoid the term niche altogether. However, Chase and Leibold (2003) revisited the concept and after clearing the decks set the stage for a new synthetic approach, framed by the recent developments of ecology. Their new,

synthetic, definition of ecological niche runs as follows.

The joint description of the environmental conditions that allow a species to satisfy its minimum requirements so that the birth rate of a local population is equal to or greater than its death rate, along with the set of per capita effects of that species on these environmental conditions.

This definition joins the two components of the niche mentioned above, and it recognizes that whenever resources and environmental conditions are altered by the organisms themselves, as in the case of predators depleting their prey or ecological engineers altering the physical structure of their environment, this aspect should be included in the niche concept. The definition by Chase and Leibold (2003) also recognizes that the difference between mortality and natality, in other words fitness, is the ultimate measure in which the requirements of the organism are to be expressed.

The niche concept has been highly instrumental in community ecology, as it allows explanations of community structure from the point of view of the placing of species along a gradient of resources or conditions according to their ecological niche. Various competition models have been derived that predict how niche overlap between species is minimized by competition and the maximum overlap between the niches of two adjacent species that will allow coexistence of both. However, less attention has been paid to the question of how the niche itself is shaped by underlying determinants founded in the physiology of the species. The science of environmental physiology, also called ecophysiology or physiological ecology, addresses these aspects. Several physiology textbooks have been written that reach out to ecology, including that by Larcher (2003) on plants, and that by Schmidt-Nielsen (1997) on animals. In this chapter we likewise aim to explore the reductionist path, from niche to genomics.

Environmental physiology has a strong focus on studies of stress. The reason for this bias is that the regulatory mechanisms allowing homeostasis of the *milieu interne* are best seen when these mechanisms are put to the test by pushing the organism to the borders of its ecological niche. This is how we will approach the issue in this chapter; by studying responses at the edges of the ecological niche we aim to reveal the regulatory mechanisms that promote fitness both within and outside the niche.

Like the niche, the concept of stress has a long and confusing semantic history in ecology. Some authors have argued that ecological stress should be defined by analogy to the physical concept of stress, which would imply that it is an external constraining or impelling force applied to an ecological system. Most biologists, however, consider stress as an internal state, brought about by a hostile environment or negative social interaction. Nowadays, there seems to be agreement on the fact that a distinction should be made between a *stressor* (an external factor), the *stress* (an internal state brought about by a stressor), and the *stress response* (a cascade of internal changes triggered by stress). Although the concept of stress can be defined at

various levels of ecological integration, stress is most commonly studied in the context of individual organisms, whereas stress responses are studied on the cellular, biochemical, and genomic levels.

We need to realize that the concept of stress is not absolute; it can only be defined with reference to the normal range of ecological function; that is, with reference to the ecological amplitude or ecological niche of the species. What is an extremely stressful condition for one organism (e.g. the absence of free air) is quite normal for another organism (a fish). A definition of ecological stress that incorporates this idea runs as follows (Van Straalen 2003).

Ecological stress is a condition evoked in an organism by one or more environmental factors that bring the organism near to or over the edges of its fundamental ecological niche.

This definition complies with the common physiological usage of the term, which is that stress is an internal condition, not an external factor. In addition, stress has the following properties: (i) it is usually transient, (ii) it involves a syndrome of specific physiological responses, and (iii) it is accompanied by the induction of mechanisms that counteract its consequences. Our niche-based definition of stress is illustrated schematically in Fig. 6.2.

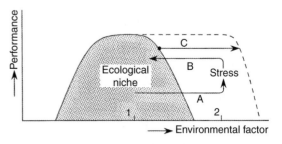

Figure 6.2 Graph illustrating a definition of stress based upon the ecological niche of a species. Ecological stress arises when the intensity of an environmental factor increases from 1 to 2 in such a way that in position 2 the organism is placed outside the niche (A). This will evoke stress and stress-response reactions, which fade away when the environmental factor relaxes and the organism returns to its niche (B). Another type of response is to move the border of the niche (C) by genetic adaptation in such a way that position 2 is not experienced as stress anymore. Reproduced from Van Straalen (2003), with permission from the American Chemical Society.

The stress response can take different forms, depending on the timescale. Calow (1989) distinguished two main types, *proximate* and *ultimate* responses. The proximate response implies induction of physiological, biochemical, and genomic mechanisms (*physiological adaptation*) that allow survival while the stress prevails. Such mechanisms cannot be maintained forever without consequences for normal cell function and so a return to the niche is necessary for long-term maintenance of fitness. The ultimate response implies that genotypes with a greater than average innate capacity to resist the stress are favoured and replace the ones with lower resistance in the next generation. Then, after some generations, the whole population consists of resistant genotypes (*genetic adaptation*). The boundaries of the niche have been shifted to include the organism's new position, and what was stress before is not stress anymore (Fig. 6.2).

The existence of genetic adaptation makes us realize that the ecological niche is not a property of a species as a whole, but may show variation between populations of a single species. In that case, a species with wide ecological amplitude (a *euryoecious* species) may consist of several local populations, each with narrow amplitude (*stenoecious* populations). Consequently, what is experienced as stress for one population is normal for another population of the same species. Such genetically determined polymorphisms in response to stress have been investigated often in evolutionary ecology and provide some fascinating examples of microevolution in real time, such as pesticide resistance in insects and metal tolerance in plants. In addition, stress may lower the threshold for expression of traits and so increase phenotypic variation and accelerate evolution (Hoffmann and Hercus 2000).

In our discussion of stress responses we will emphasize *conditions* more than *resources*. Resources are environmental factors that can be consumed and belong to the impact component of the ecological niche. Conditions are factors in the habitat, such as temperature, humidity, osmotic value, and oxygen tension, that are not consumed and can be altered only slightly by the organism. They can be plotted along an axis, as in Fig. 6.2,

and the ecological amplitude of the species can be marked by values that depend on the species' physiology. We must add that our niche-based definition of stress does not encompass all stress phenomena. For example, stress may arise in animals upon sight of a predator, or when being chased away by a group member, or when witnessing an overwhelming natural event. These kinds of stress are difficult to relate to the concept of ecological niche, but many aspects of the internal state evoked by social and mental factors are similar to the ones imposed by a harsh environment.

A considerable number of functional genomic studies have been conducted with model organisms under stress. Due to the early availability of a full genomic microarray a lot of work has been done on yeast, which has developed into a classical model for the study of stress responses. There are also a fair number of genomic studies on stress responses in *Drosophila* and *Arabidopsis*. As in previous chapters, we will discuss the studies on models even when taken out of the context of their ecology, to illustrate the principles. From there we will try to draw conclusions about the relationship between stress and the ecological niche, which may hold equally for other species.

6.2 The main defence mechanisms against cellular stress

All organisms must deploy stress-defence systems with more or less specific tasks to cope with disturbances and restore normal physiological conditions after disturbance. These mechanisms are found in any cell and several systems are conserved throughout life, from prokaryotes to eukaryotes. The widespread occurrence of stress-defence mechanisms indicates that already very early in the evolution of life the defence against disturbances was a crucial problem to solve. Stress defence is thus closely linked to the idea of *homeostasis*, the tendency to regulate the internal state at a level independent from the changeable environment. Some systems that evolved in the early days of life have maintained their tasks mostly unchanged throughout evolution. We have already glimpsed some stress-defence systems in Chapter 5, when discussing the issue of longevity. It was noted then that upregulated stress defence in the wide sense is one of the most obvious signatures of a long life. In this section we will discuss the various stress-defence systems in more detail.

Korsloot et al. (2004) reviewed the cellular stress-defence responses with an emphasis on arthropods. In the book, the authors distinguished five different systems: (i) basal signal transduction systems, (ii) stress proteins, (iii) the oxidative stress response, (iv) metallothionein and associated systems, and (v) mixed-function oxygenase. It was also noted that there are many crosslinks between the different stress-defence systems. These crosslinks help to coordinate the cellular response, which is needed to maintain integrity. In addition, many genes of the stress-defence system have promoters responding to more than one challenge. For example, the metallothionein promoter has metal-responsive elements enabling induction by metal stress, but it also has antioxidant-responsive elements and steroid hormone receptor-binding sites. Korsloot et al. (2004) even went one step further and argued that the different systems cooperate as a single, *integrated, cellular stress-defence system*. In the course of this chapter, we will meet genomic evidence that this may indeed be the case. However, before presenting the genome-wide profiling studies we will shortly discuss the five best-investigated systems separately. Later sections will show that these five are by no means the only stress-responsive systems in the cell, but they serve to illustrate the most important principles of how stress-induced gene expression is brought about.

A theme common to all stress-induced gene expression is that stress signals converge on the activation of transcription factors, which bind to specific DNA sequences in the promoters of stress-induced genes. More generally these factors are called *trans*-acting factors and the DNA sequences to which they bind are *cis*-regulatory elements. It is of considerable interest to know which sequences may act as transcription factor-binding sites and which genes have promoters with these sequences. Screening of the 5' region of a gene and identification of potential binding sites for transcription factors can help considerably to understand the

biochemical context and the function of a gene. Conversely, when groups of genes are up- or downregulated in concert, and the same transcription factor-binding site appears in their promoters, this may indicate that they are regulated by the same transcription factor. One can never be sure, however, that a certain sequence, even if it conforms to a *cis*-regulatory element consensus sequence, is acting as a transcription factor-binding site *in vivo*, because transcriptional regulation is an extremely complicated process and is very much context-dependent (Wray *et al.* 2003). TRANSFAC® is a database on eukaryotic transcription factors, their genomic binding sites, and their DNA-binding profiles (www.gene-regulation.com/pub/databases.html). An overview of consensus sequences of transcription factor-binding sites appearing in this chapter is given in Table 6.1.

The scientific literature on cellular stress covers an extensive territory of biochemistry. It would be a hopeless task to try and cover all this ground here; instead we aim to present those aspects of stress-defence mechanisms that we think are necessary to understand the genomic studies on stress responses in an ecological context. We pay special attention to the pathways by which stress is translated into gene expression.

6.2.1 Stress-activated protein kinase signalling pathways

The principle of a signal transduction pathway was introduced in Chapter 5. An extensively discussed system in that chapter was the insulin/IGF receptor, which proved to be associated with the regulation of longevity in various animals. The insulin-signalling pathway relays a hormonal signal into a cellular response, via a cascade of molecular interactions, eventually leading to inactivation of a transcription factor. We saw the same principle in the TOR and Ras pathways acting upon body size in insects and in the phytochrome signalling pathway regulating shade avoidance in plants. The principle of signal transduction is also employed in one of the most elaborate stress-responsive systems, the *mitogen-activated*

Table 6.1 Overview of stress-related transcription factors and consensus sequences of their DNA binding sites (http:www.gene-regulation.com/pub/databases.html)

Transcription factor	DNA-binding site	Consensus sequence (5′ → 3′)	Context
Activator protein 1 (AP-1)	AP-1-binding element	TGA(C/G)T(A/C)A	General stress response
Heat-shock factor (HSF)	Heat-shock-responsive element (HSE)	NGAANNGAANNG AAN	Heat shock, general stress response
Nuclear factor erythroid 2-related factor 2 (Nrf2)	Antioxidant-responsive element (ARE), electrophile-responsive element (EpRE)	TGACNNNGC	Induction of antioxidant enzymes and ROS-scavenging systems
Metal-responsive-element-binding transcription factor (MTF-1)	Metal-responsive element (MRE)	TGC(A/G)CNC	Induction of metallothionein
Aryl hydrocarbon receptor (AhR)	Xenobiotic-responsive element (XRE), dioxin-responsive element (DRE), aryl hydrocarbon-responsive element (AhRE)	TGCGTGAGAAGA (human, mouse, rat, guinea pig)	Induction phase I and II biotransformation enzymes
Centromere-binding factor 1 (CBF1)	Centromere DNA element 1 (CDE1)	(A/G)TCAC(A/G)T G (yeast)	Regulation of sulphur amino acid biosynthesis pathway, induced by exposure to heavy metals
Msn2p, Msn4p	Stress-response element (STRE)	GATGACGTGT (Msn4) (yeast)	General stress defence and carbohydrate homeostasis in yeast
DREB family transcription factors	Dehydration-responsive element (DRE)	CCGAC	Induction of defence against cold and drought in plants
Abscisic acid-responsive element binding factor (ABF)	Abscisic acid-responsive element (ABRE)	(C/T)ACGTGGC	Water stress signalling in plants
Hypoxia-inducible factor (HIF-1)	Hypoxia responsive element (HRE)	TACGTGCT (human)	Metabolic switches related to hypoxia
Oestrogen receptor (ER-α)	Oestrogen-responsive element (ERE)	AGGTCANNNTGA CCT (human, mouse, cattle, clawed frog, chicken)	Activation of processes related to oogenesis and other female functions

Notes: In some cases more than one name is given to the same factor or binding site. Some consensus sequences are specific to certain groups of organisms (where indicated). N, any nucleotide; ROS, reactive oxygen species.

protein kinase pathway (MAPK pathway; Chang and Karin 2001).

The name MAPK derives from the action of mitogens, chemical substances that stimulate cell division and often have a tumour-promoting action. Several secondary metabolites from plants are notorious mitogens, the best known being phorbol esters from the family Euphorbiaceae (spurges), such as the tropical plant *Croton tiglium*. The mitogenic action of phorbol esters is due to their activation of the MAPK pathway; some of them are used as model compounds in studies of

cell biology. It became clear later that MAPK is also involved in transducing stress signals.

The classical MAPK pathway is activated by binding of an extracellular signal molecule such as a mitogen to a receptor protein on the cell membrane (Fig. 6.3). In addition to mitogens, MAPK is sensitive to cytokines, hormones, and a variety of other molecules. These substances activate a *receptor tyrosine kinase* (RTK), a molecule with an extracellular part to which the signalling molecule can bind, and an intracellular domain with kinase activity. When the extracellular

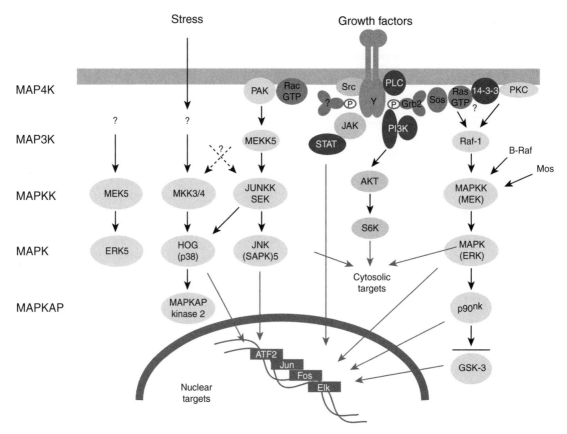

Figure 6.3 Scheme of the mammalian MAPK signalling pathway, illustrating the great complexity of reactions. The figure shows a portion of a cell with the plasma and nuclear membranes. The classical pathway (on the right) starts with binding of a growth factor to an extracellular receptor (Y, receptor tyrosine kinase), leading to three successive waves of kinase activity (MAP3K, MAPKK, MAPK) and activation of MAPKAPs. When activated, several of the downstream kinases can translocate to the nucleus and trigger transcriptional regulation of a variety of genes. Kinases can also influence cytoplasmic targets and contribute to translational control. Parts of the MAPK pathway are activated by stress signals (on the left), but the cascade of events is less well known than that triggered by cell growth factors. ATF, activating transcription factor; GSK, glycogen synthase kinase; JAK, Janus kinase; PAK, p21-activated kinase; PI3K, phosphoinositide 3-kinase; PKC, protein kinase C; PLC, phospholipase C; S6K, ribosomal protein S6 kinase; SEK, SAPK kinase; STAT, signal transducer and activator of transcription. © Sigma–Aldrich Co.

receptor is occupied, a conformational change is imposed upon the intracellular domain, which then becomes an attractive binding site for adapter proteins, and their binding triggers further molecular reactions. The cascade of reactions following on RTK activation is extremely complex. More than 100 different proteins have been described to participate in the MAPK network; many of them are *kinases*, enzymes that activate other proteins by phosphorylating amino acids critical for the three-dimensional structure, resulting in activation or

inactivation of the target. The MAPK cascade involves three successive tiers of kinase activity, which—from downstream to upstream—are designated MAPK, MAPK kinase (MAPKK, MKK, or MEK), and MAPKK kinase (MAPKKK, MAP3K, MEK kinase, or MEKK). The upstream proteins associated with binding to RTK and activation of MAP3K are sometimes called MAP4K (Fig. 6.3). Downstream of the three-tiered cascade, MAP kinases activate *effector kinases*, also called MAPK-activated proteins (MAPKAPs).

One way in which the MAPK proteins, and the effector kinases activated by them, exert their action is by translocation to the nucleus. An example is a kinase called *extracellular signal-regulated kinase* (ERK), which when activated can pass through the nuclear envelope and there activate transcription factors. Important targets for ERK are c-Jun and c-Fos, which are components of the dimeric transcription factor *activator protein 1* (AP-1), a protein which turns up as a downstream effector in various stress responses. The combination of two different proteins to a single active unity is called *heterodimerization*. Many of the transcription factors activated by MAPK are dimeric proteins, which only become active by combining the two components to one functional protein. Although transcription factors are important MAPK targets, MAPKs may also influence post-transcriptional processes in the cytoplasm, for example by contributing to mRNA stabilization.

MAPK is also activated by stress (Fig. 6.3). The part of the pathway that is specifically associated with the stress response is called the *stress-activated protein kinase* (SAPK) *signalling pathway*. However, the chain of events in SAPK is less well known than in the case of the classical pathway activated by mitogens. The stress signal may be transduced along a distinct pathway, interacting with the classical pathway, but it may also affect enzymes of the classical pathway directly. Two important kinases which are activated by stress and can translocate to the nucleus to activate transcription factors are *HOG* (product of high osmolarity gene) and *c-Jun N-terminal kinase* (JNK). Some of the MAPK proteins are known under different names in different organisms; for example HOG was first described in yeast but its orthologue in higher organisms is known as p38, whereas JNK is also known as SAPK5.

As noted above, the action of protein kinases involves phosphorylation of certain amino acids in other proteins. The two stress-related effector kinases, HOG and JNK, can only transfer to the nucleus when phosphorylated. However, they may also be dephosphorylated by the action of protein phosphatases. Phosphatases specific for MAPK are called *MAPK phosphatases* (MKPs;

Tamura *et al.* 2002). The action of these phosphatases is very important for relaxation of the MAPK signal. There are four different MKP families; some of them act on specific effector kinases (e.g. only on JNK) and others have a broader action spectrum (e.g. they dephosphorylate three effector kinases— HOG, JNK, and ERK). Interestingly, MKP genes are among the many genes activated by MAPK signalling. Their activity thus constitutes a negative feedback, which attenuates the MAPK signal after being triggered by stress or a growth factor.

One of the possible ways in which stress may activate SAPK signalling is by inhibition of MKPs. Oxidative stress in particular may lead to the oxidation of sensitive thiol groups in protein phosphatases and so allow SAPK signalling by removing the negative feedback (Korsloot *et al.* 2004). Direct activation of kinases is another possible mechanism (Fig. 6.3).

The fact that different stimuli all activate MAPK signalling, yet can activate different genes, suggests the presence of a regulatory or coordinative system within the cascade. The existence of distinct but not mutually exclusive mechanisms, and the involvement of a large number of kinases, many of them acting upon each other, could contribute to the achievement of specificity in the responses to different stimuli.

6.2.2 Heat-shock proteins

Heat-shock proteins are the best-investigated components of the cellular stress response. The traditional way of studying them is by applying a heat shock: exposure of the organism or a cell culture for a period of 30 min to a few hours to supraoptimal temperature, such as 39°C. Later research has shown that heat-shock proteins are not only induced by heat but also by many other environmental stress factors, including cold, food depletion, osmotic stress, and toxicants, and so the term *stress proteins* is actually more appropriate. However, the term heat-shock protein is now so widely used and accepted that we retain it here.

Heat-shock proteins have been found in all organisms in which they were sought (Feder and Hofmann 1999). In addition, essential features of their molecular structure are conserved over the entire tree of life, from Archaea to mammals. This extremely wide occurrence and conservation suggests that heat-shock proteins must play a very fundamental role in the cellular defence against stress. The consensus view is that this role is due to support in protein folding and unfolding, which involves four aspects: (i) stabilizing essential structural proteins, especially around the nucleus and of organelles such as ribosomes and spliceosomes, (ii) assisting transfer of proteins across membranes, by unfolding and refolding, (iii) supporting refolding of proteins denatured by stress, and (iv) supporting the degradation of aberrant proteins. On the basis of their ability to align with other proteins and facilitate changes in their three-dimensional structure, heat-shock proteins are often called *molecular chaperones*. Possibly the conservation of heat-shock proteins throughout all living beings stems from the fact that a very specific three-dimensional structure is required for conducting this chaperone function. The ability of heat-shock proteins to protect cells against the adverse effects arising from accumulation of denatured proteins can be seen as a logical extension of their normal function as molecular chaperones.

The diversity of stress proteins varies per taxonomic group. As an example we discuss here the stress proteins of *Drosophila* (Table 6.2;

Korsloot *et al.* 2004). Four families may be discerned; Sp90, Sp70, small heat-shock proteins, and ubiquitin. The proteins themselves are named after their apparent molecular mass on an electrophoresis gel, so Hsp70 has a molecular mass of about 70 kDa. Another type of classification is between *inducible* and *constitutive* proteins. Proteins of the latter category are called *heat-shock cognate proteins* (e.g. Hsc70). They are always present in the cell during normal physiological function and are not usually induced by stress factors, although under certain circumstances they may be induced a little. Proteins of the former category, especially Hsp70, are known for their very strong inducibility. Induction of heat-shock proteins was long used as a classical model of gene regulation, because the heat shock is so easily brought about and induction is very strong. Even nowadays, the promoters of heat-shock genes are often deployed in gene-expression studies, using genetic constructs in which the coding region of a gene of interest is fused with a promoter from an inducible heat-shock gene.

Organisms may have more than one representative of the heat-shock proteins indicated in Table 6.2. For instance, *D. melanogaster* has two types of Hsp70, encoded by genes at different chromosomal loci. At each locus, two or more repeats of the same gene are present, depending on the fly stock. The organization of heat-shock

Table 6.2 Overview of stress proteins in *D. melanogaster*, following Korsloot *et al.* (2004)

Family	Name	Location in unstressed cell	Location under cellular stress	Function
Sp90	Hsp83	Cytoplasm	Cytoplasm	Stabilizes specific receptor proteins and chaperones peptides to membranes
Sp70	Hsp70	Nucleus (cytoplasm)	Nucleus, nucleolus, cytoplasm	Highly inducible, rescue of aberrant proteins during stress response
	Hsp68	Nucleus, cytoplasm	Nucleus, cytoplasm	Inducible; function unknown
	Hsc70	Cytoplasm	Cytoplasm, nucleus	Protein binding in cytoplasm
	Hsc70b	Cytoplasm	Cytoplasm (nucleus)	Protein binding in cytoplasm
	Hsc71	Mitochondrion	Mitochondrion	Protein binding in mitochondrion
	Hsc72	Endoplasmic reticulum	Endoplasmic reticulum	Protein binding in endoplasmic reticulum
Small Hsps	Hsp22, Hsp23, Hsp26, Hsp27	Cytoplasm	Nucleus (cytoplasm)	Participation in development and stress response
Ubiquitin	Ubiquitin	Nucleus, nucleolus, cytoplasm	Cytoplasm (nucleus)	Histone binding, tagging proteins for degradation

genes varies both within and between species and this variation is possibly relevant for the ecological response to environmental extremes (Feder and Hofmann 1999).

Each of the stress proteins mentioned in Table 6.2 has specific functions during normal cell metabolism. Members of the Sp90 family, including *Drosophila* Hsp83, are constitutively present, but heat, anoxic conditions, and glucose deprivation increase their level. They are sometimes referred to as *glucose-regulated stress proteins*. A specific role for Hsp83 lies in stabilizing steroid hormone receptors. As long as a steroid does not occupy the receptor, Hsp83 binding ensures a receptive configuration. Upon binding of the hormone, Hsp83 detaches from the receptor, which then takes another configuration; the receptor may then enter the nucleus to act as a transcription factor. This is a mechanism by which cells sensitive to steroid hormones translate a hormonal signal into gene expression. A similar function of Hsp83 lies in stabilizing the heat-shock factor, HSF, and the aryl hydrocarbon receptor (see below). Expression of Hsp83 varies with life-stage and tissue; Hsp83 is developmentally induced in gonad tissue during oogenesis, which is understandable from the steroid hormone-receptor-binding function.

The Sp70 family includes the inducible heat-shock proteins Hsp70 and Hsp68 (Table 6.2). These proteins are abundantly expressed within minutes after commencement of a heat treatment. After induction Hsp70 is mostly found in the nucleus and on cell membranes, where it ensures protection of essential proteins against denaturation. In the recovery phase after termination of a heat treatment, Hsp70 is translocated to the cytoplasm and participates in the degradation of damaged proteins. It is likely that Hsp68 fulfils a similar role, but its precise function is less well described. When the cell is not in stress, Hsp70 is hardly detectable, so the induction has the characteristics of a rescue operation, with Hsp70 turning out only in case of emergency. However, we know from Chapter 5 that a modest continuous upregulation of Hsp70 can have beneficial effects: it increases lifespan. A large continued increase in Hsp70 levels has detrimental effects, especially in tissues

with a high rate of cell division and growth. The reason is, as we will see below in more detail, that induction of Hsp70 is accompanied by a redirection of protein synthesis in which priority is given to heat-shock protein synthesis and degradation of aberrant proteins, while normal protein synthesis is blocked.

The heat-shock cognate proteins Hsc70 and Hsc70b have roles similar to the inducible stress proteins, but they are constitutively present at high levels to assist in the folding of peptides. In particular, polypeptide chains that leave the ribosome after translation need assistance to assume the correct three-dimensional structure. It is assumed that negatively charged clusters in the Hsc bind to positively charged amino acid residues of the peptide; this binding changes the local conformation of the complex and causes a fold in the chain by forcing any neutral residues adjacent to the charged residues into the relatively hydrophobic environment inside the fold. The other two heat-shock cognate proteins, Hsc71 and Hsc72, are active in the mitochondrion and the endoplasmic reticulum, respectively. Both proteins are encoded in the nuclear genome, so Hsc71 must be transported across the mitochondrial membrane after translation in the cytoplasm.

Low-molecular-mass heat-shock proteins vary considerably in molecular mass and number across species and they form a rather heterogeneous collection with limited sequence similarity to each other. Four different small heat-shock proteins have been identified in *Drosophila* (Table 6.2). The regulation and function of small heat-shock proteins are complicated. Some small heat-shock proteins are activated by phosphorylation in response to stimuli that also activate the MAPK signalling pathway (see above). These are constitutively present heat-shock proteins that play a role in protecting and chaperoning special proteins and enzymes in normal cell metabolism. However, other small heat-shock proteins are induced in large quantities without phosphorylation and these inducible heat-shock proteins seem to be directed mainly towards nuclear structures and RNA.

The fourth family of heat-shock protein consists of the small globular protein *ubiquitin*, which like the other heat-shock proteins is extremely well conserved; only three amino acids differ between yeast and *H. sapiens*. Ubiquitin can be conjugated to other proteins by a ubiquitin protein lyase and when proteins are 'tagged' in this way they are destined for ATP-dependent cytoplasmic protein degradation. The cytoplasmic proteolytic system promotes the turnover of proteins and increases availability of free amino acids for protein synthesis. Another regulatory role for ubiquitin is due to its binding to histones. In Chapter 5 we saw that active transcription of DNA requires histones to be acetylated; if not, histones will interact with each other and cause condensation of chromatin. A similar effect is due to ubiquitin: ubiquitination of histones prevents chromatin condensation. Under stressful conditions an enzyme, ubiquitin hydrolase, is induced, which removes ubiquitin from the histones (*de-ubiquitination*) and thereby causes a general decrease of DNA processing. The liberated ubiquitin moves to the cytoplasm and participates in tagging proteins for proteolytic degradation. The system can thus be understood as a shift of priorities under stress from DNA transcription to protein degradation.

How is the activity of inducible heat-shock proteins regulated? One way in which the cell can increase the amount of heat-shock proteins is by transcriptional activation of *Hsp* genes. Inducible heat-shock protein genes all share certain conserved sequences in their promoter, so-called *heat-shock elements* (HSEs) see Table 6.1. These are the binding sites for a very important transcriptional activator, *heat-shock factor* (HSF). In yeast and *Drosophila* only one HSF has been identified, but in mammals there appear to be two, HSF1 and 2, where HSF1 is homologous to *Drosophila* HSF. Biochemical research has shown that HSF needs to undergo *homo-trimerization* to become active as a transcription factor (Fig. 6.4). This means that three HSF molecules must bind together; the trimeric HSF can then bind to the HSEs in the promoter of heat-shock genes.

In unshocked *Drosophila* cells, HSF is present in the nucleus in the monomeric form. Binding of heat-shock proteins, in particular Hsp83 and Hsp70, to HSF monomers prevents trimerization (Morimoto *et al.* 1994; Santoro 2000). When heat

shock leads to the appearance of aberrant proteins in the cytoplasm, heat-shock proteins are drawn to these proteins with an affinity greater than their binding to HSF. Consequently, HSF is no longer prevented from trimerization and the activated HSF may bind to the heat-shock elements in the promoter of *Hsp* genes. The binding itself is however not yet sufficient for maximum transcriptional activation; HSF must also be phosphorylated. This is brought about by a protein kinase, which by itself is activated by an external stimulus, transduced, for example, through the SAPK pathway. The activation of a transcription factor by phosphorylation while already bound to DNA is an example of *transactivation*. The complete model of HSF activation is shown in Fig. 6.4.

The scheme illustrated in Fig. 6.4 illustrates how environmental stress regulates the induction of heat-shock protein at three levels: (i) translocation of protein kinases to the nucleus, (ii) removal of the inactivation conferred by heat-shock proteins on HSF, and (iii) transactivation of HSF by protein kinase. Even this rather complicated scheme is still a simplification. For example, trimerization of HSF is not only dependent on the appearance of aberrant proteins and heat-shock-protein detachment, it may also be directly triggered by changes in the cellular redox state and by a peak in Ca^{2+} ions. In addition, the details are likely to differ from one species to another. Our scheme is mostly inspired by the situation in *Drosophila*, which is similar to the situation in mammals except that some heat-shock proteins have different molecular masses; for example, Hsp83 of *Drosophila* is homologous to human Hsp90. The heat-shock factor of yeast differs considerably from animal HSF. It lacks one of the domains (a leucine zipper) that allows stabilization of the monomeric configuration. Consequently, it is not stored in monomeric form in the nucleus like the animal HSF, but is already bound to the DNA in its trimeric form, requiring only phosphorylation to be activated.

The regulatory mechanisms depicted in Fig. 6.4 all relate to transcriptional processes; however, the stress response also includes regulatory mechanisms at the level of RNA processing, translation, and mRNA degradation. One of the processes that

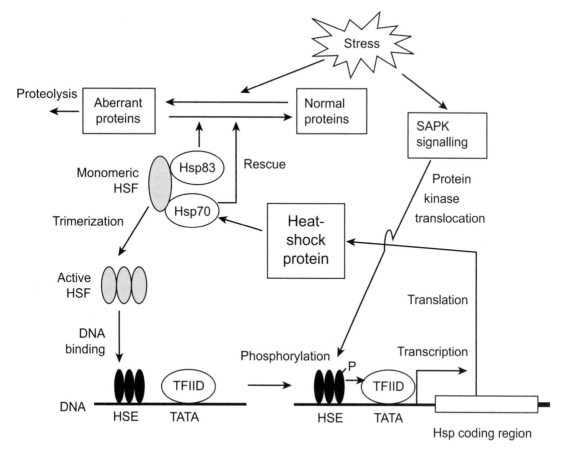

Figure 6.4 Schematic representation of the chain of events leading to induction of heat-shock proteins as a consequence of stress. TATA, TATA box; TFIID, transcription factor IID, a TATA-box-binding protein facilitating the formation of a transcription-initiation complex.

is downregulated during severe heat shock is RNA splicing: the extrusion of intron sequences from the primary transcript, occurring at specialized bodies in the nucleus, the spliceosomes. The consequence is that all pre-mRNAs that have introns in their sequence cannot be processed and are not translated into functional proteins. Interestingly, the inducible heat-shock genes themselves lack any introns and so can circumvent the splicing block. Translation itself is not inhibited: it can proceed and produce a massive amount of heat-shock protein. In addition, as we saw above, stress may cause de-ubiquitination of histones, because ubiquitin moves to the cytoplasm where it is needed for tagging damaged proteins. The loss of ubiquitin from histones causes chromatin condensation,

which is another mechanism to suppress protein synthesis. Obviously, the inducible heat-shock genes must be exempt from inactivation by chromatin condensation, which is brought about by a basal promoter-binding element (known as a GAGA factor) being permanently bound to the *Hsp* promoter; this protein displaces the histone-containing nucleosomes from the gene and so prevents chromatin condensation around the *Hsp* gene.

The heat-shock response continues to act as a model *par excellence* of homeostatic mechanisms of the living cell. It illustrates a great variety of biochemical regulatory mechanisms, of which we have discussed only the most salient features. Later in this chapter we will see how important heat-shock proteins are in the genome-wide responses to stress.

6.2.3 The oxidative stress-response system

Under the term oxidative stress come a variety of phenomena that are an unavoidable consequence of aerobic metabolism. By definition all aerobic organisms need oxygen, but at the same time they must avoid the inherently cytotoxic effects of oxygen. Toxicity of oxygen is not due to O_2 itself but to reactive oxygen derivatives, generated by cellular processes. These derivatives are jointly referred to as *reactive oxygen species* (ROS). Several of these ROS are *free radicals*; that is, molecules or elements with one or more unpaired electrons in the outer orbital. The best-known ROS are the free radicals superoxide radical ($O_2^{\bullet-}$), hydroxyl radical ($OH^\bullet$), and nitric oxide radical ($NO^\bullet$; the lack of a paired electron is indicated by $\bullet$ in the chemical formula). Non-radical ROS are hydrogen peroxide (H_2O_2) and singlet oxygen ($^1O_2$). Molecular oxygen itself has two unpaired electrons, so strictly speaking it is also a radical, but the reactivity of O_2 is limited due to the fact that the two electrons have equal spin (the so-called spin restriction). In ROS the spin restriction is lifted, which is why these forms are called activated oxygen.

ROS are produced at many places in the cell. Superoxide anion is produced abundantly in the respiratory chain of the mitochondria, the light-harvesting reactions of the choroplast, the reduction–oxidation reactions catalysed by cyto-chromes of the smooth endoplasmic reticulum, and the xanthine dehydrogenase pathway, which is involved with the degradation of purines to urate. Singlet oxygen is produced by so-called photo-sensibilization reactions, in which light energy is absorbed by molecules such as riboflavin, chloro-phyll, and retinol, and transferred to molecular oxygen. Not all ROS are equally reactive. The most reactive species are hydroxyl radical and singlet oxygen, which react immediately with a suitable molecule and so inflict injury mainly on local cellular structures. H_2O_2 and superoxide anion are more stable and can move through the cell by diffusion; H_2O_2 can even pass cell membranes. The damaging effect of these ROS is mostly due to their ability to generate hydroxyl radicals. These reactions are catalysed by transition metals, such as iron and copper. For example, $OH^\bullet$ is generated from H_2O_2 in the so-called *Fenton reaction*:

$$Fe^{2+} + H_2O_2 \rightarrow Fe^{3+} + OH^\bullet + OH^-$$

Although the concentration of free iron in the cell is very low (most iron is bound in porphyrins and ferritin and not available for the Fenton reaction), some iron is bound to low-molecular-mass chela-tors such as citrate, and this can catalyse the gen-eration of $OH^\bullet$ from H_2O_2.

In Chapter 5 we saw that ROS are implicated in aging and senescence according to the free radical theory of aging. The cellular basis for this theory is that ROS may cause damage to many macro-molecules in the cell, including proteins, DNA, and lipids. Protein damage may be caused by oxidation of free thiol (SH) groups, leading to loss of function. DNA damage may be due to thymidin dimers and strand breaks. Lipid damage is due to a process known as *lipid peroxidation*. This is a chain reaction in which an oxygen radical abstracts a hydrogen atom from an unsaturated bond in a fatty acid chain, which is followed by molecular rearrangement and uptake of oxygen, to form a lipid peroxyradical ($LOO^\bullet$), which can abstract $H^\bullet$ from another lipid. Polyunsaturated fatty acid chains of membrane lipids are particularly sensitive to lipid peroxidation. Peroxidation causes loss of membrane flexibility, loss of activity of membrane-bound enzymatic processes, and, in the most severe form, lipid destruction followed by the appearance of volatile alkanes, alkenes, and aldehydes.

Obviously, there are great advantages in the use of oxygen gas as an electron acceptor in cellular respiration; however, damage induced by ROS is an unavoidable side effect. Protection against ROS damage is an evolutionary neccessity that must compensate the disadvantage. A great variety of antioxidant systems is deployed, using two strat-egies: *scavenging* (neutralization of ROS by reaction with a reductant) and *enzymatic transformation* (dismutation or reduction) to a non-reactive form. Table 6.3 provides an overview of the major

Table 6.3   Major antioxidant systems protecting the cell against injury from radical oxygen species

Antioxidant system	Primary localization	Actions
Copper- and zinc-containing superoxide dismutase (Cu/Zn SOD)	Cytosol, nucleus	Catalyses dismutation of $O^{\bullet-}_2$ to H_2O_2
Manganese-containing superoxide dismutase (Mn SOD)	Mitochondrion	Catalyses dismutation of $O^{\bullet-}_2$ to H_2O_2
Rubredoxin oxidoreductase	Only found in anaerobic bacteria and Archaea	Catalyses reduction of $O^{\bullet-}_2$ to H_2O_2
Catalase (CAT)	Peroxisomes	Catalyses reduction of H_2O_2 to H_2O
Glutathione peroxidase (GSH-Px)	Cytosol, mitochondrion	Catalyses reduction of H_2O_2 to H_2O
Glutathione peroxidase (GSH-Px,m)	Lipid membranes	Catalyses reduction of lipid Hydroperoxides
Glutathione reductase	Cytosol, mitochondrion	Catalyses reduction of low-molecular-mass disulphides
Ascorbate peroxidase	Chloroplast (only in plants)	Catalyses reduction of H_2O_2 to H_2O
Thioredoxin (TRX)	Cytosol, mitochondrion, nucleus	Restores redox state by reducing oxidized thiols, scavenges H_2O_2 and $OH^{\bullet}$
α-Tocopherol (vitamin E)	Lipid membranes	Scavenges H_2O_2, $OH^{\bullet}$, and $LOO^{\bullet}$; interrupts lipid peroxidation
β-Carotenoid (vitamin A)	Lipid membranes	Scavenges $O^{\bullet-}_2$ and peroxyl radicals
Ascorbate (vitamin C)	Throughout the cell	Scavenges $O^{\bullet-}_2$ and $OH^{\bullet}$; contributes to regeneration of vitamin E
Glutathione	Throughout the cell	Scavenges $O^{\bullet-}_2$ and organic free radicals; substrate in enzymatic reduction reactions

Source: From various sources.

antioxidant systems. Scavenging and enzymatic systems to protect cells from oxidative stress are found in almost all organisms; however, anaerobic and microaerophilic bacteria and Archaea have an oxidoreductase not found anywhere else (Lumppio *et al.* 2001). Several antioxidant enzymes are localized in specific organelles of the cell (Table 6.3).

Among the different protective systems listed in Table 6.3 we briefly highlight the role of *glutathione*, which acts both as a scavenger on its own and as a substrate in enzymatic reactions. Glutathione is a tripeptide, γ-glutamyl-cysteinyl-glycine, which in reduced form (written GSH) has a free thiol group on the cysteine residue. In the transition to the oxidized form (GSSG) two thiol groups react with each other, losing two electrons, as in the reaction catalysed by glutathione peroxidase:

$$H_2O_2 + 2GSH \rightarrow GSSG + 2H_2O$$

Oxidized glutathione is then again reduced at the expense of reduction equivalents from NADPH by means of glutathione reductase:

$$GSSG + NADPH + H^+ \rightarrow 2GSH + NADP^+$$

Since glutathione is responsible for the majority of ROS-protection reactions and it is also involved in conjugating lipophilic substances in xenobiotic biotransformation reactions (see below), the maintenance of a large pool of reduced glutathione is very important for the vitality of the cell. Cellular stress due to ROS, altered redox state, and xenobiotic metabolism may cause *glutathione depletion*.

To be effective as protective mechanisms, the antioxidant systems listed in Table 6.3 must be upregulated when the cell perceives oxidative stress. This is indeed the case. Recent research has shown that genes encoding antioxidant protective enzymes all have a characteristic sequence in their promoter, designated the *antioxidant-responsive element* (ARE). This element is also called the *electrophile-responsive element* (EpRE), after researchers discovered that not only oxidative stress but also electrophile chemicals could activate the element. As in the case of heat-shock proteins, the sequence serves to bind a transcription factor which is activated by stress. The factor binding to AREs has been identified as *Nrf2*, also called NF-E2 (an abbreviation of nuclear factor erythroid 2-related factor 2; Nguyen *et al.* 2003, 2004; Jaiswal 2004; Kobayashi and Yamamoto 2005). Under normal physiological conditions, Nrf2 is continuously degraded under the influence of a protein known as Keap1 (Kelch-like ECH-associating protein 1). This protein supports the tagging of Nrf2 with ubiquitin, after which it is destined for cytoplasmic proteolysis. Degradation of Nrf2 is compensated by continuous synthesis, but under normal conditions the pool of Nrf2 in the cytoplasm is very small. Under conditions of oxidative stress Keap1 is inactivated and Nrf2 is stabilized (Fig. 6.5). Presumably, the oxidative-stress signal is transduced through the SAPK pathway leading to

activation of protein kinase C (PKC) and other cytosolic factors. Phosphorylation of Nrf2 by PKC results in the release of Nrf2 from its repressor. Nrf2 may then translocate to the nucleus, and will bind to an ARE. However, before becoming fully active it needs to undergo *heterodimerization*. This is comparable to the formation of activator protein AP-1 by heterodimerization of c-Jun and c-Fos (see Section 6.2.1 on SAPK signalling). In the case of Nrf2, the partner protein is suggested to be a member of the family of Maf proteins. Maf proteins are nuclear factors that like Nrf2 may bind to ARE. In the absence of Nrf2, they exert negative control over ARE-mediated gene expression. How this negative control is lifted by Nrf2 is not known

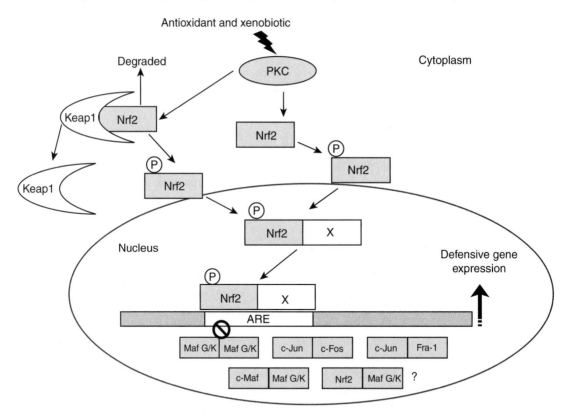

Figure 6.5 Scheme of Nrf2 activation and induction of antioxidant genes by ARE binding. Nrf2 in the cytoplasm is bound to a protein, Keap1, which promotes its degradation by ubiquitination. An oxidative-stress signal, transduced via SAPK, activates protein kinases that may phosphorylate Nrf2 and liberate it from inactivation of Keap1. In addition, Keap1 may be destabilized by reactive chemicals, for example through alkylation of cysteine residues critical for its activity. Either mechanism allows Nrf2 to translocate to the nucleus. There it undergoes dimerization with an as-yet unknown partner (X) and triggers expression of genes with AREs in their promoter. A variety of nuclear factors (Maf G/K and others) normally inhibit ARE-mediated gene expression, but are removed by the Nrf2/X heterodimer. From Jaiswal (2004), with permission from Elsevier.

precisely. Anyway, expression of the antioxidant system seems to be regulated by both negative and positive agents and this could allow fine-tuning to the physiological needs of the cell.

Our present knowledge of the regulation of antioxidant systems is almost completely limited to human cells and *Drosophila*. One of the reasons for this is that the issue raises a good deal of interest in medical research: tumor cells are characterized by an upregulated oxidative-stress response. How the knowledge generated in the medical sector translates to an ecological context of animals and plants in the wild is difficult to evaluate at the moment; however, it may be expected that, like the stress-activated signalling pathways and the stress proteins, most of components of the oxidative-stress response are evolutionarily conserved. In the genome-wide studies discussed later in this chapter we will discuss some evidence supporting this statement.

6.2.4 Metallothionein and associated systems

Metallothionein is a peculiar protein with an extremely high affinity for free metal ions such as Zn^{2+}, Cd^{2+}, and Cu^+. In the presence of sufficient metallothionein the free concentrations of these metals in solution are reduced to the picomolar range. Interestingly, iron is not bound by metallothionein but instead has its own pathways of uptake and intracellular transport, involving ferrotransferrin and ferritin. As we will see below, changes in iron trafficking are part of the general stress response, but the regulation of these changes does not appear to interact with metallothionein and therefore we leave iron out of consideration here.

As the name indicates, metallothionein contains not only a high amount of metal but also of sulphur. This is due to an extraordinarily high percentage of cysteine residues (around 30%). These cysteines participate in the formation of *metal-thiolate clusters*, in which the thiol groups of several cysteines coordinate with a group of metal atoms. In the metallothionein of mammals there are two clusters, one binding four metal ions using eleven cysteines, the other binding three metal ions with nine cysteines. These two clusters appear as two separate protein domains, a C-terminal α-domain (four-metal cluster) and an N-terminal β-domain (three-metal cluster), which are separated by a short linker sequence (Fig. 6.6). There are no aromatic amino acids in metallothionein; the whole protein is very hydrophilic.

The two-cluster structure of metallothionein has also been found in other vertebrates, invertebrates, and plants; however, in some species the α and β clusters are in reversed configuration with respect to their N- and C-terminal positions. Other species have two three-metal clusters. The metallothionein genes (*Mt*s) of *Drosophila* are an oddity among animals. *Drosophila* has four *Mt* genes, each encoding a small metallothionein with one metal-binding domain (Valls *et al.* 2000; Egli *et al.* 2003). Such single-cluster *Mt*s have not been found in other animals up to now, but they are present in fungi. In plants the *Mt* genes encode proteins in which the two metal-thiolate clusters are connected by a very long linker of non-cysteine amino acids (Cobbett and Goldsbrough 2002). So it appears that the evolution of metallothionein, quite unlike the stress proteins and the components

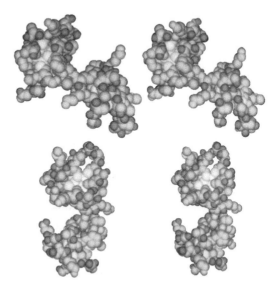

Figure 6.6 Model, based on crystallography data, of rat liver metallothionein with five cadmium and two zinc atoms, bound in two metal-thiolate clusters. The molecule can be viewed in three dimensions by stereoscopy. From Robbins *et al.* (1991), with permission from Elsevier.

of signalling pathways discussed above, has come with a considerable reshuffling of the molecule (Van Straalen *et al.* 2006).

When metallothionein was first described, the function attributed to it was to regulate the cellular concentrations of essential metals. This was assumed to involve donating metals to specific metal-requiring ligands (enzymes, zinc fingers, structural proteins), while preventing aspecific binding to macromolecules by keeping the free concentrations of metals very low. In addition some metallothioneins turned out to be highly inducible by non-essential heavy metals (e.g. Cd) and this suggested a detoxification role. This classical, dual role of metallothionein has come under fire recently. The following issues describe the more complicated situation today.

First, it turned out that metallothionein is induced not only by metal ions, but also by a variety of other stresses, including changes in redox state, oxidative stress and stress hormone signals. This suggests that the protein might be a member of the integrated stress response and has other roles as well. A function as a scavenger of free radicals is often suggested. Second, metallothionein should not be considered a single protein. Many organisms have more than one *Mt* gene; the human genome has no less than 16 *Mt* genes. Most invertebrates investigated so far have two genes, one strongly inducible by cadmium and encoding a cadmium-binding protein, the other not inducible and encoding a copper-binding protein. The presence of a specific zinc-binding metallothionein in invertebrates is doubtful; the copper- and cadmium-binding metallothioneins of *Drosophila*, nematodes, earthworms, and snails are not inducible by zinc. Third, other metal-chelating substances have been found; this happened initially in plants, hence the name *phytochelatins*. They are the main zinc-binding ligands in plant cytoplasm and could play a similar role in invertebrates. Phytochelatins are peptides of variable length with the general formula $(\gamma\text{-Glu-Cys})_n\text{-Gly}$, where n varies from 2 to 5. The peptides are synthesized from glutathione by the enzyme *phytochelatin synthase*. The gene encoding this enzyme is not only found in plants but also in nematodes,

earthworms, and chironomids (Cobbett and Goldsbrough 2002). Phytochelatin synthase is also present in the genome of the tunicate *Ci. intestinalis*, as mentioned in Section 3.3.5, but it is absent from vertebrates.

Induction of metallothionein by exposure to heavy metals has been studied extensively, but the mechanism is not yet clear. As in the case of antioxidant enzymes, inducibility is due to the presence of specific sequences in the promoter of the gene, which bind a transcription factor activated by metals. In the case of metallothionein these sequences are called *metal-responsive elements* (Table 6.1). Such elements are not only present in the promoters of *Mt* genes, but also in those of genes encoding membrane-bound zinc transporters and enzymes associated with glutathione synthesis (Andrews 2001). The best-characterized transcription factor binding to these sequences is *metal-responsive element-binding transcription factor* (MTF). Induction by metals takes place by activation of MTF in the cytoplasm, followed by translocation to the nucleus; however, how MTF is activated by metals is not clear. One model suggests that in uninduced circumstances MTF is inhibited by a factor called metallothionein transcription inhibitor (MTI). This MTI has possible binding sites for zinc, which if occupied would result in the release of MTF (Palmiter 1994; Roesijadi 1996; Haq *et al.* 2003). Under this model, activation of MTF by cadmium is explained by an increase in the free zinc concentration in the cell, brought about by cadmium displacing zinc from cellular binding sites (Fig. 6.7).

The validity of the model for MTF activation by free zinc may be limited to mammals and cannot be extrapolated without modification to fish or invertebrates. In rainbow trout it was shown that silver could activate metallothionein expression without mediation by zinc (Mayer *et al.* 2003). In *Drosophila*, zinc itself does not activate MTF, although cadmium does. In earthworms and nematodes induction must involve an entirely different mechanism because metal-responsive elements seem to be completely absent; yet induction of the *Mt* gene by cadmium is possible (Stürzenbaum *et al.* 2001). Another, non-MTF/metal-responsive element-like induction mechanism

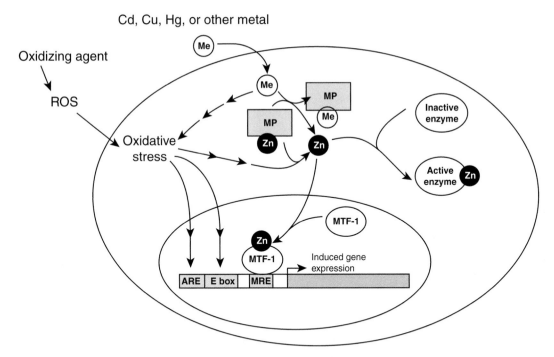

Figure 6.7 Model for induction of metallothionein in mammals. Heavy metals such as cadmium, copper, and mercury can displace zinc from metal-containing ligands (MP), thereby increasing the free zinc concentration in the cell. Excess free zinc then activates metal-responsive element-binding transcription factor-1, MTF-1, which binds to metal-responsive elements (MREs) in the promoter of the metallothionein gene. Additional transcriptional activation is recruited from oxidative stress signals (activating AREs and adenomajor late transcription factor (MLTF), binding to the E box. After Haq *et al.* (2003), with permission from Elsevier.

is found in yeast. Here metallothionein (*CUP1*) is induced primarily by copper and the system is more direct than the one described above. A transcription factor has been isolated which binds to upstream activation sequences only when it contains copper; this, probably in combination with other transcription factors, results in enhanced transcription (Thiele 1992).

There is a great deal of tissue specificity in metallothionein induction. Dallinger *et al.* (1997) isolated two metallothioneins from the snail *Helix pomatia*. One of these binds only copper and is present mainly in the mantle while the other binds only cadmium and is present in the midgut gland. In zebrafish, *Mt2* is induced by methylmercury but only in the liver—not in muscle or brain—despite the fact that methylmercury accumulates mainly in the brain (Gonzalez *et al.* 2005). The metallothioneins of mammals are also tissue-specific. Among the four groups of iso-enzymes (MTI–IV), MTI and

MTII are inducible and expressed in nearly every cell; these proteins seem to qualify best as members of the stress-response. MTIII and MTIV are expressed constitutively and only in specific tissues: MTIII is expressed only in the brain and MTIV is found only in squamous epithelium cells.

Metallothionein has many links with the stress-activated systems discussed above, especially with SAPK signalling and the antioxidant stress response. One of the interactions is due to the fact that after induction by metals MTF must still be activated by kinase activity (Zhang *et al.* 2001). Yu *et al.* (1997) showed that in mammalian cell lines metallothionein induction by metals can be suppressed by protein kinase C inhibitors, suggesting that MTF must be phosphorylated by stress signalling. In addition, sequencing of metallothionein promoters has revealed that they contain not only metal-responsive elements, but also antioxidant responsive elements. This explains why metallothionein is also induced by oxidative stress. A model of metallothionein induction in mammals, involving both MTF and antioxidant signalling, is represented in Fig. 6.7.

Finally, we note that the metal-scavenging function of metallothionein should be supplemented by other metal-handling systems, if it is to make any sense in the metabolism of the cell. Exactly how excess metal is removed after being bound by metallothionein is not completely known. One of the mechanisms suggested is that metallothionein donates excess metals to vesicles of the lysosomal system. Especially in invertebrates, tissues with an intestinal and hepatic function are full of these vesicles, which in electron microscope images become visible as electron-dense granules (Hopkin 1989). These granules may be removed from the cell by exocytosis or apoptosis of the cell. This scheme is supported by Liao *et al.* (2002), who identified a cadmium-responsive gene (*Cdr-1*) in the genome of *C. elegans*, encoding a putative metal pump associated with the lysosomal membrane. A remarkable feature of this gene was that it was cadmium-specific and not induced by oxidative stress.

6.2.5 The mixed-function oxygenase system

Aromatic endogenous compounds such as steroid hormones, signalling molecules, vitamins, feeding deterrents, and anti-herbivore toxins are all metabolized in the cell by enzyme-mediated reactions covered under the general term *biotransformation*. Also, compounds coming from the environment, such as plant toxins, environmental pollutants, and drugs are subject to biotransformation. One of the most important enzyme families participating in these reactions is known by the name of the *cytochrome P450s*. The genes encoding these enzymes, which are designated with the prefix *Cyp*, can be found in all genomes, including those of most prokaryotes. About 200 different families of *Cyp* genes have been described; they represent probably the most diverse superfamily of enzyme systems known. Their evolutionary success story is due to frequent gene duplication, as well as gene conversion, genome duplication, gene loss, and

lateral transfer (Werck-Reichhart and Feyereisen 2000; Werck-Reichhart *et al.* 2002). Comparative genomic analysis has shown that cytochrome P450 diversification has occurred independently in different lines (Ranson *et al.* 2002). Many species have several tens of *Cyp* genes (*D. melanogaster* has 84 and *C. elegans* has 74), but their number has exploded into the hundreds in plants (*A. thaliana* has 249 and *O. sativa* has 323). The extreme diversity in plants is thought to result from the increased need for versatile defence mechanisms in sessile organisms, which cannot avoid stress factors by moving away. A complicated nomenclature is used for distinguishing the various cytochrome P450 families and higher groupings, designated as clans (http://drnelson.utmem.edu/CytochromeP450.html).

In eukaryotes cytochrome P450 is anchored in the membrane of the smooth endoplasmic reticulum, which can be isolated by differential centrifugation as a fraction containing microsomal vesicles; that is why P450 activity is also called *microsomal monooxygenase*. Other designations are aryl hydrocarbon hydroxylase, aromatase, and mixed-function oxidase. The latter term obviously relates to the huge diversity of substrates that can be attacked by cytochrome P450 enzymes. Most of the biotransformation reactions are oxidations introducing a hydroxyl group on to an aromatic ring, according to the following overall reaction:

$$RH + O_2 + NADPH + H^+$$
$$\rightarrow ROH + H_2O + NADP^+$$

where R is any substrate. This scheme shows that molecular oxygen is split into two oxygen atoms, one of which is introduced into the substrate and the other of which reacts with two hydrogen atoms to form water. The reactive centre, which is responsible for the binding of oxygen, contains an iron atom in a porphyrin ring structure. The introduction of a hydroxyl group on to an aromatic ring makes the substrate more polar and therefore more water-soluble, which is important if it has to be excreted.

Cytochrome P450 usually conducts its biotransformation reactions in conjunction with conjugation enzymes such as glucuronyl transferase, sulphotransferase, and glutathione S-transferase. These enzymes transfer a polar endogenous compound (glucuronic acid, sulphate, or glutathione, respectively) to the product of the oxidation catalysed by cytochrome P450. Because of the sequence of events, the initial oxidation by cytochrome P450 is called *phase I biotransformation*, and the conjugations thereafter are called *phase II biotransformations*. Interestingly, the phase II enzymes seem to have diverged to nearly the same degree as cytochrome P450; for example the *Drosophila* genome has 27 annotated genes encoding a glutathione S-transferase.

The importance of biotransformation reactions is well recognized in ecological biochemistry (see Harborne 1997). In plants no less than 15–25% of protein-encoding genes may be involved with biotransformation reactions of secondary metabolism. The evolution of novel secondary compounds presents a fascinating illustration of how metabolic diversity may be generated by gene duplication, repeated evolution, and convergence (Pichersky and Gang 2000; Wittstock and Gershenzon 2002). In animals, biotransformation is often deployed to detoxify plant toxins and excrete them in water-soluble form. Some specialist herbivores sequester plant-derived toxins in their bodies to support their own defence against predators; others metabolize them to reproductive pheromones (Nishida 2002). It is often assumed that the great diversity of plant secondary compounds to which herbivores are exposed was an important selective force in the evolution of biotransformation mechanisms. In pharmacology and toxicology biotransformation is studied intensively because it determines the half-life of drugs and environmental pollutants in the body.

In the context of the stress response we must pay special attention to biotransformation reactions directed towards plant toxins and xenobiotics. Such reactions are usually considered as detoxifications, which increase the water solubility of the compound and assist its elimination. However, depending on the chemical, phase I reactions may give rise to intermediate compounds that are more reactive than the parent compound. Such compounds may react with macromolecules (DNA,

proteins) before phase II metabolism can detoxify them. This process is called *bioactivation*. A well-known example is the activation of certain polycyclic aromatic hydrocarbons (a group of chemicals occurring in crude oil, diesel exhaust, and tar) to very reactive intermediates that are highly mutagenic and carcinogenic. The production of such very reactive compounds with obvious negative metabolic effects can be seen as an unavoidable evolutionary trade-off against the capacity to metabolize toxins and xenobiotics. In addition, with some substrates cytochrome P450 produces large amounts of ROS as a by-product. These two negative side effects of biotransformation may explain why upregulated monooxygenase is often accompanied by upregulation of antioxidant enzymes and heat-shock proteins. Fig. 6.8 provides an overview of the different possibilities for the fate of a foreign compound that undergoes biotransformation.

Many biotransformation enzymes are greatly inducible; that is, their activity can increase by several orders of magnitude when they are exposed to certain chemical compounds. However, not all isozymes of cytochrome P450 are inducible to the same degree and not all chemical compounds induce the same set of enzymes. Traditionally, toxicologists have made a discrimination between two types of inducer: *phenobarbital-type inducers* (PB-type inducers) and *3-methylcholanthrene-type inducers* (3MC-type inducers). The two compounds, phenobarbital and 3-methylcholanthrene, are used as model substrates. The distinction is important because only the 3MC-type inducers act according to a mechanism of receptor-mediated transcriptional regulation. The common property of 3MC-type inducers is a planar molecular structure, such as is present in certain dioxins, certain polychlorinated biphenyls and polycyclic aromatic hydrocarbons such as benzo(*a*)pyrene and 3-methylcholanthrene itself. 3MC-type more than PB-type induction is associated with cytotoxic effects. The most potent inducer in this class, and at the same time the most toxic anthropogenic chemical, is 2,3,7,8-tetrachlorodibenzo(para)dioxin (TCDD), a compound which arises as a by-product of the manufacture of chlorinated pesticides and other organochlorines.

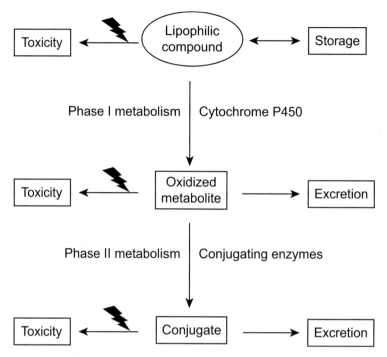

Figure 6.8 Overview of the fate of lipophilic compounds subjected to two phases of biotransformation. Phase I metabolism is conducted by cytochrome P450; phase II metabolism consists of conjugation enzymes such as sulphotransferase (transferring a sulphate moiety to the activated product of phase I metabolism), uridinediphosphate glucuronyl transferase (transferring glucuronic acid), and glutathione S-transferase (transferring glutathione). In several cases the product of phase I metabolism is very reactive and more toxic than the original compound. This also happens sometimes with phase II reaction products.

Induction of cytochrome P450, especially the genes known as *Cyp1a1* and *Cyp1a2*, is initiated by binding of the inducer with a cytosolic receptor, the *aryl hydrocarbon receptor* (Ah receptor). In the uninduced state this receptor is stabilized by heat-shock protein Hsp83 (Hsp90 in mammals). This role of Hsp83 is very similar to its stabilization of the heat-shock factor HSF and the steroid hormone receptor (see Fig. 6.4 and Table 6.2). If a 3MC-type inducer binds to the Ah receptor, the protein is activated and can translocate to the nucleus, where it is phosphorylated by protein kinase C and forms a transcriptional activator complex with another protein, Ah receptor nuclear translocator (also known as ARNT). The complex then binds to sequences known as *xenobiotic-responsive elements*. The elements are also called dioxin-responsive elements because of the use of dioxin as a model compound; the term Ah receptor elements is also used. Such sequences are present in the promoters of both phase I and phase II genes. Genes activated by the Ah receptor are jointly referred to as the *Ah battery* (Nebert *et al.* 2000). The group involves at least two P450 genes (*Cyp1a1* and *Cyp1a2*) and four genes involved with phase II biotransformation and the antioxidant stress response. Interestingly, the promoters of phase II biotransformation enzyme genes in the Ah battery contain not only xenobiotic-responsive elements but also AREs.

There is a close link between xenobiotic biotransformation and oxidative stress (Nebert *et al.* 2000; Kong *et al.* 2001; Fig. 6.9). Some xenobiotics such as dioxins and polychlorinated biphenyls are very potent inducers of *Cyp1a1*, but are themselves hardly metabolized by the cytochrome P450 enzymes. Instead, upregulated enzyme activity generates a lot of ROS and induces prolonged oxidative stress. In addition, some metabolites generated by P450 activity are very *electrophilic*, which means that they react easily with other compounds to compensate their shortage of electrons. Electrophiles and oxygen radicals induce antioxidant enzymes by the mechanisms discussed above. The presence of AREs in the promoters of phase II biotransformation enzymes ensures that these genes are also induced. The chronic toxicity of compounds such as dioxin is ascribed to a situation of sustained oxidative stress in the whole organism.

Compounds of the PB-type predominantly induce cytochrome P450s of the IIB group and to a certain extent also members of the III family, which are normally induced by steroid hormones. PB-type induction is not as specific as 3MC-induction and

it does not depend on the Ah receptor. The precise mechanism is not known. One possibility is that PB-type inducers activate cytochrome P450 by binding to a cytosolic repressor, causing derepression of *Cyp* genes. However, the great structural variety of PB-type compounds makes it unlikely that this mechanisms holds for all inducers. Another possibility is that BP-type inducers introduce a change in redox state, upon which SAPK signalling is triggered, leading indirectly to transcriptional upregulation of *Cyp* genes. A comprehensive summary of the various pathways that may lead to induction of biotransformation activities, based on Nebert *et al.* (2000) and Korsloot *et al.* (2004), is given in Fig. 6.9.

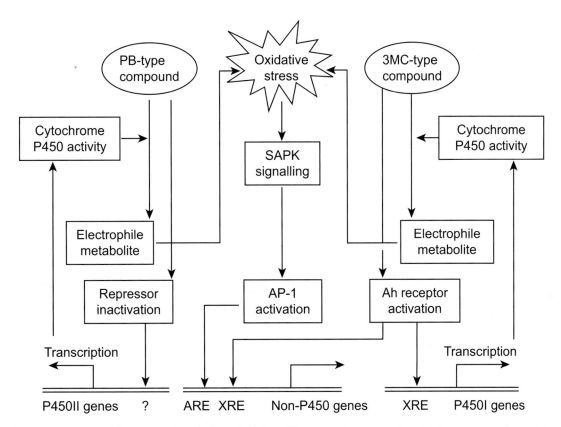

Figure 6.9 Summary of the various pathways leading to induction of biotransformation enzymes. Three inducing agents can trigger activity, PB-type inducers, oxidative stress, and 3MC-type inducers. The first type of inducer presumably activates transcription of cytochrome P450 II genes by binding to a repressor. The latter type of substance induces cytochrome P450 I genes via activation of the Ah receptor, which binds to a xenobiotic-responsive element (XRE) in the promoters of P450 I genes. In addition, P450 activity generates electrophilic metabolites and ROS and this may, through SAPK signalling, induce antioxidant genes and non-P450 genes from the Ah battery.

6.3 Heat, cold, drought, salt, and hypoxia

Most of the genome-wide stress analyses concern model organisms exposed to stress factors in the laboratory. The aim of such studies is usually to reveal biochemical regulatory mechanisms in the cell or to identify genes that act as targets of signalling cascades. The most extensively studied organism in this respect is baker's yeast, *S. cerevisiae*. There is also a lot of work on stress responses in mammalian cell lines conducted in the context of tumor cell biology. Hardly ever are cellular stress studies aimed at explaining the performance of organisms in the wild. It must be noted, however, that genomic studies in ecologically relevant systems are only just beginning. It is to be expected that essential insights obtained from the work on model organisms can be translated to ecologically relevant contexts, since, as we have seen above, several aspects of the stress response are conserved over large parts of the tree of life.

In this section we address a number of physical factors of the environment that elicit stress responses when they attain extreme values. Among these factors, temperature stands out as a major determinant of the niche, because the great majority of species—prokaryotes, ectothermic animals, and plants—cannot regulate the temperature of their internal environment. In these organisms the rate of all metabolic processes is ultimately determined by the ambient temperature. The importance of temperature is easily demonstrated by the fact that numerous species exhibit a distribution range that is bounded by some aspect of temperature, for example a winter frost isotherm. Other major conditions that determine niche boundaries are humidity, salinity, oxygen tension, and redox potential. Responses to drought have received a lot of attention in plant studies, because water deficit is often the most severe limiting factor for crop productivity. In this section we will explore what kind of stress responses are triggered by abiotic factors at the niche edge.

6.3.1 Responses to abiotic stress factors in yeast

S. cerevisiae has been called the vanguard of a truly integrative biology, because with its limited number of protein-encoding genes (around 6000), the availability of mutants deleted for each of these genes, and the early development of tools such as microarrays, it seemed possible to capture all interactions, including the transcriptome, proteome, and metabolome, into an integrative approach of the living cell. Stress-response studies are an important part of yeast integrative biology, because by removing the cell from its normal operating range the various compensatory and regulatory mechanisms are forced to reveal themselves.

Two pioneering studies of whole-genome responses to stress in yeast were published at around the same time: Gasch *et al.* (2000) and Causton *et al.* (2001). These authors studied the transcriptome of yeast under a variety of stress factors: temperature shocks, chemicals generating ROS, osmotic shock, and nutrient depletion. The profiles were studied in time and for some agents as a function of the dose. Interestingly, no two expression programmes were precisely the same in terms of the genes affected, the magnitude of expression alteration, and the changes in time. Gasch *et al.* (2000) introduced the term *choreography of expression* to describe the sequence of events occurring after a specific stimulus. Each stress factor seemed to trigger its own choreography; the uniqueness of each programme highlights the precision by which yeast responds to environmental change.

Despite the specific choreographies, the studies also showed that a large fraction of the yeast genome responded to stress in a stereotypical manner. Gasch *et al.* (2000) identified two clusters, one upregulated and one downregulated, of ~900 genes in total—more than 14% of the yeast genome—which demonstrated similar responses across the various stress factors. These genes together were designated as the *environmental stress response*. In a similar manner, Causton *et al.* (2001) identified 499 genes, corresponding to ~10%

of the yeast genome, which were common to most of the transcriptional changes observed when yeast cells were exposed to a number of different environmental changes. They called these genes the *common environmental response* (CER). Table 6.4 lists the functional categories to which the genes of the environmental stress response belong.

The genes of the environmental stress response fit into a syndrome of changed priority from protein synthesis to protective mechanisms. Almost all repressed genes have something to do with translation at the ribosomes, whereas many upregulated genes relate to stress-defence mechanisms, such as scavenging of ROS, antioxidant defence, and repair of aberrant proteins. A major fraction of the upregulated genes is due to heat-shock proteins and other stress proteins discussed in Section 6.2. Another group of upregulated genes with a less obvious stress-defence function involved enzymes of sugar metabolism, especially in the pathways of trehalose and glycogen. The genome-wide regulation of carbohydrate metabolism under stress seems to be specific to yeast and may reflect the pervasive importance of carbohydrates in the natural environment of yeast. Interestingly, both synthetic and catabolic enzymes of carbohydrate metabolism were upregulated under stress. Gasch *et al.* (2000) suggest that these apparently conflicting functions may reflect the need for the cell to increase its capacity for regulated flux of carbohydrates so as to rapidly buffer energy reserves and manage osmotic instability.

The coherent induction and repression of genes belonging to the environmental stress response would suggest that they are all regulated by a single master process. Earlier research had suggested a key role for the yeast transcription factors Msn2p and Msn4p. These important transcription factors are associated with changes of nutritional state and diauxic shift; they regulate many genes related to carbohydrate metabolism, but are also involved in the stress response. Translocation of Msn2p to the nucleus under the influence of SAPK and other signalling pathways is a characteristic feature of the yeast stress response (Görner *et al.* 1998). Msn2p and Msn4p exert transcriptional control of stress-responsive genes by binding to a specific *stress-response element* in the promoters of these genes. However, the genomic work by Gasch *et al.* (2000) showed that not all stress responses were dependent on these transcription factors. For example, genes of the thioredoxin cluster were induced in *msn2; msn4* mutants to the same degree as in the wild type, suggesting that there must be alternative regulators of environmental-stress-response gene expression. In fact, it seems to be more a rule than an exception that genes are regulated by more than one transcription factor, depending on specific environmental conditions. A possible additional group of stress-responsive gene regulators are the *yeast activator protein* (Yap) factors. There are eight *Yap* genes in the yeast genome and for five of them a function in some aspect of the stress response has been established (Rodrigues-Pusada *et al.* 2004).

Table 6.4 List of functional categories containing genes regulated in the environmental stress response of yeast

Functional categories of genes repressed	Functional categories of genes induced
Growth-related processes	Carbohydrate metabolism
RNA processing, RNA splicing	Cellular redox reactions and antioxidant defence
Translation initiation and elongation	Protein folding
Nucleotide synthesis	Protein degradation and vacuolar functions
Secretion	DNA-damage repair
Ribosomal proteins	Intracellular signaling

Source: From Gasch *et al.* (2000).

Numerous other studies have been published on genome-wide responses to stress in yeast (see Gasch and Werner-Washburne 2002 for an overview); however, since their focus was mainly biochemical, discussing them all would lead us too far away from the ecological focus of this chapter. From the sample studies discussed above we may conclude that yeast cells respond to environmental stress factors by means of a number of essentially independent pathways that are integrated in an overall genomic expression programme. When cells are exposed to two or more stress factors simultaneously the resulting expression programme largely approximates the sum of each individual stress response. Coherence is brought about by plenty of crosstalk between the various pathways and the presence of different transcription factor-binding sites in the promoters of stress-responsive genes. It is also obvious from the yeast studies that stress responses elicited by sudden changes of abiotic conditions are essentially transient. The remodelling of the transcriptome reflects an adaptation phase in which the cell adjusts its metabolic machinery to the new conditions. Transcript levels turn back to normal even when the stress factor persists. In accordance with this model, the time over which the genome shows altered transcription is correlated with the seriousness of the disturbance. Finally, the discovery of a common set of genes (the environmental stress response) being involved with defence against a variety of physical and chemical stressors is an important lesson that is equally applicable to other organisms.

6.3.2 Plant responses to drought, cold, and salt

Plants have a remarkable ability to cope with environmental stress factors, including extremes of temperature, humidity, and salinity. The genomic responses to such stresses are of potential importance to agriculture, because a better understanding of abiotic stress tolerance may improve the basis for breeding of crop plants. In some parts of the world cultivation of stress-resistant crops under marginal conditions is the only way to increase food production. Whereas people have selected plant species for 10 000 years to grow under a variety of climatic conditions, breeding for stress tolerance has proven difficult because the traits involved are determined by multiple genes. Using a genome-wide approach it might be possible to identify factors upstream in a stress signalling cascade and so increase the likelihood that determinants of genome-wide stress tolerance can be identified, manipulated, and possibly introduced into crop plants. It may be assumed that the fundamental aspects of stress tolerance are present in all plants, but what distinguishes species, it seems, is how fast and how persistent the stress-tolerance machinery is engaged (Bohnert *et al.* 2001).

Extremes of temperature, humidity, and salinity are most often studied in plant stress-response studies, but a shortage of nutrients (nitrate, sulphate, etc.) is an equally important factor that may limit the ecological niche of a species. Responses to nutritional stress are, however, more specific than responses to physical stress and the two hardly interact with one another. Therefore we leave nutritional stress out of consideration here. A recent thorough overview of the various molecular aspects of plant responses to drought and salt stress is given by Bartels and Sunkar (2005).

Genomic studies of *Arabidopsis* have shown that there is a great deal of commonality in the responses to drought, cold, and salt. Seki *et al.* (2001, 2002) developed a cDNA microarray of *Arabidopsis* genes and monitored the expression of 7000 genes when plants were desiccated, exposed to 4 °C, or grown in hydroponic solution with 250 mM NaCl. In total, 277 genes were upregulated more than 5-fold by drought, 53 were cold-inducible, and 194 were induced by high salinity (Fig. 6.10). Among these genes, 22 responded to all three stress factors. A large number of genes overlapped between drought and salinity, and fewer between either of these factors and cold. The fact that the greatest number of genes is induced by drought suggests that tolerance to drought requires the largest transcriptional alteration in plant cells and that water deficit may be considered the most severe limiting factor of

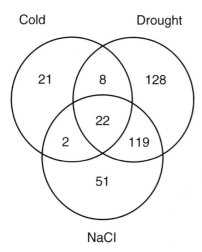

Cold   Drought

21   8   128

22

2   119

51

NaCl

Figure 6.10 Venn diagram of gene expression in *A. thaliana* Columbia exposed to three stress factors: cold, drought, and high salinity (NaCl). In each intersection the number of genes is specified whose expression ratio showed more than 5-fold upregulation compared to unstressed plants. From Seki *et al.* (2002), by permission of Blackwell Science.

plant growth. However, as Seki *et al.* (2002) admit, this result may also be due to the intensity of the stress factors applied; dose-dependence of gene expression was not investigated in this study.

The various *Arabidopsis* genes induced by drought, cold, and high salinity may be classified into two functional groups. The first group includes proteins that play a direct role in combating stress. These genes include heat-shock proteins, osmoprotectants, water-channel proteins, sugar transporters, and potassium transporters. Each of these proteins is targeted to solve a specific aspect of the stress condition. For example, KIN proteins induced by cold have a unique ability to prevent freezing of fluids by neutralizing ice nucleators. Aquaporins (members of a larger family of major intrinsic proteins, MIPs) regulate the flux of water across the membrane. A group of proteins called late embryogenesis-abundant (LEA) proteins have a role similar to heat-shock proteins and protect macromolecules from denaturation. Table 6.5 provides an overview of the various upregulated genes involved directly in stress tolerance. The second group of drought-, cold-, and high-salinity regulated genes contains mainly regulatory proteins.

These are transcription factors, protein kinases, protein phosphatases, and genes associated with plant hormones and signalling molecules. No fewer than 40 genes, which is 11% of all stress-regulated genes in the study of Seki *et al.* (2002), encoded transcription factors. In a similar study Chen *et al.* (2002) found 57 transcription factors in the *Arabidopsis* genome to be regulated by one or more stress factors (cold, salt, wounding, pathogens). Both studies illustrate the importance of transcriptional control in the tolerance to stress.

One specific group of stress-inducible transcription factors in plants is the *DREB family*, which belongs to the larger group of AP-2/ERF-type transcription factors. DREB proteins bind to the so-called *dehydration-responsive element* (DRE), a 9-bp conserved DNA sequence which is found in promoters of several stress-responsive plant genes (see Table 6.1). The element is assumed to exert an important *cis*-acting influence on stress-responsive gene expression. DREB1s are involved in cold-responsive gene expression, whereas DREB2s are associated mainly with drought. Altering the expression of such master genes can be a simple lever for increased stress tolerance. For example, overexpression of DREB1A in transgenic *Arabidopsis* results in enhanced tolerance to drought, cold, and salt (Kasuga *et al.* 1999).

Another control mechanism regulating the stress response in plants goes via the plant hormone *abscisic acid* (ABA). ABA is an important stress-responsive plant hormone that triggers a signalling pathway ultimately converging on a *cis*-acting DNA sequence known as *abscisic-acid-responsive element* (ABRE). Transcription factors binding to ABREs belong to the large group of *basic leucine zipper* (bZIP) proteins. The seemingly awkward name derives from a DNA-binding domain rich in basic residues adjacent to a leucine zipper domain which supports dimerization of the protein (required for DNA binding). Among the 22 genes that were upregulated by all three stress factors in the study of Seki *et al.* (2002), 16 contained a DRE in their promoter and 15 contained an ABRE, illustrating the importance of both abscisic acid-dependent and -independent gene regulation. The presence of different transcription

Table 6.5 List of gene categories associated with genomic responses to drought, cold, or high salinity in *A. thaliana*

Group of genes	Functional significance
Proteins involved directly with stress tolerance	
Late embryogenesis-abundant (LEA) proteins and heat-shock proteins	Protect macromolecules from denaturation
Cold-inducible (KIN) proteins	Inhibit ice-crystal growth and neutralize ice nucleators
Osmoprotectant biosynthesis-related proteins	Production of sugar and proline as osmolytes protecting cells from dehydration
Carbohydrate metabolism-related proteins and sugar transporters	Transport of sugars through plasma membrane and tonoplast to adjust osmotic pressure
Water-channel proteins (aquaporins)	Regulate water flux over plasma membrane and tonoplast in relation to osmotic homeostasis
Potassium transporters	Control potassium and sodium uptake in relation to salinity tolerance
Detoxification enzymes	Protection against ROS and electrophiles
Proteases, protease inhibitors, and senescence-related genes	Increase protein turnover and availability of amino acids; accelerate leaf scenescence
Ferritin	Protects cells from iron-catalysed oxidative damage through the Fenton reaction
Lipid-transfer proteins	Repair of stress-induced membrane damage, alteration of lipid composition of membranes
Regulatory proteins	
Transcription factors	40 different genes encoding DNA-binding proteins regulating stress-inducible genes
Protein kinases and protein phosphatases	Transducing stress signals and regulating stress-inducible genes
Plant hormone-related genes	Biosynthesis, regulation, and action of ethylene, jasmonic acid, and auxin

Notes: Two main categories are shown; one group that is involved in direct combating of stress, another that is involved in signal transduction and transcriptional regulation.
Source: After Seki *et al.* (2002).

factor-binding sites in promoters of the same gene may explain the partial overlap between transcriptional profiles of different stress factors.

In section 6.2 it was mentioned that the expression of several stress-related genes (e.g. metallothionein) shows a high degree of tissue-specificity. This is also applicable to the transcription profiles of plants exposed to abiotic stress factors. In particular, roots and leaves may show diverging transcriptional profiles. Kreps *et al.* (2002) monitored expression of 8100 *Arabidopsis* genes using an oligonucleotide gene chip and identified 2409 genes with a greater than 2-fold change over controls when plants were exposed to salt, osmotic, and cold stress. However, expression of many genes was specific for either roots or leaves, especially in the response to cold. Less than 14% of the cold-specific changes were shared between roots and leaves. Some transcripts had different tissue-specific temporal dynamics; for example, a gene was expressed initially in both roots and leaves but the transcript in the root disappeared quickly and only in the leaves was it observed as a sustained and consistent change. The importance of time-specific responses was also underlined in a study by Kawasaki *et al.* (2001) on salt stress in rice: genes for general stress defence were induced within 15 min, and most genes reached a peak after 1 h, but some only subsided after 7 days.

In the real life of plants, abiotic stress factors often occur simultaneously. This is especially valid for drought and heat in semi-arid or desert environments. The question is, can the common environmental stress response provide protection against two such factors at the same time? Interestingly, this does not seem to be the case, at least not for heat and drought in plants. Rizhsky *et al.* (2004) showed that to combat both heat and drought, plants deploy a partial combination of two multigene defence pathways, plus an additional 454 genes that are expressed specifically during a combination of drought and heat. As shown in Fig. 6.11, there is actually very little similarity between the responses of *Arabidopsis* to drought and heat. Only 29 genes were found to overlap. The largest overlap was between the responses to heat and a combination of heat and drought. This suggests that large portions of the defence programme against heat are also turned

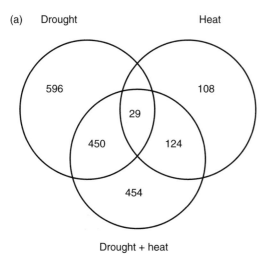

(a)   Drought   Heat

596   108
                        29
            450   124
                        454

Drought + heat

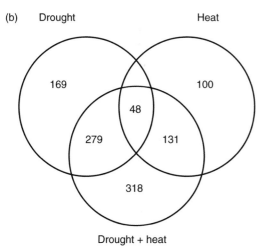

(b)   Drought   Heat

169   100
                        48
            279   131
                        318

Drought + heat

Figure 6.11 Venn diagrams showing genes regulated by drought (decrease over 6 days of plant water content to 70–75%), heat (38°C for 6 h), and a combination of the two treatments in *A. thaliana* Columbia analysed using an oligonucleotide gene chip. (a) Upregulated genes, (b) Downregulated genes. After Rizhsky *et al.* (2004). Copyright American Society of Plant Biologists.

out in defence against drought, but in addition new genes are activated to deal with the combination. The combined response is characterized by enhanced respiration, suppressed photosynthesis and accumulation of sucrose and other sugars. It was particularly striking that the amino acid proline, which is a common osmoprotectant accumulating under drought stress, was not accumulated when plants were exposed to both drought and heat. This suggests that the combination of drought and heat imposes a different kind of stress to plant cells compared to drought alone. Perhaps proline is avoided and sucrose favoured because heat ameliorates the toxicity of drought-induced proline.

The study of Rizhsky *et al.* (2004) suggests that there is an element of 'collision' in the defence pathways to different stress factors. The presence of antagonism in gene-expression programmes was also suggested by Tamaoki *et al.* (2003). These authors investigated the role of three plant hormones—ethylene, jasmonic acid, and salicylic acid—in the regulation of gene expression of *Arabidopsis* exposed to ozone. By studying the stress responses of mutants disturbed in each of the hormone signalling pathways, it became obvious that ethylene regulated the ozone response of 73 genes, jasmonic acid regulated 62 genes, and salicylic acid regulated 24 genes; however, there was a considerable over lap between these genes. Many defence genes induced by ethylene and jasmonic acid signals were suppressed by salicylic acid signalling, suggesting that the salicylic acid pathway acts as an antagonist to the other two pathways. Such interactions in gene-expression programmes of plants are in contrast with the additive nature of stress responses assumed for yeast (Gasch and Werner-Washburne 2002; see above).

Although most of the mechanistic knowledge on stress tolerance comes on the account of *A. thaliana*, plant biologists are also very interested, for obvious reasons, in the stress responses of crop species such as rice and barley (Kawasaki *et al.* 2001; Ozturk *et al.* 2002). In addition, the study of species growing naturally under extreme conditions might add insights that cannot be attained from mesophilic plants. Models for the study of halotolerance are the ice plant, *Mesembryanthemum crystallinum*, and the green alga *Dunaliella salina* (Cushman and Bohnert 2000; Bohnert *et al.* 2001). The resurrection plant, *Craterostigma plantagineum*, which shows a remarkable ability to restrict cell damage during desiccation and rehydration of its tissues, is a promising model for xerotolerance. Gene-discovery programmes in such naturally tolerant models have focused on EST sequencing of stressed and unstressed libraries. Such studies have demonstrated that ESTs related to stress are under-represented in the current genomic databases, which suggests that there may still be unknown mechanisms involved in plant stress tolerance. Comparative genomics of the type discussed in the context of aging in Section 5.2 has an important role to play here. Comparisons among stress-tolerant species from different evolutionary lineages may help to identify the universal gene complement underlying stress tolerance in plants.

6.3.3 Abiotic stress responses in animals

In addition to yeast, fruit flies have been widely used as a model for studying genome-wide responses to elevated temperature. As expected, supraoptimal temperature induces heat-shock proteins, but it also causes a great variety of other changes in the transcriptome. As an example, consider the study of Leemans *et al.* (2000). These authors found that 74 genes were affected significantly in *D. melanogaster* embryos exposed to a mild heat shock (36 °C for 25 min). Among these 74 genes, 36 had increased and 38 had decreased expression levels (Fig. 6.12). A very strong induction was seen for the small heat-shock proteins *Hsp22*, *Hsp26*, *Hsp27*, and *Hsp23*; this induction was many times greater than the fold regulation found for other genes. No induction was seen for the heat-shock cognate proteins *Hsc70-1*, *Hsc70-4*, and *Hsc70-5*; however, a small degree of upregulation was noted for *Hsc70-3*. Two signal transduction genes were also upregulated, *Shark* and *Rabgap1*. *Shark* encodes a protein kinase involved in the JNK cascade and *Rabgap1* is a Ras GTPase activator; both proteins are part of signalling

transduction pathways regulating cell growth (see Fig. 6.3). Finally, many changes were seen in genes encoding proteins involved with transcriptional regulation and metabolism, and these changes included both up- and downregulation (Fig. 6.12).

The study of Leemans *et al.* (2000) illustrates that the transcriptional change induced by heat shock may be more complicated than suggested by the biochemical work discussed in Section 6.2. Although the strong upregulation of heat-shock proteins is in accordance with the earlier theory, the many changes seen in other functional categories suggest that there is no simple downregulation of overall protein synthesis, but a more complicated adjustment to the metabolic needs of the cell. Some of the responses documented by Leemans *et al.* (2000) may be specific to the life stage (embryos); for example, the abundance of small heat-shock proteins, which are known to be developmentally expressed. Also we must remember that the heat-shock response involves post-transcriptional regulatory mechanisms, so not all transcriptional changes depicted in Fig. 6.12 need to be expressed at the protein level.

Although the heat-shock response has become the most widely used model for studies on stress-induced transcriptional change, other stress responses may sometimes be more closely linked to an environmental context. An example illustrating this point is the response to *hypoxia* (low oxygen levels) in fish. Dissolved oxygen is a limiting factor for many fast-swimming, active fish species. Fish species inhabiting sediment burrows in estuarine ecosystems are particularly tolerant of hypoxic conditions and the study of such naturally tolerant species is expected to shed light on cellular responses to hypoxia in general.

From mammalian research it is known that gene expression induced by hypoxia is controlled by *hypoxia-inducible factor 1* (HIF-1) in a manner which is very comparable to the action of Nrf2 in the oxidative stress response (see Section 6.2). HIF-1 comes in two subunits, HIF-1α and HIF-1β, which are both expressed constitutively at high levels (Wenger 2002; Schulte 2004). HIF-1β is identical to the Ah receptor nuclear translocator, a protein that we met in Section 6.2 as a dimerization partner of

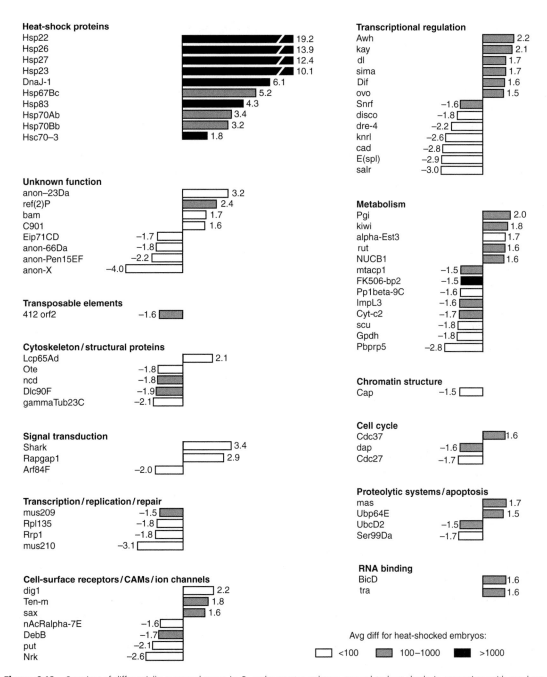

Figure 6.12 Overview of differentially expressed genes in *D. melanogaster* embryos exposed to heat shock, in comparison with non-heat shocked embryos. The genes are grouped according to functional class; bars represent fold regulation. CAM, cell-adhesion molecules. In the key, the darkness of the bar indicates the extent to which expression in heat-shocked embryos differs (on average) from the standard condition. Avg Diff is the ratio of the two expressions. After Leemans *et al.* (2000) by permission of the National Academy of Sciences of the United States of America.

the Ah receptor. Bound to HIF-1α, it promotes translocation of HIF-1 to the nucleus. However, under normal physiological conditions HIF-1α is degraded rapidly through ubiquitin-mediated cytoplasmic proteolysis. Degradation is initiated by enzymatic hydroxylation of two proline residues, a process that is sensitive to the oxygen concentration in the cell. Under low oxygen conditions prolyl hydroxylase is inhibited, with the consequence that HIF-1α is no longer degraded and the heterodimer can translocate to the nucleus. A second hypoxia-dependent regulatory mechanism is located in the nucleus. To act as a transcription factor, HIF-1α must bind to nuclear factor p300, which is prevented by another enzymatic hydroxylation, in this case of an asparagine. Under hypoxic conditions this hydroxylation is also impeded, allowing HIF-mediated transcription to occur. Interestingly, HIF-1 is also required for heat acclimation and contributes even to metal resistance (Katschinski and Glueck 2003; Treinin et al. 2003). Such cross-tolerance phenomena present another illustration of the interconnectedness between the various stress-response pathways.

In an pioneering genomics study Gracey et al. (2001) documented transcription profiles of the long-jawed mudsucker, *Gillichthys mirabilis* (Perciformes, Gobiidae), exposed to hypoxic conditions (0.8 mg/l at 15°C). SSH was used to generate cDNA libraries enriched with hypoxia-regulated genes (see Section 2.1). More than 5000 PCR-amplified cDNA clones were printed on an array, which was then used to monitor transcriptional change in liver and muscle tissue of fish under hypoxic conditions. Clones that were differentially expressed by 2.5-fold or greater were sequenced and their putative function was established by homology to database sequences. A total of 126 distinct hypoxia-regulated cDNAs were found, of which 75 could be identified by homology.

Gracey et al. (2001) were able to interpret many of the changes in gene expression in terms of an ecophysiological strategy employed by the fish to allow its survival under hypoxic conditions (Table 6.6). The transcription profile suggested a metabolic switch in which very rapidly after the onset of hypoxia the major energy-requiring processes like protein synthesis and locomotion were repressed. Then, after about 24 h, the metabolic machinery was directed towards anaerobic ATP production and synthesis of glucose from non-carbohydrate sources (*gluconeogenesis*). The changes in genes for amino acid metabolism indicate that amino acids are the main source for gluconeogenesis. At the same time, cell growth and proliferation is repressed by means of binding circulating IGFs and by attenuation of MAPK signalling.

Among the genes with unclear function was an inducible pseudogene encoding an antisense mRNA matching the 5' end of retinoblastoma-binding

Table 6.6 Summary of transcriptional change observed in long-jawed mudsucker, *G. mirabilis*, exposed to hypoxic conditions

Functional category	Tissue	Examples of genes	Possible functional significance
Energy metabolism	Liver	Lactate dehydrogenase (+), enolase (+), trisephosphate isomerase (+)	Maintenance of glucose homeostasis by gluconeogenesis
Locomotion and contraction	Muscle	α-Tropomyosin (−), myosin heavy chain (−), myosin regulatory light chain 2A (−)	Decreased locomotory activity
Translation, protein synthesis	Muscle	Elongation factor 2 (−), several ribosomal proteins (−)	Reduced protein synthesis
Iron metabolism	Liver	Haem oxygenase-1 (+), ferritin (+), transferrin (−)	Increased production of erythrocytes
Cell growth and proliferation	Liver	IGF binding protein 1 (+), MAPK phosphatase (+)	Suppression of cell growth by attenuating MAPK signalling
Amino acid metabolism	Liver	S-Adenosylmethionine synthase (+), tyrosine aminotransferase (+)	Catabolism of amino acids used for gluconeogenesis

Notes: The direction of change is indicated with + or −.
Source: After Gracey et al. (2001).

protein 2 (RBP2). An *antisense RNA* is a sequence complementary to a certain mRNA, the translation of which is suppressed by binding to the 5′ end. The production of antisense RNA in response to stress is a means to antagonize the expression of other genes, which may contribute to the fine-regulation of metabolic processes. That this type of mechanism was involved in the response to hypoxia was not known before.

The long-jawed mudsucker study is remarkable because it illustrates several of the points raised in Chapter 1 as dilemmas of ecological genomics: (i) it is possible to explore the transcriptional profile of an organism about which no sequence data existed before, (ii) cDNA microarrays can be developed from SSH libraries without prior knowledge of the genome, (iii) microarrays are useful as exploratory instruments, (iv) even with a limited sequencing effort, a great deal of insight into the ecophysiology of a species can be reconstructed, and (v) the study of species adapted to extreme conditions can reveal new insights into homeostatic mechanisms that may be relevant for many other species.

6.4 Herbivory and microbial infection

Not only abiotic conditions, but also biotic factors may limit the ecological niche and can elicit specific stress responses in plants and animals. Such biotic factors can be other organisms that decrease the fitness of plants or animals by consumptive action. Ecologists distinguish herbivores (animals consuming plants or parts of them), predators (animals consuming other animals after catching and killing them), parasites (organisms living inside or on the surface of plants or animals, diverting resources from the host to themselves), and parasitoids (animals living inside other animals but killing the host to complete the life cycle). The initial contact between the two players in such interactions is invariably accompanied by stress and specific defence responses, especially in the victim being preyed upon, being consumed, or acting as a host. In this section we will review studies dealing with stress responses associated with attack by herbivores and pathogenic

microorganisms. The importance of both processes in regulating the abundance of species in the wild stands beyond doubt; however, herbivory has received much more attention from ecologists than parasitism and disease. Genomic responses to microbial infection of plants is a major topic in plant pathology, but this is mostly studied in an agricultural context and therefore not discussed here. Predation, although a popular subject among ecologists, has not been studied at the genomic level and so is also not considered.

6.4.1 Plant defence against insect herbivory

Plants respond to herbivore attack with a wide array of defence mechanisms. One of the strategies employed is to synthesize *secondary compounds* such as alkaloids and terpenoids that are toxic to herbivores and pathogens. Such compounds are called secondary because they do not belong to the metabolism of primary cell constituents (carbohydrates, proteins, and lipids). Secondary compounds can be synthesized after initial damage by the herbivore (*damage-induced defence*), or they may be synthesized at all times (*constitutive defence*). The first strategy has the advantage that the costs associated with biosynthesis only burden the plant when actually under attack; however, the disadvantage is that the defence may not be rapid enough, or the initial damage may be too severe (Wittstock and Gershenzon 2002). The second strategy is more effective, but obviously costs of biosynthesis are incurred for as long as the plant grows. The costs of anti-herbivore defences can indeed be considerable, especially under competitive growing conditions, and so inducibility is assumed to have evolved as a cost-saving mechanism (Zavala *et al.* 2004). We will focus on damage defence induced by insects in this section.

Induced defence against herbivores can be direct or indirect. Direct responses aim at preventing further feeding of the herbivore by some kind of toxic action such as sensory irritation, gut convulsion, or paralysis. Indirect responses involve the use of volatile alarm chemicals to attract predators and parasitoids of the herbivore. There is a great deal of specificity in these responses, involving chemical communication between plants, herbivores, and natural enemies, which has led to the concept of *tritrophic interaction* (Price 1980). A better understanding of the genomics of herbivore defence by plants may thus have important ramifications outside the plant proper and benefit community ecology (Dicke *et al.* 2004).

Over the last decade, significant progress has been made in identifying the nature of volatile chemicals that are produced by plants when attacked by insects (Kessler and Baldwin 2002; Dicke *et al.* 2003). Three groups of chemicals can be distinguished: (i) so-called *green leaf volatiles*, C_6 alcohols and aldehydes which are synthesized quickly after damage and are not very specific to the plant species or the type of leaf damage, (ii) terpenoids from the octadecanoid pathway, commonly called *oxylipins*, which are emitted slowly, typically 24 h after the damage and are more specific, and (iii) derivatives of the aromatic amino acid precursor shikimate, such as *methyl salicylate*, which are emitted after herbivore damage but not after mechanical wounding. Volatile chemicals are not only released locally from damaged leaves but also from undamaged parts of the plants in distinct temporal patterns (a *systemic response*). Some constituents of the volatile releases are just a passive consequence of damage to cell compartments, notably vacuoles and trichomes; however, many are due to *de novo* biosynthesis under control of three plant hormones: ethylene, jasmonic acid, and salicylic acid. The question is, how is the signal emitted from herbivore-induced damage transduced into regulation of these biosynthetic pathways?

Studies on *Arabidopsis* provided the first cues on the nature of transcriptional changes induced by herbivory in plants. Reymond *et al.* (2000) prepared an array with 150 EST probes of genes that were known to be involved in stress defence. Temporal change of gene expression was analysed in response to mechanical wounding and feeding by cabbage white caterpillars, *Pieris rapae* (Lepidoptera, Pieridae). Mechanical wounding of the leaf induced a clearly recognizable stress response in *Arabidopsis*. Upregulation was seen for many genes from the stress-responsive pathways

discussed in Section 6.2, including several MAPKs, a metallothionein, two glutathione S-transferases, and a cytochrome P450. A cluster of 17 genes was regulated in a coherent fashion; this cluster included general stress genes as well as genes implicated in the synthesis of the plant hormone *jasmonic acid*, and known to be induced by this hormone (Fig. 6.13). Earlier research had shown that three key enzymes in the jasmonate biosynthetic pathway (lipoxygenase (LOX), hydroperoxidaselyase (HPL), and allene oxide synthase (AOS)) can be used as indicators for wounding and these three enyzmes were also responsive in the microarray study (Fig. 6.13). There was a very good correlation between expression of the inducible gene cluster and leaf concentrations of jasmonate, while the metabolic precursors of jasmonate, OPDA and dnOPDA, rose more slowly, peaking ~6 h after wounding. The observations suggest strongly that the transcriptional response to wounding is triggered by a burst of jasmonate within 1 h, followed by jasmonate synthesis.

The availability of *Arabidopsis* mutants insensitive to jasmonate allowed the role of this trigger to be assessed in more detail. About one half of the genes regulated by wounding were no longer induced or repressed in a jasmonate-insensitive mutant, demonstrating that for a significant number of genes the plant response to wounding depends strictly on the action of jasmonate. The other half of the genes responded independently of jasmonate and this group included many genes that were also activated by water stress. So it seems that the transcriptional profile of wounding includes signatures triggered by jasmonate signalling, which are specific to wounding, as well as water stress signalling, which, as we have seen above, is regulated by abscisic acid.

Further experiments by Reymond *et al.* (2000) have demonstrated that the transcriptional profile of wounding is not the same as the profile induced by herbivory. Feeding by cabbage white caterpillars induced many transcripts that were also induced by mechanical wounding, but additional genes were induced that mainly responded to feeding. Interestingly, there was an under-representation of water stress-induced

(a)

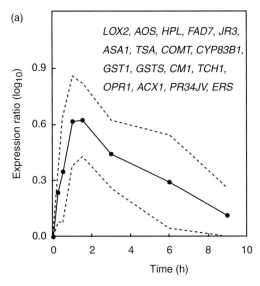

(b)

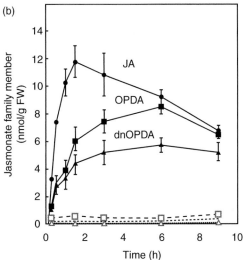

Figure 6.13 Temporal changes of gene expression in *Arabidopsis* after mechanical wounding of leaves. (a) Average expression of 17 genes with similar temporal change; dashed lines indicate the standard deviation. LOX2, lipoxygenase; AOS, allene oxide synthase; HPL, hydroperoxide lyase; FAD7, fatty acid desaturase; JR3, aminohydrolase; ASA1, anthranilate synthase α subunit; TSA, tryptophan synthase α subunit; COMT, O-methyltransferase; CYP83B1, cytochrome P450; GST1 and GST5, glutathione S-transferases; CM1, chorismate mutase; TCH1, calmodulin; OPR1, OPDA reductase; ACX1, acyl-CoA oxidase; PR3AIV, chitinase; ER5, late embryonesis-abundant (LEA)-like protein. (b) Temporal change of leaf concentrations of jasmonate (JA), 12-oxo-phytodienoic acid (OPDA) and dinor OPDA (dnOPDA). The open symbols represent control measurements without wounding. After Reymond *et al.* (2000). Copyright American Society of Plant Biologists.

genes in the herbivore-induced transcriptional profile. These observations suggest that insects in some way or another can minimize the dehydration stress experienced by the plant while damaging the leaf. Behavioural mechanisms can possibly explain these observations. Many herbivorous insects remove tissues only from the edge of the leaf and make semicircular holes without cutting the midvein. In doing so they will cause less damage than most mechanical-wounding operations.

It must be noted that transcriptomic changes following wounding and herbivory are not restricted to those triggered by jasmonate and abcissic acid; they involve many other effects that indicate a major metabolic switch, as in the case of abiotic stress responses. Transcripts related to photosynthesis and ribosomal processes are found to be downregulated by wounding, whereas expression of genes associated with protein turnover, carbohydrate metabolism, cell-wall modification, and antimicrobial defence was increased (Moran *et al.* 2002; Hui *et al.* 2003; Zhu-Salzman *et al.* 2004). The genome-wide nature of these changes suggests that in addition to jasmonic acid signalling, which is specific to wounding, other signalling pathways are also activated, for example those triggered by salicylic acid, ethylene, and abscisic acid. The changes can be summarized as a metabolic switch from growth priority to stress-defence priority.

The fact that some aspects of the transcriptional changes to herbivory are specific to insect feeding and do not occur after mechanical wounding suggests that certain chemical cues emanating from the insect are recognized by the plant. Two classes of these so-called *stress elicitors* have been identified (Kessler and Baldwin 2002; Korth 2003). The first class includes digestive enzymes present in the saliva of the herbivore, such as β-glucosidase, glucose oxidase, and alkaline phosphatase. The second class comprises amino acid conjugates of fatty acids, fatty-acid–amino acid conjugates (FACs), which are present in the digestive tract, frass, and regurgitate of insects (Schittko *et al.* 2001; Roda *et al.* 2004). The fatty acid moiety in these compounds derives from the plant itself; FACs probably represent products of phase II bio-transformation in the insect, aimed at solubilizing and excreting lipophilic compounds (see Section 6.2). Recognition of chemical cues that are specifically associated with insect feeding allows plants to fine-tune their defence to certain herbivores and to attract natural enemies of the herbivore by releasing specific volatiles.

An interesting complication is that some herbivores are able to express detoxification enzymes of the cytochrome P450 family even before the plant has been able to produce defensive toxins. Herbivores do this by responding to the jasmonate and salicylate burst of the plant. This would provide protection in the critical window between accumulation of plant defence compounds and production of enzymes that can metabolize them. This *eavesdropping* on plant defence signals is especially advantageous in polyphagous herbivores such as the noctuid *Helicoverpa zea*, for which this mechanism was first described (Li *et al*. 2002).

The interaction between herbivore-induced damage and chemical signalling has been investigated in great detail in a system consisting of wild tobacco, *Nicotiana attenuata* (Solanaceae), and its specialist herbivore, tobacco hornworm, *M. sexta* (Lepidoptera, Sphingidae; Fig. 6.14). Wild tobacco is an interesting model because it is diploid, exhibits a large amount of phenotypic plasticity, and evolved in a habitat that can be considered primordial to the agricultural niche, the environment created by wildfires in woodlands (Baldwin 2001). *N. attenuata* uses two different defence mechanisms against herbivores; the production of nicotine and the production of terpenoid volatiles. The level of *nicotine*, a potent neurotoxin, is greatly induced by mechanical wounding under the influence of jasmonate signalling. Nicotine production seems to act mainly as a direct defence against browsing mammalian herbivores. Specialist herbivores such as *M. sexta* can store nicotine in their tissues without any toxicity and may even use it to defend themselves against avian or mammalian predators. Consequently, the plant must rely on other chemical defences, such as oxylipins, to combat such specialist herbivores. Oxylipins are also assumed to support indirect

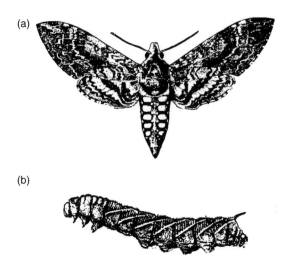

(a)

(b)

Figure 6.14 Tobacco hornworm, *M. sexta* (Lepidoptera, Sphingidae), a specialist herbivore of wild tobacco and used as a model for genomic studies of plant responses to insect herbivory. (a) Adult; (b) larva. From Gillot (1980), with permission from Springer.

defence by attracting natural enemies of the herbivore. The ecological importance of jasmonate-induced oxylipins in the defence against insect herbivores was demonstrated elegantly by Kessler *et al*. (2004). These authors used transformed lines of *N. attenuata* which were planted outdoors in an experimental field. Mutants in which key enzymes of the jasmonate pathway (LOX, HPL, and AOS) had been silenced were not only more vulnerable to damage by sphingid caterpillars but also attracted new herbivores such as leaf hoppers and a chrysomelid beetle.

Genomic investigations into the tobacco–herbivore interaction have used a microarray which was developed from cDNAs identified by pre-genomic differential screening techniques such as differential display, cDNA-AFLP, and subtractive hybridization (see Chapter 2 for a discussion of these techniques; Halitschke *et al*. 2003; Hui *et al*. 2003). Despite the small number of genes on the array (initially 241), which precludes a truly genome-wide inventory of expression change, the use of a targeted approach proved to be quite succesful. Development of small *boutique arrays*, encompassing a focused selection of cDNAs from genes relevant in a certain context, is a good strategy for ecological laboratories working with incompletely sequenced organisms (Held *et al.* 2004). Later the herbivore stress array was extended to 789 probes, represented by 50-mer oligonucleotides (Heidel and Baldwin 2004).

The studies on the tobacco herbivore community have revealed interesting patterns of herbivore-specific gene expressions (Heidel and Baldwin 2004; Voelckel *et al.* 2004; Voelckel and Baldwin 2004). Comparisons were made between transcription profiles induced by different species of chewing lepidopteran: *M. sexta* (Sphingidae), a specialist herbivore, and two generalists, *Heliothis virescens* and *Spodoptera exigua* (both Noctuidae). Chemical analysis had shown that the composition of stress elicitors (FACs) in the regurgitate of these species is varied. The *M. sexta* regurgitate is dominated by a FAC named *N*-linolenoyl-L-glutamate. This is absent in the other two species, which both have a compound known as volicitin, or *N*-(17-hydroxylinolenoyl)-L-glutamine. The reason for these specific-specific FAC profiles is unknown; maybe they relate to different substrate specificities of biotransformation enzymes of the insects' xenobiotic metabolism. The FAC profiles were correlated with herbivore-induced gene expression in the plant; there was a large overlap between the transcriptional profiles of the two noctuids, whereas the overlap between either of the noctuids and *M. sexta* was much smaller (Voelckel and Baldwin 2004).

Other differences between herbivore-elicited transcriptional profiles were observed in a comparison of diverging herbivore-feeding guilds. Chewing herbivores such as sphingid and noctuid caterpillars consume pieces of leaf tissue completely; however, mirid bugs (Heteroptera, Miridae) puncture holes in the tissue and feed on the cell contents, while aphids insert their stylet between the cells and suck only the phloem. Comparing *M. sexta*, *Tupiocoris notatus* (Heteroptera, Miridae), and *Myzus nicotianae* (Homoptera, Aphididae) Voelckel *et al.* (2004) observed that the aphids elicited only weak responses, both in a qualitative sense (fewer genes affected) and in a

quantitative sense (fold regulations were lower). An overview of the differences in terms of numbers of genes is given in Fig. 6.15. Thus the herbivore-induced transcriptomic changes could be proportional to the degree of damage caused by the herbivore. It is also relevant to note that FACs or other elicitors have never been isolated from aphids (Moran *et al.* 2002). Interestingly, Voelckel *et al.* (2004) found a plant gene encoding the enzyme glutamate synthase to be induced by aphid feeding, but not by the other two herbivores. This could indicate an alteration of the plant amino

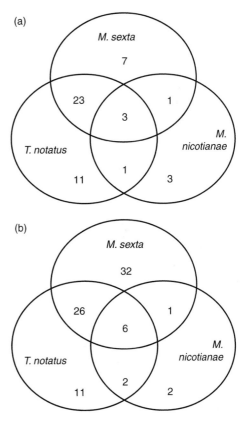

Figure 6.15 Venn diagram of the number of cDNAs showing differential expression in *N. attenuata* in response to three herbivorous insects, *M. sexta* (Lepidoptera, Sphingidae), *Tupiocoris notatus* (Heteroptera, Miridae), and *Myzus nicotianae* (Homoptera, Aphididae). The different feeding modes employed by these herbivores induce distinct but overlapping gene expression, with the largest overlap between *M. sexta* and *T. notatus*. (a) Upregulated genes; (b) downregulated genes. After Voelckel *et al.* (2004), by permission of Blackwell Science.

acid metabolism imposed by the aphid. Amino acids are a limiting resource for aphid growth and any upregulation of their concentration in the phloem would greatly benefit the aphid. These data added to an earlier study by Moran *et al.* (2002) suggest that plant responses to aphids are fundamentally different from responses to chewing herbivores. The response to aphids has less of a wounding signature and includes facilitation of the host plant by the herbivore.

Our short account of the defence mechanisms associated with plant–herbivore interactions demonstrates that there are still important gaps in our knowledge. On the mechanistic side, the source of stress elicitors in the plant-feeding insect is not known and also the plant receptor upon which these elicitors act remains to be elucidated. On the ecological side, the tritrophic aspect of herbivore defence needs more attention, as well as the way in which the different stress responses interact with different feeding groups of herbivores. These questions indicate that genomic analysis of plant–herbivore interaction represents a highly promising research area, where genomics meets ecology, with pay-offs to both sides.

6.4.2 Genomics of the immune response in *Drosophila*

In Chapter 3 we saw that the genomes of invertebrates, including the tunicate *Ci. intestinalis*, do not encode proteins of the *adaptive immune system*, which only evolved in the vertebrate lineage, to supplement the older *innate immune system*. The adaptive immune system is to be considered as a major evolutionary innovation: it is specific to particular antigens, it shows an extremely large diversity, partly inherited and partly acquired during maturation of the system, and it builds up a memory of previous antigen encounters. The latter property implies that when an antigen reacts with a clone of cells with specificity for that antigen, these clones expand greatly and adapt to give the highest possible specificity for the antigen. All these properties are lacking in the innate immune system, which constitutes a general first-line defence with low specificity.

The organization of the innate immune response of invertebrates shows many similarities with the vertebrate innate response, suggesting that they have a common origin, and that defence against microorganisms was already a priority in the first metazoans (Hoffmann and Reichhart 2002). Research on the innate immune response has greatly benefitted from the use of *Drosophila* as a model and some important molecules involved, such as the antifungal compound *drosomycin*, were first isolated from *Drosophila*. Studies in *Drosophila* can reveal aspects of the human innate immune response that may otherwise be obscured by the adaptive response (Govind and Nehm 2004). In addition, the evolutionary conservation of the innate immune system implies that the principles discovered in *Drosophila* have a general validity for all animals.

The link between between immunology and ecology is only weakly developed at the moment. Without doubt, pathogens are a very important aspect of ecological functioning, in both plants and animals. Diseases can limit the distribution of species or prevent their establishment in newly colonized habitats. Lee and Klasing (2004) therefore called for a role of immunology in invasion biology. Disease and parasitism are also potent evolutionary driving forces. Continued adaptation to parasites was implicated in Van Valen's (1973) *Red Queen hypothesis*, which holds that evolution is very much in line with the remark made by the Red Queen, whom Alice met in Lewis Carroll's famous book *Through the Looking Glass*, saying 'Here you see, it takes all the running you can do to keep in the same place'. Still, the number of studies that bridge the two fields, ecology and immunology, is limited.

An interesting target of investigation in ecological immunology is the *major histocompatibility complex* (MHC), a large cluster of genes encoding proteins involved in the adaptive immune response of vertebrates. MHC proteins bind to specific recognition sites of antigens and present them on the surface of leucocytes. A part of the molecule, a cup-like structure responsible for the recognition of a specific topographical structure of the antigenic molecule (the *epitope*), shows an extremely high degree of sequence polymorphism. The MHC genes are therefore interesting markers for resolving population structure and dispersal (Beebee and Rowe 2004). Beebee and Rowe (2004) also discuss the possibility that the striking polymorphism of the MHC complex is maintained partly by selective mate choice. Obviously, genetic variation of MHC genes may be the cause of differential susceptibility to disease, although not many studies have actually demonstrated this in an ecological context. A case in point is a study on Atlantic salmon, *Salmo salar* (Salmoniformes, Salmonidae), in which an association has been found between allele frequencies at the MHC IIB locus and susceptibility to bacterial infection (Langefors *et al.* 2001). One of the alleles had a particularly high frequency among fish resistant to the enteric pathogen *Aeromonas salmonicida* (Deltaproteobacteria).

The innate immune system, although less specific than the adaptive system, nevertheless shows considerable transcriptional change upon infection. This has become obvious from studies in *Drosophila*, which we discuss here to illustrate the genome-wide nature of immune responses and the signalling pathways involved (De Gregorio *et al.* 2001; Irving *et al.* 2001; Dionne and Schneider 2002; Govind and Nehm 2004). Microbial infection in *Drosophila* activates a number of processes. One cascade leads to blood coagulation at the site of infection followed by the production of melanin, which is toxic to microorganisms. This process of *melanization* is accompanied by encapsulation of the pathogen. Another reaction is a massive synthesis of *antimicrobial peptides* by the fat body; these peptides bind to specific surface molecules of bacteria and other invaders. Third, blood cells analogous to mammalian *macrophages* become active as microbe engulfers.

The responses of the innate immune system are controlled by two signalling pathways, *Toll* and *immune deficiency* (Imd; Fig. 6.16). The Toll protein is a membrane-bound system with an extracellular receptor that recognizes a cytokine called Spätzle. However, the Spätzle protein must first be cleaved to become active, and this is achieved by means of a proteolytic cascade originating from the

infection. Activation of Toll triggers an cytoplasmic signalling pathway converging on two transcription factors of the nuclear factor-κB family called DIF and DORSAL, which translocate to the nucleus and promote transcriptional activation of a number of antimicrobial peptides. The Toll cascade is mainly directed towards pathogenic fungi and Gram-positive bacteria and it includes the antifungal protein drosomycin. Compared to Toll, the Imd pathway is less well described and the extracellular cascades triggering the receptor remain undefined at the moment. Imd signalling may activate caspases leading to apoptosis of the cell, but it may also act upon a protein complex called IKK signalosome, where a transcription factor called RELISH is activated. RELISH promotes transcription of genes encoding peptides directed towards Gram-negative bacteria (Fig. 6.16).

Infection of *Drosophila* with the Gram-negative bacterium *E. coli*, the Gram-positive *Micrococcus luteus*, and the entomopathogenic fungus *Beauveria bassiana* (Hyphomycetes) leads to a transcriptional response in genes from many different functional classes, including actin-associated proteins, calcium-binding proteins, cell-adhesion proteins, heat-shock proteins, and many others. Also the Toll and Imd genes were induced. In total 543 genes were found to be differentially expressed (Irving *et al.* 2001), which nevertheless seems a modest number in comparison to some of the abiotic stress responses discussed above. Irving *et al.* (2001) could not find specific signatures for infection by Gram-positive bacteria, Gram-negative bacteria, or fungi; however, bacterial infection seemed to regulate a larger number of genes than fungal infection. De Gregorio *et al.* (2001) selected a set of 400 *Drosophila* genes as immune-regulated, of which 230 were induced and 170 repressed. Among these genes a large number had not previously been associated with the immune response and only 34% could be assigned a designated immunological function.

An overview of functional classification of immune-regulated genes of *Drosophila* is provided in Table 6.7. Several genes had already been identified as immune-regulated in previous studies, but many more were added to the list.

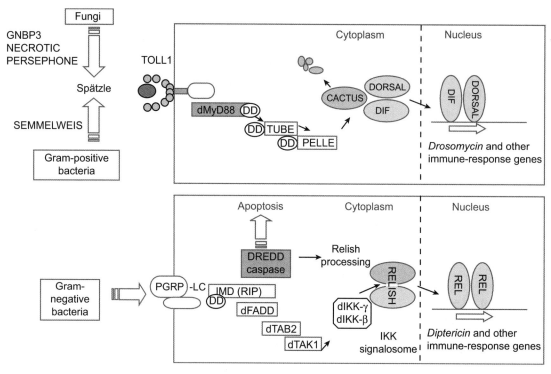

Figure 6.16 Scheme of two signal transduction pathways of the innative immune response in *Drosophila*. (a) The Toll pathway, which is characterized by a proteolytic cascade, cleavage of Spätzle, activation of transcription factors DIF and DORSAL, and production of peptides directed against fungi and Gram-positive bacteria. (b) The Imd pathway, converging on the transcription factor RELISH and the production of peptides directed against Gram-negative bacteria. DD, death domain; this is a heterodimerization domain present in several proteins involved in signal transduction, originally described for proteins involved with cell death. PGRP-LC, peptidoglycan-recognition protein LC, where LC stands for low complexity, a certain class of bacterial surface proteins. d before a protein name indicates *Drosophila*; e.g. dMyD88 is the *Drosophila* homologue of MyD88, a macrophage differentiation marker. For more information on the other protein identifiers, the reader is referred to Flybase (http://flybase.bio.indiana.edu). Reproduced from Govind and Nehm (2004).

A surprising aspect of the genomic inventory was the large number of trypsin-like serine proteases and the many antimicrobial peptides of unknown function. Maybe these induced proteins represent new classes of antimicrobial action. Another unexpected property of the genomic immune response was that it involved sequestration of extracellular iron. What the immune response has to do with iron metabolism is difficult to see, but maybe the role of iron as a catalyst of ROS is relevant in this respect. A change in cellular iron trafficking was also seen in the response of plants to abiotic stress (see Section 6.3).

The two pioneering genomic studies on the immune response of *Drosophila* leave many questions unanswered (Dionne and Schneider 2002). For example, it is unclear to what extent the responses are due to wounding alone. The plant studies discussed above have demonstrated that damaging a tissue can already activate some 50 genes. In addition, the large number of genes with unclear function in the immune response makes interpretation difficult. Further work on mutants may help to resolve these issues. It is also expected that the knowledge on innate immune responses in *Drosophila* will be an important guide to explore the immune system of vectors of human disease, such as the malaria mosquito, *An. gambiae* (Dimopoulos *et al.* 2000; Osta *et al.* 2004). In addition, the conclusion regarding the ecological relevance of immunogenomics must come from field studies assessing genome-wide immune responses in wild animals.

Table 6.7 List of functional categories of genes induced by septic injury (Gram-positive and Gram-negative bacteria) and fungal infection in *D. melanogaster*

Gene categories	Functional significance
Recognition and phagocytosis	
Peptidoglycan-binding proteins	Bind to cell envelope of Gram-positive bacteria
Imaginal-disc growth factor proteins	Stimulate cell growth required for wound healing
Thiolester proteins	Forming a complement by binding to invader surface
Serine protease cascades	
Trypsin-like serine proteases	Extracellular signalling molecules of the Toll pathway
Serpins	Inhibition of trypsin-like serine proteases
Serine protease inhibitors of the Kunitz family	Possibly similar to serpins, but not previously implicated in immune response
Melanization and coagulation	
Pro-phenoloxidase activating enzymes	Proteolytical activation of phenoloxidase from its precursor, conversion of dopamine to melanin
Fibrinogen-like protein	Possible role in blood clotting
Antimicrobial peptides	
Drosomycin	Protein toxic to fungal metabolism
IM-2	Small antimicrobial protein, precise function unknown
Signalling pathways	
Cytokine-like small peptides	Activation of stress signalling pathways
Genes of the Toll pathway	Regulation of antifungal peptide production
Genes of the Imd pathway	Regulation of antibacterial peptide production
Proteins of JNK signalling pathway	General stress response (see Section 6.2)
Iron metabolism	
Transferrins and iron transporters	Sequestration of extracellular iron

6.5 Toxic substances

The study of ecological effects of toxic substances in the environment is designated as *ecotoxicology* (Walker *et al.* 2001). This multidisciplinary science is a meeting place of environmental chemists, toxicologists, and ecologists. Chemists determine the concentration of substances in the environment and study their distribution over environmental compartments and chemical ligands; toxicologists analyse uptake kinetics, biotransformation, and metabolic effects of toxicants; ecologists study the effects of toxic insults at the population, community, and ecosystem levels. Ecotoxicology traditionally has a strong link with environmental policy. Through this link, scientific support is provided for decisions about issues like standard setting, remediation of contaminated sites, and pesticide registration. In industrialized countries new substances are produced continuously, but their application in society is regulated by safety requirements concerning human and environmental health. Most legislatory systems require that new substances must be tested for their possible adverse effects on ecological receptors before they are admitted to the market.

A well-known principle in toxicology is that toxicity is not an absolute property of a substance, but that the effect of a substance depends on the dose given to the organism. The classical poison is effective in very low amounts, but in principle all substances can be toxic if dosed highly enough. This was recognized already by the Austrian alchemist and physician A.P.T.B. von Hohenheim (1493–1541), better known as Paracelsus, who wrote (Koeman 1996):

Alle Ding sind Gifft . . . allein die Dosis macht das ein Ding kein Gifft is (Everything is a poison . . . it is only the dose that makes it not a poison).

Following the Paracelsus principle, an important activity of toxicologists is the establishment of *dose–effect relationships*, which in ecotoxicology usually take the form of a graph in which some aspect of the performance of a tested organism (e.g. growth or reproduction) is plotted as a function of the concentration of a given toxicant in water, soil, or air. From such a graph two important benchmarks are estimated, the exposure concentration at which a 50% effect is observed (EC_{50}) and the highest exposure at which still no effect is seen (NEC, no effect concentration). Ecotoxicologists are usually concerned with effects that show up after chronic (long-term) exposure and, unlike human toxicologists, study end points that are important for the ecological functions of an organism. In the case of animals, many ecological functions are associated with feeding and behaviour—for example, macroinvertebrates grazing to suppress algal blooms, or earthworm burrowing to improve soil structure—that is why end points in ecotoxicology can be different from those in human toxicology.

The application of genomic technology in toxicology is called *toxicogenomics* (Lovett 2000; Pennie *et al.* 2000; Burczinsky 2003; Waters and Fostel 2004), and its ecological counterpart as *ecotoxicogenomics* (Snape *et al.* 2004). An important aim of toxicogenomics is to characterize the mode of action of toxicants on the basis of expression profiles. When two toxicants induce the same set of genes in a target organ, they are likely to have the same mode of action (Hamadeh *et al.* 2002). New substances, such as drugs, industrial chemicals, or pesticides, can be screened for their transcription profile and when the profile is compared with a database of earlier-investigated chemicals any similarities may provide an indication of the hazardous properties of the compound. The first commercial microarrays, designed to screen induction of enzymes in the human liver, were developed at the end of the twentieth century. It is likely that such tools will also be developed for environmental applications, but standardized assays are not yet available.

In this section we will address the question of how genomic technology can improve our insight into ecotoxicity of environmental chemicals. Out of the huge number of chemicals that may cause environmental problems we have selected three classes of toxicant: heavy metals, pesticides, and endocrine disrupters. These three groups have very different environmental effects and serve to illustrate the principles of ecotoxicogenomics.

6.5.1 Heavy metals

Under the term heavy metal fall all elements in the Earth's crust with a density of greater than $5\,g/cm^2$ in their metallic form. Thus defined, a large proportion of the periodic table of elements belongs to this category; however, many heavy metals are very rare or extremely unavailable and are of no environmental concern. The toxicity of heavy metals is not due to the metal itself, but to ionic forms and other chemical species (e.g. Pb^{2+}, $HgCH_3^+$, and $Cr_2O_7^{2-}$). The active and toxic form of a metal usually constitutes only a small proportion of the total concentration in an environmental compartment, and depends on properties of the environment as well as the metal. One of the most important influences is due to environmental pH: a low pH promotes dissociation of metal complexes and may increase the fraction of metal present in ionic form without changing the total concentration. The dynamic processes occurring at the interface of environmental ligands and biotic surfaces are studied under the heading of *bioavailability*. Unfortunately, bioavailability is usually ignored in laboratory toxicity experiments, and many of the toxicity data reported in the literature refer to artificial media in which the speciation of metals is biased towards a high fraction of free, ionic metal forms. Total concentrations in such studies cannot be extrapolated easily to the environment. This also holds for most of the toxicogenomic studies using heavy metals.

Studies in yeast were the first to reveal the genome-wide effects of exposure to heavy metals (Gross *et al.* 2000; Momose and Iwahashi 2001; Vido *et al.* 2001; Eide 2001). Many yeast genes are induced by cadmium, including obvious genes such as heat-shock proteins, but also unexpected genes, such as genes related to the synthesis of methionine. In fact, the whole sulphur-salvage pathway, including sulphate uptake, sulphate reduction, methionine synthesis, and glutathione synthesis, was upregulated (Fig. 6.17). This is all understandable since cadmium is a sulphur-seeking metal and its detoxification requires the sulphur-containing amino acid cysteine. In addition, cadmium introduces oxidative stress, which,

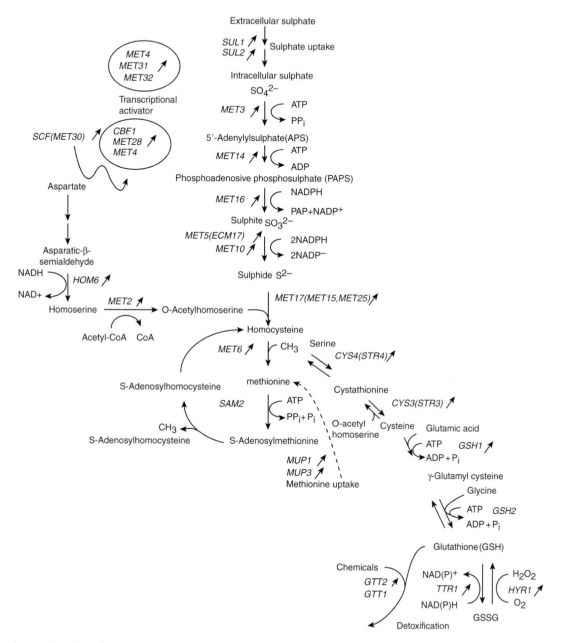

Figure 6.17 Effects of cadmium on the sulphur-salvage pathway of yeast (*S. cerevisiae*). The pathway starts with uptake of sulphate, followed by the formation of phosphoadenosine phosphosulphate (PAPS), sulphate reduction, synthesis of cysteine, methionine, and glutathione (GSSG). Genes found to be upregulated by cadmium are indicated with an oblique arrow next to the name. After Momose and Iwahashi (2001) by permission of the Society of Environmental Toxicology and Chemistry.

as we have seen above, is counteracted by means of enzymes such as glutathione S-transferase, which again requires reduced sulphur.

Interestingly, Momose and Iwahashi (2001) as well as Vido *et al.* (2001) did not find induction of metallothionein when yeast cells were exposed to cadmium. Yeast has a copper-binding metal-lothionein (*CUP1*) that is induced by copper, but not by cadmium. Instead, cadmium in yeast follows a glutathione-dominated pathway, comparable to the phytochelatin pathway of zinc in plants and invertebrates. This is consistent with the finding that among the strongly upregulated genes was a transporter of metal-glutathione, which translocates metals to the vacuole. Metallothionein in yeast is more strongly induced by oxidative stress than by cadmium; however, at higher doses cadmium also causes oxidative stress, so the two effects are difficult to separate in practice.

When analysing the promoter sequences of genes affected by cadmium, Momose and Iwahashi (2001) noted that many of them had a sequence motif known as centromere DNA element I, CDEI. This element binds a transcription factor Cbf1, which is known to regulate the sulphur amino acid-biosynthesis pathway. The results of Momose and Iwahashi (2001) suggest that this pathway is also activated under cadmium stress, maybe through the depletion of glutathione. It is well known among toxicologists that glutathione depletion is a common consequence of oxidative stress and metal stress (see Section 6.2). All in all, the yeast study is a nice example of how the results of a transcription-profiling exercise can be understood from mechanistic knowledge of the underlying processes.

6.5.2 Pesticides

Pesticides, also called crop-protection products, represent a wide range of chemical substances, most of them organic, often with very complex structures, designed for a specific target in the pest organism. Despite the intention of combating only pests, no pesticide is 100% selective against the pest and almost all pesticides have greater or smaller side effects on non-target organisms. Such side effects can be manifested in places far away from the site of application if the pesticide is persistent and transported easily by air or water. Modern pesticides are short-lived and effective in very low doses. It is expected that genomic analysis will contribute to a more sensitive assessment of possible side effects and to an ever increasing precision in the development of new products. This expectation is reflected by the interest in toxicogenomics shown by the pesticide industry.

In contrast to heavy metals, pesticides are designed to affect a specific biochemical target, such as a single photosynthesis protein, a specific ion channel in the nervous system, or a key enzyme in a biosynthetic pathway. The biochemical damage caused by reacting with such a specific target is called the *primary lesion*. An example is the reaction of organophosphate insecticides with the enzyme acetylcholinesterase, which when inhibited will cause accumulation of acetylcholine, a neurotransmitter, in the synaptic cleft, leading to uncoordinated behaviour, spasms, and mortality in insects. From this fundamental toxicological principle one would expect that gene-expression profiles induced by pesticides would be limited to a small fraction of the genome. This appears not to be the case. Most pesticides, when administered to organisms outside of the context of agriculture, are less selective than expected. We discuss a few examples to illustrate this point.

Paraquat   (1,1'-dimethyl-4,4'-dipyridilium dichloride) is a herbicide used to kill weeds prior to emergence of the crop and to destroy foliage of crops such as potatoes in preparation for harvesting. Under the influence of UV light paraquat enters into a redox-cycling process, producing large amounts of ROS that destroy the surface of leaves in an aspecific manner. Such redox cycles are also triggered easily inside organisms once paraquat is activated by cytochrome P450. Because of this bioactivation, paraquat is rather toxic to humans, with lung damage being the first apparent effect. Due to its ability to generate ROS, paraquat is often used as a model agent to impose oxidative stress on animals in the laboratory.

Girardot *et al.* (2004) exposed *D. melanogaster* to paraquat and two other oxidative stress agents,

H_2O_2 and tunicamycin, and assessed gene expression using the Affymetrix Drosophila Genome Array. No fewer than 1111 genes (12% of the probes) were found to be up- or down-regulated in flies exposed to the highest dose, illustrating the genome-wide nature of oxidative-stress defence. The genes included many representatives of the general stress response, such as cytochrome P450s, glutathione S-transferases, peptidases, and triacylglycol lipases. These expressions are all indicative of increased effort towards detoxification, removal of damaged proteins, and repair of membrane lipids. A notable aspect was that iron-binding proteins were specifically induced by paraquat. As noted above, effects on iron metabolism are observed in many different contexts (drought stress in *Arabidopsis*, hypoxia in fish, and microbial infection in *Drosophila*), suggesting that changes in iron metabolism are a part of the general stress response.

A striking feature revealed by the paraquat study was that some of the expression was highly specific to a family of isoenzymes. This was very obvious in the cases of the cytochrome P450s and the glutathione S-transferases. As illustrated in Fig. 6.18, Cyp9b2 was hardly affected by paraquat, Cypb18a1 was repressed, and Cypb4e3 was greatly induced. Similarly, glutathione S-transferase E10 was downregulated, but glutathione S-transferase D5 was upregulated. A similar gene-specific inducibility was found among the cytochrome P450s of *C. elegans* (Menzel *et al.* 2001). Among the 10 *CYP35* genes of the nematode, four were found to be induced strongly by a few specific substrates, four were induced weakly by a wide range of substrates, and two were not inducible at all. This variation reflects the different modes of cytochrome P450 induction, as highlighted in Fig. 6.9.

Monitoring gene expression in organisms in the wild is often proposed as a strategy to evaluate exposure to pollutants in the environment (e.g. Snell *et al.* 2003). The observations of Girardot *et al.* (2004) make it very clear that such measurements need to be specific to a particular isozyme to be of

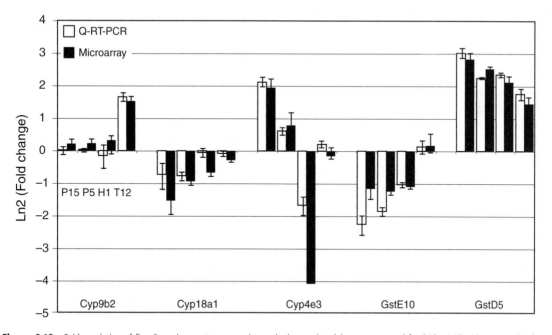

Figure 6.18 Fold regulation of five *D. melanogaster* genes observed when male adults were exposed for 24 h to 15 mM paraquat in the medium (P15), 5 mM paraquat (P5), 1% H_2O_2 (H1), or 12 µM tunicamycin (T12). Data are shown for two methods of differential expression screening: quantitative PCR applied to reverse-transcribed mRNA (Q-RT-PCR) and microarray hybridization. Cyp, cytochrome P450 (three isoenzymes are shown); Gst, glutathione S-transferase (two isoenzymes). After Girardot *et al.* (2004) with permission from BioMed Central.

any indicative value. This means that the probes on a microarray must be able to discriminate between the various isoforms within gene families. Such specificity is easier to achieve with quantitative PCR (see Section 2.3) than with cDNA microarrays; in the study of Girardot et al. (2004) there was a very good correspondence between Q-RT-PCR measurements of expression and microarray hybridizations (Fig. 6.18).

Whereas the study on paraquat was a typical mechanistic laboratory analysis without direct relevance to environmental exposure, similar transcription-profiling protocols are being developed with the aim of applying them to animals exposed to pesticides or other pollutants in the wild (Sultan et al. 2000; Fredrickson et al. 2001; Bultelle et al. 2002; Miracle et al. 2003; Perkins and Lotufo 2003; Azumi et al. 2004; Sansone et al. 2004; Straub et al. 2004; Miracle and Ankley 2005). Most of these studies are still at the stage of method development: the major activities are collection of differentially expressed clones and checking of the reproducibility of microarray screening. Studies in the coastal marine environment have progressed furthest at the moment, because the application of physiological and molecular markers in monitoring programmes, such as 'mussel watch', was started earlier there than in other environmental compartments. We will discuss one marine study to illustrate the direction that the field is taking.

Pacific oyster, *Crassostrea gigas* (Bivalvia, Ostreidae), live in coastal marine habitats and are exposed to pesticides from surface run-off and river discharge. In particular water-soluble and persistent herbicides such as atrazine, diuron, and isoproturon (Fig. 6.19) can reach estuarine and coastal environments from agricultural run-off and river discharge. To assess the stress response of oysters exposed to herbicides, Tanguy et al. (2005) developed forward and reverse SSH libraries (see Section 2.1) from the gills and digestive glands of animals exposed to a cocktail of the three herbicides shown in Fig. 6.19. Some 137 sequences were retrieved as showing differential expression and the genes could be assigned to six major metabolic functions: (i) xenobiotic detoxification, (ii) nucleic acid and protein regulation,

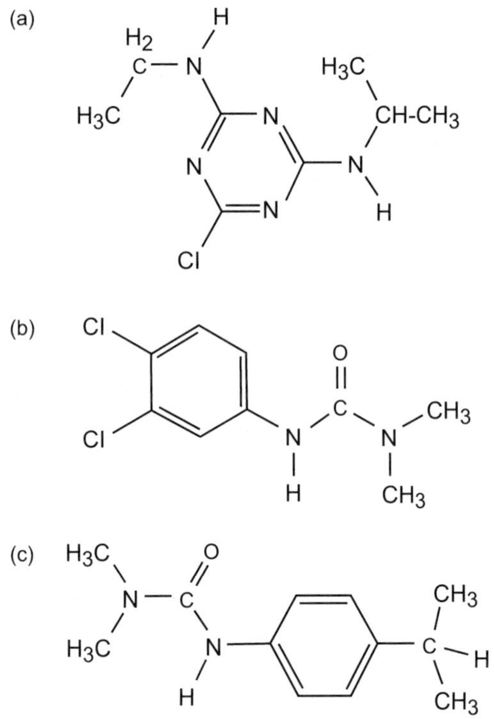

Figure 6.19 Structural formulae of three water-soluble and relatively persistent herbicides: (a) atrazine (6-chloro- N-ethyl-N-isopropyl-1,3,5-triazine-2,4-diamine), (b) diuron (3-(3,4-dichlorophenyl)-1,1-dimethyl ureum), and (c) isoproturon (3-(4-isopropyl)phenyl-1,1-dimethyl ureum).

(iii) respiration, (iv) cell communication, (v) cytoskeleton maintenance, and (vi) energy metabolism. Among the genes consistently upregulated by herbicide exposure was a glutamine synthetase. The enzyme encoded by this gene plays an important role in detoxification of ammonia, synthesis of glutamine, and clearance of glutamate as a neurotransmitter. How these functions relate to pesticide exposure is unclear; however, induction by a wide variety of compounds suggests that glutamine synthetase is part of a general stress response rather than a specific detoxification pathway.

It is also interesting to note from this study that even at relatively low and environmentally relevant exposure concentrations (0.5–2 µg/l) many genes are regulated by herbicides, even though herbicides are not known for their great toxicity to animals. The altered transcriptional profile indicates a broad metabolic effect including a general upregulation of energy production. What the long-term effects of this alteration may be remains uncertain; however, the gene-expression changes seen by Tanguy *et al.* (2005) are consistent with earlier physiological measurements on mussels, in which energy metabolism, often expressed as *scope for growth*, is one of the most sensitive indicators of pollution (Bayne 1989). The data suggest that bivalves in the marine environment possess great flexibility to respond to xenobiotic exposure with adequate transcriptional change, but that this change could alter their filtration capacity and so their ecological functioning in the long term.

6.5.3 Endocrine disrupters

Towards the end of the 1990s environmental scientists began to realize that a wide variety of chemicals in the environment could disrupt endocrine functions of animals. The initial discovery was made by Soto *et al.* (1991), who reported that breast cancer cells sensitive to oestrogen responded to a then-unknown compound leaching in very low amounts from laboratory plasticware made of polystyrene (Colborn *et al.* 1996). The compound was identified as *p*-nonylphenol, one of a family of synthetic chemicals called alkylphenols, which are added to polystyrene and polyvinylchloride to improve stability of the plastic. Alkylphenols are also produced as biodegradation products of alkyl polyethoxylate detergents, and are therefore found in sewage effluent and wastewater from septic tanks. The effect of nonylphenol on oestrogen-sensitive cells appeared to be due to its binding to the *oestrogen receptor*, one of several steroid hormone-binding proteins present in the cytoplasm. When activated by oestrogen the receptor undergoes homodimerization, translocates to the nucleus and binds to *oestrogen-responsive elements*, regulating transcription of a great variety of genes. Although endogenous steroid hormones such as 17β-oestradiol have the greatest affinity for the oestrogen receptor, numerous compounds, of which some are very abundant in the environment, have been shown to

interfere with steroid hormone receptors, either as *agonists* (mimicking the effect of natural hormones) or *antagonists* (suppressing the natural action of a steroid by blocking its receptor). In addition, several environmental chemicals influence hormone metabolism by inhibiting certain forms of cytochrome P450 that metabolize steroids, for example the conversion of testosterone into oestradiol, an activity known as *aromatase*.

Endocrine-disrupting chemicals may be pesticides or metabolites of pesticides, as well as industrial and household chemicals. Oestrogen-active compounds are also found naturally in plants, and are called *phytoestrogens*. A well-known case of phytoestrogen action is the occurrence of isoflavonoids in Australian clover, *Trifolium subterraneum* (Leguminosae), which was identified as the cause of impaired sexual performance in sheep. Phytoestrogens from soybean (*Glycine max*, Leguminosae) are considered as an alternative to oestrogen therapy for menopausal symptoms. Why plants would produce compounds that affect the endocrine system of vertebrates is not clear. Maybe it is just a side effect, while the main function of these compounds lies elsewhere. For example, isoflavonoids are also implicated in signalling between plants and microorganisms in the rhizosphere. Still, it is often assumed that plants have evolved phytoestrogens as a defence strategy against herbivory. If this is the case, one would expect that effects of endocrine disruptors to be more severe in carnivores than in herbivores, because herbivores have had ample opportunity to adapt to plant-derived endocrine disrupters. This prediction, forwarded by Wynne-Edwards (2001), still needs to be evaluated.

Endocrine disruptors are associated with a range of adverse effects observed in terrestrial and aquatic wildlife (almost exclusively vertebrates), varying from developmental disorders to decreased fertility. One of the effects which has attracted a lot of attention is the occurrence of intersex fish due to the feminization of males. An indicator of feminization is expression of the gene *Vtg*, which encodes an egg-yolk protein, *vitellogenin*, which is normally only expressed in the female gonad. The levels of vitellogenin in male trout, caged in rivers at several distances from sewage-treatment facilities in the UK, were found to decrease with increasing distance from the sewage outlets (Harries *et al.* 1997). Following the discovery of such effects, many industrialized countries have started programmes to screen all existing chemicals for their possible endocrine-disruptive properties.

That many chemicals can be labelled as potential endocrine disrupters is now beyond doubt. Laboratory studies have demonstrated that anticonception oestrogens, alkylphenols, phthalates, and some organochlorine pesticides can cause reproductive disorders in fish in the nanogram and low-microgram per litre range (Mathiessen 2000; Mills and Chichester 2005; Sumpter 2005). However, evidence that these chemicals actually impair populations of fish in the wild is less convincing. One of the problems is the lack of reliable and sensitive methods to assess the reproductive status of wild fish. Against this background, endocrine disruption recently became a favourite model for studies in ecotoxicogenomics. Because steroid hormone receptors are regulators of gene expression, any agonistic or antagonistic impact on such receptors should be clearly visible in transcription profiles.

Larkin *et al.* (2002) developed a screening method for profiling 132 genes in largemouth bass, *Micropterus salmoides* (Perciformes, Centrarchidae). Their 'boutique' gene array was used subsequently to assess transcriptional profiles in fish exposed to *p*-nonylphenol and DDE (2,2-bis(*p*-chlorophenyl)-1,1-dichloroethylene), a degradation product of the insecticide DDT (2,2-bis(*p*-chlorophenyl)-1,1,1-trichloroethane) and a suspect hormone disrupter. Oestradiol was used as a positive control. As Fig. 6.20 shows, there was considerable overlap between the expression profiles of nonylphenol and oestradiol. Four vitellogenin genes were induced, plus two choriogenins. These effects were expected because it is known that oestradiol induces the production of yolk proteins as part of the process of oogenesis. Induction of aspartic protease may also be related to this process, since work on zebrafish has suggested that this enzyme plays a role in post-translational processing of vitellogenin in the liver prior to secretion into the

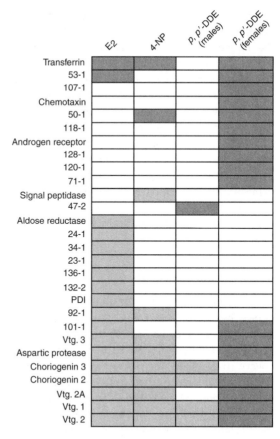

Figure 6.20 Summary of gene-expression changes in largemouth bass (*Micropterus salmoides*) exposed to oestradiol (E$_2$), *p*-nonylphenol (4-NP), and *p,p'*-DDE, a stable degradation product of DDT. Genes upregulated are shown by dark-grey shading; genes downregulated are shown by light-grey shading. Vtg, vitellogenin; PDI, protein disulphide isomerase. Genes with unknown function are indicated by codes. After Larkin *et al.* (2002) with permission from Elsevier.

bloodstream. The role of the other genes in the steroid hormone response is less clear.

Interestingly, some genes were regulated by nonylphenol but not by oestradiol. This suggests that nonylphenol cannot be considered a pure oestrogen mimic but must have additional modes of action that are independent of the oestrogen receptor. The same conclusion was reached in a proteomics study on zebrafish by Shrader *et al.* (2003; see Fig. 1.8). The compound DDE downregulated several oestrogen-responsive genes and a number of others (Fig. 6.20). It was striking that DDE had a much stronger effect on females than on males. This compound obviously interferes with fish reproduction but its mode of action requires further investigation.

A synthetic view of the mode of action of endocrine disrupters such as nonylphenol is not yet available. It may be expected that the picture will be quite complicated. There are several different steroid hormone receptors in the cell that can be activated by steroids such as testosterone, oestradiol, cortisol, progesterone, and others. Endocrine disrupters may activate but also inactivate these receptors; in addition, some endocrine disrupters also interact with the receptors for thyroid hormone and retinoids and the cascades triggered by these receptors may interact with steroid hormone transcriptional activation. This suggests that hormone disruption is not to be considered as a simple linear process. A genome-wide understanding of endocrine disruption in fish will probably only come from species with completely sequenced genomes, such as zebrafish.

This example of endocrine disruption completes our short overview of ecotoxicogenomics. It is obvious that this subdiscipline of ecological genomics is lagging behind considerably in comparison with other areas of stress ecology, such as abiotic stress in plants. A question specific to ecotoxicogenomics is what would be the role of genomics techniques in the risk assessment of chemicals? Bishop *et al.* (2001) analysed the situation and made three recommendations that we endorse: (i) risk assessors must be proactive and involved in identifying research issues that ecotoxicogenomicists should address, (ii) risk assessors must assist genome investigators in finding ways of interpreting gene-expression data that support insights into hazards and dose–effect relationships, and (iii) there must be a continuing and effective dialogue in both directions, so as to maximize information exchange in a way that can inform policy decisions.

6.6 Genomic approaches to ecological stress: an appraisal

We started this chapter by framing stress responses in the context of the ecological niche. Have the genomic insights reviewed in this chapter deepened our insight into what constitutes the boundaries of the ecological niche? It seems fair to admit that, more than in the case of community ecology (Chapter 4) and life-history analysis (Chapter 5), genomics and niche theory still seem to exist on different planets. It may be that our charge for this chapter was a bit too ambitious; however, some general principles have emerged that can help us to connect the two disciplines.

We have seen that stress factors in the environment can impinge on many aspects of the metabolism of cells. Outlines were given for several pathways that translate stress signals into gene expression. Although the details vary per pathway, some common properties emerge: (i) in the case of physical stress, deviation from normal conditions is noted by a stress-specific sensing system, (ii) in the case of chemical stress, there is a specific interaction with a cytosolic receptor protein, (iii) some stress pathways have a system of double-negative control—that is, stress removes the degradation of an activator, (iv) the stress signal is transduced via a more or less complicated network, often involving protein kinases, (v) there is a lot of interaction (crosstalk) between the pathways triggered by different stress signals, (vi) most stress-transduction systems converge on a transcription factor that is translocated to the nucleus, (vii) additional nuclear factors are often needed for activation of the transcription factor, (viii) gene expression is promoted by binding to specific DNA sequences, which are present in a battery of genes, and (ix) many genes have more than one transcription factor-binding site and thus are activated by more than one stress signal.

We have also seen that there is both commonality and divergence in the stress response. Some aspects of the stress response, for example induction of heat-shock proteins, are very general and can be found in nearly every cell under nearly every stress. Other aspects are specific for the stressor and trigger a limited number of genes addressing the stress factor; for example, the induction of proteins preventing the growth of ice crystals in body fluids.

Despite the great increase in knowledge represented by the genomic studies reviewed in this chapter the biological significance of stress-induced transcription profiles still needs further examination. It is not impossible that environmental stress induces groups of genes that are not functionally related. In Chapter 1 we discussed a study by Spellman and Rubin (2002), who noted the presence of transcriptional territories in the genome of *Drosophila*: sets of physically adjacent but functionally unrelated genes that are expressed jointly in association with chromatin remodelling. Some of the unusual gene expressions observed in transcription profiles may be due to these effects. The lesson could be that it is always necessary to frame gene expression in terms of a biological scheme, such as the stress-transducing systems and transcription factor cascades discussed in this chapter.

Ecological studies applying genome-wide approaches to assessing physiological stress in animals or plants under natural conditions have not yet been published. The most commonly analysed single-gene indicator of stress is Hsp70 (Feder and Hofmann 1999). Expression of heat-shock proteins is used to obtain an indication of physiological stress and this may explain ecological interactions in the field. For example, Burnaford (2004) showed that a species of chiton, *Katarina tunicata* (Mollusca, Polyplacophora), was experiencing temperature stress during low tide on a rocky shore under unshaded conditions, and this could explain its association with a species of kelp, *Hedophyllum sessile* (Phaeophyta, Laminariales), under which it finds shelter. The author showed that it was abiotic stress, not predation risk or lack of food, that limited the animal's habitat use on exposed rock surface. In this study stress was measured by Hsp70 expression; one can easily imagine how transcription profiling could further deepen the insight into explaining habitat choice and niche dimensions of organisms in the wild.

The question may be asked, does the transcription profile of an organism under stress contain all the information needed to identify the nature of the stress factor? That is, can we read the cause of stress from the transcriptome? This *inverse approach* to transcription profiling seems particularly relevant in an applied context, when genomics is used to assess the quality of the environment. Although it seems reasonable to assume that this question can be answered in the affirmative, up to now a proof of principle is lacking.

Problems facing the inverse approach are 2-fold. In the first place, antagonism in mixtures may mask gene expressions. If one stress factor suppresses the inducing action of another a combination of the two will not be noted in the transcription profile. Studies on plant responses to abiotic stress reviewed in this chapter have shown that such antagonistic interactions may be realistic. Secondly, the inverse approach only works if there is a monotonic relationship between the intensity of the stress and the degree of gene expression. At low stress intensities this may indeed be the case, but under severe stress some genes induced by low stress may be repressed due to toxicity or other disturbing factors. In that case a low level of gene expression found in a transcription profile does not have an indicative value because it may have two different causes.

In general, the student of stress responses can learn from toxicology that biological responses are always dose-dependent. Toxicologists are trained to characterize intensity and duration of exposure carefully and to always include several exposure levels in an experiment. A similar attentiveness to the exposure side of stress is often lacking in biochemical stress studies.

In conclusion we note that the large gap between niche theory and genomics is fed by the fact that almost all genomic studies on stress responses are conducted in the safe environment of the laboratory. We may expect that genomic studies will now be expanded quickly to include profiling of organisms under natural conditions, in gradients of environmental stress, in extreme environments, or as a function of their distributional range. This chapter has illustrated that the genomic tools and their biochemical interpretation are ready for ecologists to take and apply.

CHAPTER 7

Integrative ecological genomics

As indicated in Chapter 1, a large part of genomics can be qualified as a science of discovery. An impressive amount of new and often unexpected information comes from sequencing genomes and analysing transcriptomes. In this book we have seen many examples of surprises that have arisen from discovery. There is, however, another side to pure discovery science: sequencing of genomes and cataloguing gene expressions are basically descriptive processes and not hypothesis-driven. We do not argue that descriptive activities are always unscientific. Every science includes an inductive phase; unprejudiced exploration is a wealthy source of new hypotheses. In fact, a large part of ecology can be characterized as mainly descriptive, especially the painstaking work conducted by field biologists who document the whereabouts of species in the wild. The point is, in good ecology this type of research is only one phase in the scientific cycle, and is followed up by hypothesis-driven research, either in the laboratory or in the field again. In the same spirit, molecular biologists are convinced that the discovery science of genomics must be followed up by integrative, hypothesis-driven, new activities. In this chapter we discuss some of the postgenomic, integrative approaches and evaluate their relevance to ecology.

7.1 The need for integration: systems biology

Systems analysis has been a part of ecology for a long time. It originates in the work of Eugene P. Odum, who in his classical book *Fundamentals of Ecology* in 1953 made a plea for considering the ecosystem as a unit in its own right, with characteristics that refer to the whole and that are amenable to investigation in the same way as a physiologist investigates a single organism. Odum also recognized that an ecosystem had properties that were not measurable at the levels of populations or communities but were characteristic of the system as a unitary whole. Such properties were called *emergent properties*. From the beginning, systems ecology relied strongly on mathematical analysis of energy fluxes and nutrient cycles, using differential equations and computer simulation. Systems ecology remained an integrative but limited part of ecology, next to the much larger fields of community and population ecology, but in the 1990s it saw a renaissance, spurred by research programmes on acid rain and climate change, which called for large-scale approaches (Schindler 1987).

It is interesting to see how around the year 2000 the developments in ecology were mirrored by similar developments in molecular biology, leading to the birth of a new field, systems biology (Ideker *et al.* 2000; Kitano 2002). In molecular biology, the immediate cause was a feeling that the trend towards more and more reductionism was reaching its limits and in some respects was even impeding progress (Van Regenmortel 2004; Strange 2005). According to the reductionist paradigm, understanding is to be gained from studying processes underlying, and thus determining, the phenomenon of interest. Therefore biological processes should be deconstructed into their component parts and the physicochemical properties studied to achieve understanding. Reductionism argues that any biological process ultimately finds its foundation in the laws of chemistry and physics. What is missing in this description is that many biological phenomena do not only depend on the properties of the component parts, but also on other biological phenomena. Gene expression takes place in a certain cellular context, which is moulded by expressions of other genes. Even the expression of a single gene may require gene products from at least 10 other genes; for example, components of the polymerase complex, transcription factors, enhancers, and modifiers. Recognizing the complexity of gene expression, molecular biologists started to analyse the network of interactions between the expression of different genes.

Despite the explosive use of the term systems biology in the biochemical literature since 2004, according to Cornish-Bowden and Cárdenas (2005) not all activities grouped under this heading can be regarded as true systems biology. Systems biology is not to be considered as a way of integrating information from diverse components into a model of the system as a whole, it must be seen as a view on the whole to understand the parts. So systems

biology argues from the whole to the components, not the other way around. Seen in this way, systems biology and reductionism are not necessarily in conflict with each other.

Westerhoff and Palsson (2004) give an historical perspective on the origin of systems biology. These authors recognized two independent lines of development: one originating in the discovery of DNA, followed by recombinant technology, and large-scale sequencing, the other originating in non-equilibrium thermodynamics, followed by Jacob and Monod's work on the *lac* operon and molecular kinetics (Fig. 7.1). The second timeline was dominated by modelling of molecular processes but this was often seen as rather theoretical and not based in 'real' biology due to the lack of adequate data. The quantum leap of data acquisition in the first timeline, brought about by the genomics revolution, caused a sudden convergence of the two developments and marked the birth of systems biology.

The structure of systems biology can also be viewed as an activity in which the two lines of

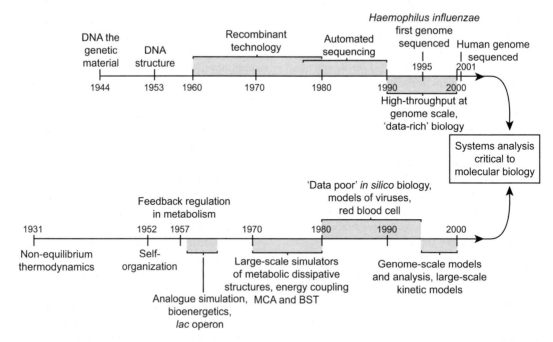

Figure 7.1 Scheme showing how systems biology emerged at the end of the twentieth century from the convergence of two lines of enquiry that had remained separate for several decades. MCA, metabolic control analysis. From Westerhoff and Palsson (2004), by permission of Nature Publishing Group.

inquiry delineated in Fig. 7.1 are fused into a circular process (Fig. 7.2; Kitano 2002). In this cyclic view, hypothesis-driven modelling, computer simulation, and predictions follow up genomic experiments. This part of the cycle is also called *in silico* biology (Palsson 2000). A tremendously important role is played by systematic perturbation of the genome using high-throughput genetic manipulation. For instance, a collection of deletion mutants is now available for essentially all genes in the genome of *S. cerevisiae* and tools exist for systematic manipulation of numerous genes via different constructs in many strains simultaneously. Analysing the phenotypic consequences of such genetic manipulations provides a vast and rich experimental basis on which hypotheses from computer simulation can be tested.

One way in which the complexity of genomic interactions can be analysed is by considering gene expression as a network. Indeed, a great deal of systems biology is concerned with *network*

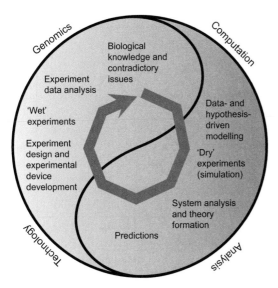

Figure 7.2 Illustrating the structure of systems biology. The cycle of research is characterized by construction of models based on genomics data, followed by computer simulation (*in silico* or dry experimentation), which leads to prediction about the function of specific parts of a genomic network. These predictions are then tested using 'wet' experiments, for example genetic manipulations in the network of interest, which leads to renewed biological knowledge and refinement of models. Reprinted with permission from Kitano (2002). Copyright 2002 AAAS.

analysis. A first logical step in such an analysis is to group genes, proteins, and metabolites into functional units. According to the concept of *modularity*, the functions of a cell can be partitioned into a number of modules where membership of a module is defined by a specific task, which is separable from those of other modules. Four different types of functional module may be discerned: (i) physically delineated molecular machines such as the ribosome or the flagellum, (ii) signalling cascades, in which membership of the module is defined by initial binding of a signal molecule such as in insulin/IGF-1 signalling (Section 5.2) or MAPK (Section 6.2), (iii) collections of genes that are all regulated by the same transcription factor, such as the Ah battery (Section 6.2; these are called *transcription modules*), and (iv) networks defined by the processing of a substrate or a group of metabolites, for example glycolysis, Krebs cycle, or nutrient salvage (see Fig. 6.17).

The greatest advances in metabolic network analysis have been made in *E. coli* and *S. cerevisiae*. One of the lines of investigation focuses on classification of genes according to regulatory modules defined by common responses to environmental conditions. Segal *et al.* (2003) analysed a large number of microarray experiments and proposed a classification of the yeast genome into 50 distinct modules, where each module showed a specific response to a set of conditions, while all the genes of a module were regulated in concert. For example, a 'respiratory module' was defined as consisting of 55 genes, of which 39 encoded respiratory proteins and six encoded enzymes of glucose metabolism. The module was primarily regulated by the transcription factor Hap4, and 29 of the 55 genes had a known Hap4-binding site in their promoter. In addition, 32 genes had an STRE element, which is recognized by the stress-responsive transcription factor Msn4 (see Section 3.2). This type of classification is an important step towards an integrative understanding of responses to environmental conditions in yeast and eventually also in models of ecological relevance.

In addition to classifying genes according to their transcriptional regulation, systems biologists also map entire sets of genes in a genome (Featherstone and Broadie 2002; Tong *et al.* 2004; Zhang *et al.* 2005). To decide whether two genes interact with each other a technique known as *synthetic genetic array* (SGA) analysis is used. The availability of deletion mutants for many genes in the genome is a crucial element in this technique and this condition is satisfied in yeast. An SGA screen starts with a viable mutant that has a query gene deleted. That mutant is crossed into an array of viable deletion mutants for many other genes to generate double mutants which are isolated automatically and scored for loss of fitness or lethality. Double mutants with reduced fitness are indicative of interaction between the two genes, because if each of the single mutants is viable while the double mutant is not, this suggests that the gene products buffer each other in the wild type. Interactions thus defined are often found between genes belonging to the same gene ontology category, which confirms the validity of the interpretation.

Analysis of genetic networks in yeast has revealed that most of them have a structure characterized as *scale-free topology* (Featherstone and Broadie 2002; Tong *et al.* 2004). This structure implies that the distribution of links over the nodes of a network is not random but biased towards a few highly connected nodes ('hubs'), which participate in a large number of metabolic reactions, while the majority of nodes is hooked on only via a single link. Technically speaking, in a scale-free network the probability $P(k)$ that a node interacts with k other nodes (the *degree distribution* of the network) follows a power law, where $P(k) \sim k^{-\gamma}$. The parameter γ approximates a value of 2.2 for all organisms. When this relationship is plotted on a graph (Fig. 7.3), the horizontal axis represents the degree of interaction that a gene has in the genome (highly interactive genes are on the right), whereas the vertical axis represents the number of genes that is found for each degree. In an SGA screen conducted by Tong *et al.* (2004) 132 different yeast

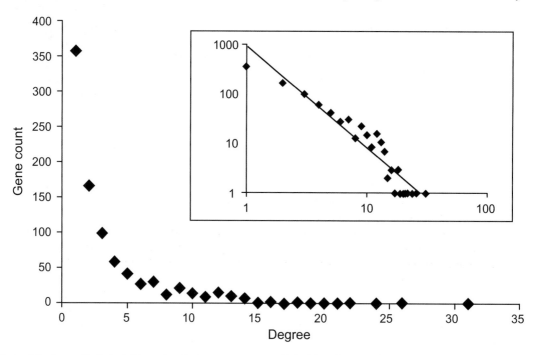

Figure 7.3 Degree distribution of interactions between yeast genes established through SGA analysis. A frequency distribution is shown of the number of interactions (degree) that yeast genes have with other genes. The shape of the distribution is suggestive of a power law, which is confirmed by the linear relationship obtained when the two axes are expressed logarithmically (inset). Networks with a power law distribution of degrees are called scale-free. An example of the topology is shown in Fig. 7.4a. Reprinted with permission from Tong *et al.* (2004). Copyright 2004 AAAS.

genes were queried and each of these genes was tested for interactions with 4700 other genes. As expected, genes with only one interaction were most abundant but some had interactions with 32 other genes (Fig. 7.3).

The scale-free property of genetic-interaction networks is in conflict with the idea of modularity referred to above. The existence of modules implies some degree of organization and a more or less uniform distribution of links over nodes. These two seemingly opposing tendencies are reconciled in the concept of *hierarchical modularity*, introduced by Ravasz *et al.* (2002). In a hierarchical modular network there are many small, highly connected modules that combine into larger units according to a power law. The hierarchical network has a scale-free topology with embedded modularity (Fig. 7.4c). Ravasz *et al.* (2002) conducted an extensive analysis of the metabolic networks of 43 different organisms and found that in all species investigated there was significant evidence for hierarchical modularity.

Network analysis in yeast is becoming more and more comprehensive. Drawing together information from SGA screens, protein–DNA interaction, and protein–protein interaction, Zhang *et al.* (2005) developed a network involving no less than 5831 nodes (genes, proteins) and 154 659 links between them. Such extremely complicated networks can be analysed for the occurrence of *network motifs*: patterns of interconnectedness between three or more genes that occur more frequently than expected from a random combination of links. Overlying a collection of network motifs is a *network theme*: an interconnected cluster of motifs reflecting a common organizational principle. Network themes can often be linked to a biological process and there are only a few of them in the cell. At a still higher organizational level is defined a *thematic map*, which captures the dynamic relationships between themes.

We have dwelt on network theory to illustrate the approaches that systems biologists are applying in order to understand the metabolism of the living cell. The question arises, should ecological genomics follow a similar path to understand ecological systems and link genomics to ecology?

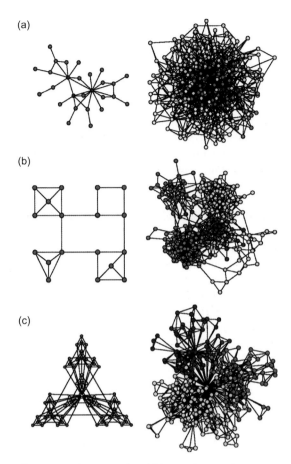

(a)

(b)

(c)

Figure 7.4 Three types of model for complex networks. (a) A scale-free model, characterized by a few, highly connected nodes; (b) a modular network, consisting of four highly interlinked units, connected to each other by a few links; and (c) an hierarchical network, which combines a scale-free topology with hierarchical clustering. Reprinted with permission from Ravasz *et al.* (2002). Copyright 2002 AAAS.

Network analysis is not uncommon in community ecology. Food webs can be considered as networks and ecologists are interested in what properties of food webs contribute to their stability. One such property, discussed by Neutel *et al.* (2002), is the pattern of *interaction strengths* across the web. Interaction strength of a trophic couple is a measure of the influence of one species on the population increase of another species when both are near equilibrium. The negative effect of a predator on its prey is usually larger than the positive effect of a prey on its predator; therefore, interactions strengths are greater for top-down relationships than for bottom-up relationships. Interaction strengths are particularly important when there are loops in the web; that is, paths that return on the same species without visiting other species more than once. The mean interaction strength of all the pairs in a loop is called the *loop weight*. Examination of 104 published food-webs revealed that the longer the loop, the lower its loop weight. Mathematical analysis showed that this property contributes greatly to the stability of the network. So, loop weight in a food web can be considered a network motif of a significance comparable to the ones identified by Zhang *et al.* (2005) for metabolic networks in the cell.

We believe that approaches similar in spirit to systems biology should ultimately be adopted to enable genomic answers to ecological questions. That is, a systems approach is needed to link genomics to ecosystem function, to life-history pattern, and to the ecological niche. Chapters 4, 5, and 6 of this book have shown that such links are currently far from complete. When ecological processes are governed by a limited number of signal transduction pathways, as in some of the stress responses discussed in Chapter 6, a network analysis of interactions seems to be a suitable option (see the figures in Chapter 6 showing induction of gene expression by stress). Also, in the cases of nutrient cycles catalysed by well-characterized genetic complexes in microorganisms, a systems approach seems to be feasible. However, in general it is difficult to forecast which type of systems biology is required as an integrative framework for ecological genomics. In addition, we see three main issues that lack in the present systems-biology approaches but which are nevertheless crucial for ecology.

Spatial considerations. Space is a very important aspect of ecological analysis. Many processes in communities require some degree of proximity between different organisms (e.g. in the case of syntrophy). Reproduction in animals usually requires physical contact between the sexes; in flowering plants pollen has to travel from anther to pistil and the distance between these organs

matters. Stratification of the environment, or other types of heterogeneity, are often crucial for species to coexist with each other. Colonization events can alter the functioning of ecosystems dramatically if the colonizer is an invader and outcompetes local species. All these issues, so obvious for an ecologist, are completely absent from present-day systems biology. It is unclear how genomics and systems biology can contribute to spatial ecology.

Temporal considerations. Like space, time is another important dimension of ecological analysis. Any ecosystem bears all kinds of traces from its prior development; the way in which an ecosystem functions is determined *partly* by what went before. Historical issues are, for example, reflected in the build-up of the soil profile and the presence of peat accumulated from previous plant growth. If birds for some reason in the past decided to use *one* piece of land rather than another, this situation may continue for a long time because habitat use is often culturally transmitted or imprinted in the offspring. So, ecologists are accustomed to the fact that some phenomena in nature can only be understood by referring to past events, and that there is a more or less predictable succession of events during the development of an ecosystem. It is unclear how such temporal phenomena can be reconciled with a systems-biology perspective.

Bi-directional interaction with the environment. In the systems-biology treatment of the living cell, the environment of the cell is considered as given; expressed in ecological terms, it is a set of external conditions that are not altered by activities of the cell. In real ecological systems, the environment is altered significantly by organisms, for example by gas exchange, by consuming resources, or by altering physical structure (e.g. burrowing by earthworms). Including such interactions in a systems-biology model would imply that resources and nutrients are considered dynamic variables that can be depleted or otherwise altered by the organism. Such bi-directional interactions are not yet part of systems-biology analyses.

7.2 Ecological control analysis

In addition to network theory, which emphasizes topology and structure, systems biology also deploys a set of quantitative tools, aiming to capture the dynamics of transcriptome, proteome, and metabolome in mathematical terms. One such approach developed in the 1970s goes under the name of *metabolic control analysis (MCA)*. The foundations of MCA theory were laid by Henrik Kacser in 1973 (Fell 1992; Kacser and Burns 1995). The theory focuses on the flux of metabolites through a biochemical pathway, where a certain substrate undergoes successive modification catalysed by enzymes. An example is the well-known glycolytic pathway, which involves 10 different catalytic steps starting with phosphorylation of glucose and ending in the formation of pyruvate. Under normal physiological conditions, the concentration of each intermediary metabolite will be constant, while there is a constant flux of glucose down to pyruvate. It is the flux of substrate that drives the metabolism of the cell, not the concentration. The question asked in MCA is, which enzyme exerts the greatest control over the flux?

In the formularium of MCA, the flux of substrate through the pathway is designated by J. The control of an enzyme i over the flux is expressed as the increase of J that would be brought about by a small increase *in* enzyme concentration, e_i. If an enzyme exerts a large control, the flux will increase strongly with a small increase of e_i; that is, the derivative of J with respect to e_i will be large. The *flux-control coefficient* C_i is therefore defined as the partial derivative of J with respect to e_i, relative to J and e_i themselves, under steady-state conditions:

$$C_i = \frac{\partial J}{\partial e_i}\frac{e_i}{J} = \frac{\partial \ln J}{\partial \ln e_i}$$

Flux-control coefficients are dimensionless constants that characterize the action of an enzyme in the pathway. However, they should not be considered as properties of an enzyme as such, because their value depends on the state of the whole pathway. Control coefficients can only be measured while the whole pathway is intact; this can be done by measuring the flux as accumulation of the end product in cells with experimentally manipulated enzyme concentrations, for example using specific inhibitors. Another approach is to use genetic means, for example by comparing homozygotes with heterozygotes, by mutating a gene to downregulate its activity, by introducing a plasmid carrying an extra copy of the gene, or by replacing the promoter of the gene by an artificial promoter that can be modulated by an external factor. The rise of DNA technology has greatly expanded the possibilities for manipulating enzyme activities and has allowed measurements of flux control that were difficult to achieve with earlier biochemical methods (Jensen *et al.* 1995).

An interesting property of flux control coefficients in a metabolic pathway is that, leaving aside some special cases, the sum of their values over the whole pathway equals unity:

$$\sum_{i=1}^{n} C_i = 1$$

where n is the number of enzymes in the pathway. This so-called *summation theorem* of MCA holds independent of the structure of the pathway; it is valid for linear pathways such as glycolysis, but also for non-linear pathways, including those with feedback, coupling, and branching.

The summation theorem, which was proven by Kacser and Burns (1995), implies that control is not necessarily attributable to a single enzyme. The idea that there is always one enzyme in a metabolic pathway that is rate-limiting is not supported by theory or biochemical experiments. Instead MCA leads to the concept of *distributed control*: all enzymes in a metabolic pathway exert some degree of control over the flux. It is possible that all enzymes take an equal share in the control, but the most common situation is that several enzymes in a pathway contribute only little to flux control. This is supported by evidence from heterozygotes that carry a defective mutant allele; in such genotypes the activity of the enzyme involved may be reduced to 50% in the case of full recessiveness, but still the flux of metabolites through the pathway is often hardly affected.

A second important concept in MCA concerns the way in which enzyme activities depend on the substrate concentration. This is expressed in the *elasticity coefficient* ε, which is defined for any enzyme i as the partial derivative of its reaction rate v_i with respect to the substrate, S, again normalized to the substrate concentration and the rate:

$$\varepsilon_i = \frac{\partial v_i}{\partial S} \frac{S}{v_i} = \frac{\partial \ln v_i}{\partial \ln S}$$

Elasticity coefficients are determined by the kinetic properties of the enzymes. In contrast to the flux-control coefficient, they are local properties and can be studied using the enzyme when it is isolated from the pathway. Flux-control coefficients lose their meaning when the enzyme is isolated, but elasticity can still be studied *in vitro*.

MCA theory further learns that there is a relationship between elasticity coefficients and flux-control coefficients. This is a consequence of the fact that under steady-state conditions the flux neither accumulates nor oscillates in the pathway. If we consider two adjacent steps, 1 and 2, in a metabolic pathway, we have two elasticities, ε_1 and ε_2, and two flux-control coefficients, C_1 and C_2. The relationship between these quantities is:

$$C_1\varepsilon_1 + C_2\varepsilon_2 = 0$$

This relationship is known as the *connectivity theorem*. For a formal proof the reader is referred to Kacser and Burns (1995). The relationship can be generalized from two adjacent steps to the whole metabolic pathway, in which case the connectivity theorem takes the form of a matrix equation.

The connectivity theorem is one of the strongest results of MCA, since it implies a formal link between enzyme kinetics and flux control. If the elasticity coefficients of all enzymes in a pathway are known, their flux control coefficients can be derived by applying the connectivity theorem. Derivation of flux control from elasticities is called *forward control analysis*. Equally interesting is *reverse control analysis*, which argues from control coefficients to elasticities. Westerhoff *et al.* (1994) showed that through a single matrix inversion step the *in-situ* elasticities of the enzymes in a pathway can be obtained from properties of the pathway as a whole, the flux-control coefficients.

MCA is interesting because it offers a powerful framework with potential application to ecology. A link between MCA and ecology is developed in

trophic control analysis (Getz *et al.* 2003). In this approach the MCA methodology is applied to food chains rather than biochemical pathways. Trophic chains are different from biochemical pathways in a one crucial aspect: the flux process is not conservative. Due to growth, respiration, excretion, and mortality in food chains, biomass is lost from the flux of matter; control analysis has to take this into account. Getz *et al.* (2003) developed an MCA-like food-web model by which it is possible to investigate which trophic groups exert the largest control over the biomass flux. The aim of the study was to test the trophic cascade hypothesis, which holds that predators exert control over their prey at the top of the food chain in such a way that the top-down control trickles down to lower trophic levels and so also limits populations at the base of the food web. Trophic control analysis demonstrated, however, that control is distributed over several different levels of the trophic chain, rather than residing in one particular place.

Another type of application of MCA to ecology goes under the name *ecological control analysis*. This type of analysis stays close to MCA by focusing on the flux of some material through a community in the absence of significant biomass production from the flux. This may hold as an approximation for microbial communities under anaerobic conditions in which growth is extremely slow and almost all of the carbon is dissimilated. As a consequence enzyme concentrations and reaction stoichiometries can be assumed to be in steady state and are conserved. W.F.M. Röling *et al.*, in work not yet published, considered a simple syntrophic community consisting of three functional groups of microorganisms, degrading a specific organic substrate by fermentation (Fig. 7.5). The fermenting microorganisms produce acetate and hydrogen but they are inhibited by these products and cannot grow when acetate and hydrogen accumulate in their environment. In a syntrophic community, fermentation products are utilized by terminal electron-accepting microorganisms such as iron reducers, sulphate reducers, and methanogens (see Section 4.3). It is assumed in the model that acetate and hydrogen are used by different microorganisms (Fig. 7.5).

(a)

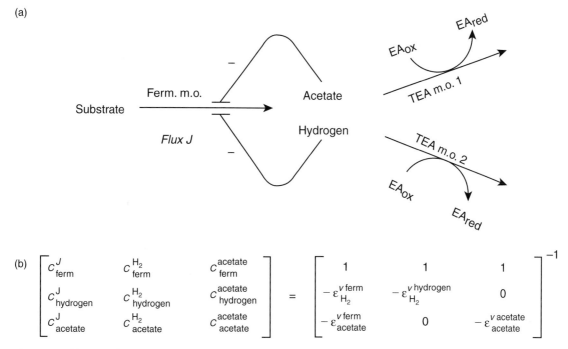

(b)

$$
\begin{bmatrix}
c_{\text{ferm}}^{J} & c_{\text{ferm}}^{H_2} & c_{\text{ferm}}^{\text{acetate}} \\
c_{\text{hydrogen}}^{J} & c_{\text{hydrogen}}^{H_2} & c_{\text{hydrogen}}^{\text{acetate}} \\
c_{\text{acetate}}^{J} & c_{\text{acetate}}^{H_2} & c_{\text{acetate}}^{\text{acetate}}
\end{bmatrix}
=
\begin{bmatrix}
1 & 1 & 1 \\
-\varepsilon_{H_2}^{v\,\text{ferm}} & -\varepsilon_{H_2}^{v\,\text{hydrogen}} & 0 \\
-\varepsilon_{\text{acetate}}^{v\,\text{ferm}} & 0 & -\varepsilon_{\text{acetate}}^{v\,\text{acetate}}
\end{bmatrix}^{-1}
$$

Figure 7.5 (a) Model of a syntrophic microbial community consisting of three members, one fermenting an organic substrate and producing acetate and hydrogen, the other two consuming the intermediates. Arrows indicate material transfer; the curved lines ending with bars indicate negative control of the products. Ferm. m.o., fermenting microorganisms; TEA m.o., terminal electron-accepting microorganisms; EA_{ox}, EA_{red}, oxidized and reduced electron acceptors, respectively; J, flux of substrate. (b) Control matrix (left) obtained as the inverse of the elasticity matrix (right) that corresponds to the scheme shown in (a). C, flux-control coefficient; the superscript indicates which flux (J, hydrogen, acetate) is controlled, the subscript indicates the controlling functional group. ε, Elasticity coefficient; the superscript indicates the functional group, the subscript indicates the substrate. From W.F.M. Röling, B.M. Van Breukelen, H.W. Van Verseveld, and H.V. Westerhoff, unpublished work.

Röling *et al.* estimated numerical values for the elasticity coefficients in this system and showed that in contrast to current views, degradation fluxes are not always limited by a single functional group, but can be controlled by several groups simultaneously. The model was used to explore under what conditions degradation of a substrate is controlled by which group of microorganisms. It turned out that under denitrifying and iron-reducing conditions flux control by terminal-electron accepting microorganisms is generally negligible and control is exerted mainly by the fermentors; however, under less favourable redox conditions flux control by terminal electron-accepting microorganisms can strongly increase. The redox potential is therefore a dominant driver that determines where flux control is concentrated.

The results have very important practical implications, for example in the remediation and management of polluted groundwater systems.

For the moment, ecological control analysis seems to fit best with problems of food-web ecology and biogeochemical cycles. However, we see no reason why application should be limited to these fields. We have seen several examples in this book that can be phrased in terms of fluxes (contaminants, oxygen radicals, resource allocation) and that could be analysed with some kind of control analysis. There is also a link between metabolic flux and population genetics, as explored in an early paper by Dykhuizen *et al.* (1987). These authors analysed enzymes of the *lac* operon of *E. coli* and showed that the β-galactosidase enzyme had only a small control coefficient with regard to fitness. Therefore, genetic variants of this enzyme in which enzyme activity is changed mildly in comparison to the wild type will mostly be selectively neutral. This was in contrast with the β-galactosidase permease locus, which encodes an enzyme with strong control over fitness. Thus it was predicted that mutations in the latter locus, even if they change enzyme activity by only a little, could be under directional selection. The evolutionary fate of an allele is thus determined by the flux-control coefficient of the enzyme it encodes. There are many examples where alternative variants of enzymes exert a different degree of flux control. Ecologists can utilize the natural variation in such enzymes to test the evolutionary implications of MCA.

7.3 Outlook

For the final section of this book we will discuss a some emerging issues that we think will become important for ecological genomics in the near future. Whereas genomics develops at staggering speed, it is difficult to predict how much attention each of these issues will get; the ones we have selected logically follow from the topics addressed in this book.

7.3.1 Organization of model species communities

One of the most frequently experienced obstacles of ecological genomics at the moment is the availability of just a few fully sequenced genomes of ecologically relevant species. However, we believe that this situation will improve soon and that within a few years many species on which more than a few ecological laboratories are working will have a sufficiently large genomic database to allow genome-wide analyses of ecological questions. Transcription profiling using microarrays is expected to be the major activity for ecological genomics in the near future, and although the use of microarrays is only useful when preceeded or followed up by major sequencing efforts, a complete genome sequence is not really necessary to work with microarrays. In

the transition phase the use of small, tailor-made arrays (so-called boutique arrays) can be a good compromise, as we have seen in several sections of this book (Chapters 4 and 6). Such arrays can be extended gradually as more genomic information on the species becomes available.

The real obstacle for ecological genomics could very well be the creation of sufficiently large collaborative networks around model species. As we have noted above, ecologists are fascinated more by the biodiversity of species than by one representative species, and have difficulty in accepting the idea of model species. Still, it is unavoidable that ecological laboratories intending to specialize in genomics will collaborate and agree on a single model. Benefit from genome information can also be obtained by working on a wild, close relative of a genomic model, as we have seen in the case of Brassicaceae related to *Arabidopsis* (Chapter 3).

7.3.2 Large-scale sequencing of the environnment

In Chapter 4 we saw that complete genomes can be reconstructed from DNA extracted from the environment, allowing analysis of metabolic potentials of organisms that have never been cultured in the laboratory. This strategy seems to be especially fruitful when applied to simplified communities in extreme environments. Such large-scale sequencing projects produce a hardly imaginable amount of new information and without doubt new discoveries are in store. It seems that an order-of-magnitude increase in sequencing effort is still needed to apply this approach effectively to mesophilic communities including eukaryotes. If technology and computing power allows for such an increase in the near future, large-scale sequencing of the environment could make an enormously important contribution to mapping the Earth's biodiversity and archiving sequences of species that risk extinction in the near future.

In addition to sequencing, functional screening of metagenomic libraries is also expected to provide a lot of new information. As we have discussed in Chapter 4, metagenomic studies at the moment are focused on discovering new genes or biosynthetic pathways for products that have potential application in medicine, biotechnology, or agriculture. We expect that in the near future metagenomics will also benefit ecology. Many functions important for nutrient cycling, mutualism, quorum sensing, plant–microbe signalling, etc. are still hidden in the genomes of partly unknown microorganisms. New technological developments that combine genomics with localization and the use of tracers seem to be particularly promising.

7.3.3 Wild transcription profiling

The great majority of genomic studies reviewed in this book were conducted with organisms in a laboratory environment. The microbial studies reviewed in Chapter 4 are an exception and that is why we have argued that the process of merging ecology with genomics has shown the greatest progress in microbiology. However, microbial ecological genomics has to deal with challenges that plague botanists and zoologists to a lesser degree, the *terra incognita* of biodiversity.

As the genomes of more and more ecologically relevant model species are sequenced, transcription profiling of organisms collected directly from their natural environment will come within reach. Such studies have not yet been conducted, but we expect that they will be crucial to understanding the dynamics and ecological relevance of transcription profiles. It could very well be that the profiles that have been observed until now in laboratory-cultured organisms will differ greatly from those collected in the wild. This may especially hold for species that face conditions in the field that are very different from their laboratory environment. For example, most ecologists are convinced that disease and parasitism are important limiting factors in the field. Do transcription profiles of plants and animals in the wild contain signatures of frequently upregulated immune responses? We don't know.

7.3.4 The mechanistic framework

Sequencing and transcription profiling run the risk of only scraping the surface of what ecological genomics has to offer. Indeed, collecting endless lists of genes and gene-expression profiles is not an aim in itself. In this chapter we discussed the necessity of linking genomics with hypothesis-driven research. In our opinion, gene-expression profiling makes sense only if the genes sooner or later can be positioned in an analytical framework that is grounded in physiological or biochemical knowledge. Even ecologists, if they take ecological genomics seriously, will have to get to grips with the mechanisms of the processes they study. We have seen that the most successful stories of ecological genomics discussed in Chapter 5, on longevity in *C. elegans* and flowering time in *Arabidopsis*, came from previous painstaking work in genetics and biochemistry. We argued in Chapter 2 that microarray-based transcription profiling should perhaps be viewed as just an exploratory instrument, or as only a stop on the way to some basic question, not as a goal in itself. We should avoid the impression, sometimes made, that genomics is a very advanced form of stamp collecting.

7.3.5 New methods of data analysis

The field of bioinformatics is developing a tremendous number of new analytical tools to match the high-throughput pipelines of comparative and functional genomics. Most of these methods are aimed at dimension reduction using various types of multivariate statistics, cluster analysis, pattern recognition, etc. Another avenue is the application of univariate models such as analysis of variance in a gene-by-gene manner, which raises statistical issues about false-positive results in large data-sets. Ecologists are traditionally acquainted with statistics as a necessary tool for the analysis of noisy data; however, the sheer size of genomics data-sets adds a new dimension to statistical analysis in ecological genomics.

We argued in Chapter 2 that the statistical approaches presently developed for microarray-derived transcription profiles are not necessarily optimal for ecological genomics. In particular, we argued that clustering of conditions may better fit with ecological problems than clustering of genes.

Another challenge for bioinformatics in ecology is how to define the normal operating range of species; that is, the set of expression profiles that reflects routine physiological variation and the absence of stress. Finally, the development of integrative approaches of the type visited earlier in Chapter 7, linking genomics and transcriptomics to questions of community structure, life-history, and ecological niche, may be the greatest challenge of all.

7.3.6 Comparative genomics

Whereas this book focuses on problems of ecology we have discussed evolutionary issues only marginally. In the last few years the field of comparative genomics has expanded its scope tremendously and is responsible for many exciting discoveries. Using techniques such as phylogenetic footprinting and phylogenetic shadowing information is obtained about the content of a genome in such a way that maximal use is made of homologies elsewhere in the tree of life. Using comparative genomics, organismal groups with unclear phylogenetic affiliation will soon be positioned solidly. Interestingly, new species are being sequenced not for the sake of these species but for the sake of a more or less distant model (usually humans). This development can actually benefit ecology. An example is the availability of a full genome sequence for the sea squirt, *Ci. intestinalis*, which was sequenced for comparative reasons, but is an equally good model for ecological studies in the coastal marine environment. We expect that the science of molecular evolution, supported by comparative genomics, is facing a glorious future.

7.3.7 Focus on natural variation

Natural variation in gene expression among and within field populations can be an extremely powerful tool for revealing evolutionary mechanisms, as illustrated in an elegant study on killifish, *Fundulus heteroclitus* (Oleksiak *et al.* 2002). This study and others suggest that inter-individual variation of transcriptional regulation is very common in natural populations. Such variation,

which is ultimately due to polymorphisms in promoter sequences or variation in *trans*-acting factors or signal transduction pathways, can be an important template for natural selection (Wray *et al.* 2003). The recent literature seems to suggest that, at least in animals, evolution acting upon transcriptional regulation could be at least as important as evolution of coding sequences (see the discussion in Stearns and Magwene 2003).

The use of natural variation is also important because it allows ecological genomics to get away from the present focus on laboratory mutants (see the work presented in Chapter 5). Knocking-out genes to analyse their function is a crucial tool in mechanistic studies, but the relevance of the results for understanding functions in the environment is limited. The point is, we need to know the fitness consequences of gene mutations under ecologically relevant conditions. Where nature is already providing this variation, it is best to make optimal use of it.

7.3.8 Genetical genomics

The by now classical approaches of quantitative genetics and QTL analysis are badly integrated with ecological genomics at the moment. We foresee a renewed interest in quantitative character analysis when ecological genomics develops further. There is an increasing realization that even the expression of a single gene behaves as a quantitative character because several other gene products (transcription factors, enhancers, repressors, ribosomal factors, and splice factors) are involved with gene expression. Expression of single genes, when measured with quantitative methods such as real-time PCR, can be subjected to the same statistical analysis as traditional quantitative characters like body size and development time. To give an example, Roelofs *et al.* (2006) measured expression of a metal-binding protein, metallothionein, in a species of soil-living insect, *Orchesella cincta*, and estimated heritability of metallothionein expression from parent–offspring regression. The expression was extremely variable between individuals of this species, which was partly due to significant polymorphism in the promoter of the metallothionein gene. Such studies can potentially link gene-expression data to ecologically relevant quantitative characters. Several authors have expressed similar views (Walsh 2001; Gibson and Mackay 2002).

The link between quantitative characters and genomics has developed further into a new field, called *genetical genomics* (Jansen and Nap 2001). The idea is to exploit the added value of genetic segregation by studying transcription profiles in large pedigrees of recombinant inbred lines. In principle it would be possible to treat each transcription profile as a quantitative trait and to use QTL methodology to dissect the expression profile into its underlying genetic components. This approach is potentially of enormous importance for ecological genomics, because it can forge a link between transcription profiling and QTL mapping.

7.3.9 Epigenetics

We know now that regulatory information that is not defined in the DNA can still be transmitted between cells, and from one generation to another, using DNA methylation, histone acetylation, etc. By means of epigenetic processes cells may canalize their potential expression repertoire and adopt specific functions or states. In addition, an epigenetically regulated state may be transmitted in cell lineages. Such epigenetic processes seem to be especially important when transmitting information that must be remembered in the daughter cells in order to respond in an adequate way to some environmental factor. We have seen several examples of epigenetic processes in this book, for example the regulation of flowering time in *Arabidopsis* by temperature (Chapter 5). Still, epigenetics has had little influence on ecology (Jablonka and Lamb 2002). We call for more attention to this important phenomenon, for example by studying the frequency of epigenetic variants in natural populations.

One of the mechanisms of epigenetic regulation is due to RNAi, a phenomenon by which double-stranded RNAs specifically suppress a target protein by degradation of its mRNA or by transcriptional gene silencing (Novina and Sharp 2004). Following the discovery of RNAi in nematodes and plants in the 1990s, realization has come that it is a widespread natural phenomenon and an evolutionarily ancient mechanism of genome defence. It may be expected that RNAi is also involved in regulating ecologically important processes, but this has not yet been demonstrated.

7.3.10 The unification of biology

The enormous scientific success of biology in the course of the last century is evident from the fact that many other disciplines have adopted biological epithets. This is not limited to biochemistry and biophysics, but extends to disciplines such as biogeology, bioarchaeology, and biopsychology. As a consequence of this spreading movement, biology has become such an extensive scientific field that it is hardly possible to view the whole. What was just zoology, botany, and microbiology before is now scattered over a wide variety of subdisciplines. The same diversifying trend is present in ecology itself, which has to cover the widest span of all: from molecule to ecosystem. How sustainable is this situation? Should we fear for the future of ecology as an homogeneous science? Not denying the tendency to diversify, which is typically seen in any growing field, we also expect an opposite trend, the unification of biology. Genomics could very well turn out to act as a new point of crystallization, bringing biologists of various plumages to the same core. As demonstrated by the recent establishment of a Gordon Research Conference series on Evolutionary and Ecological Functional Genomics (Feder and Mitchell-Olds 2003), there is an increasing interest among molecular biologists in issues of evolution and ecology. Likewise, more and more ecologists are becoming interested in the molecular biology of ecological phenomena. We hope that this book has contributed at least a little to this admittedly ambitious perspective.

References

Adam, D. (2000) Now for the hard ones. *Nature* **408**: 792–793.

Adamcyk, J., Hesselsoe, M., Iversen, N., Horn, M., Lehner, A., Nielsen, P.H., Schloter, M., Roslev, P., and Wagner, M. (2003) The isotope array, a new tool that employs substrate-mediated labeling of rRNA for determination of microbial community structure and function. *Applied and Environmental Microbiology* **69**: 6875–6887.

Adams, M.D., Celniker, S.E., Holt, R.A., Evans, C.A., Gocayne, J.D., Amanatides, P.G., Scherer, S.E., Li, P.W., Hoskins, R.A., Galle, R.F. *et al.* (2000) The genome sequence of *Drosophila melanogaster*. *Science* **287**: 2185–2195.

Allen, E.A. and Banfield, J.F. (2005) Community genomics in microbial ecology and evolution. *Nature Reviews Microbiology* **3**: 489–498.

Altschul, S.F., Gish, W., Miller, W., Myers, E.W., and Lipman, D.J. (1990) Basic Local Alignment Search Tool. *Journal of Molecular Biology* **215**: 403–410.

Amasino, R.M. (2003) Flowering time: a pathway that begins at the 3′ end. *Current Biology* **13**: R670–R672.

Amasino, R. (2004) Take a cold flower. *Nature Genetics* **36**: 111–112.

Amores, A., Force, A., Yan, Y.-L., Joly, L., Amemiya, C., Fritz, A., Ho, R.K., Langeland, J., Prince, V., Wang, Y.-L. *et al.* (1998) Zebrafish *hox* clusters and vertebrate genome evolution. *Science* **282**: 1711–1714.

Andersen, G.L., DeSantis, T.Z., Murray, S.R., and Moberg, J.P. (2004) *Phylogenetic chip for microbial detection*. 10th International Symposium on Microbial Ecology, Cancun, Mexico. International Society of Microbial Ecology, Geneva: 110.

Andrews, G.K. (2001) Cellular zinc sensors: MTF-1 regulation of gene expression. *BioMetals* **14**: 223–237.

Aparicio, S., Chapman, J., Stupka, E., Putnam, N., Chia, J., Dehal, P., Christoffels, A., Rash, S., Hoon, S., Smit, A. *et al.* (2002) Whole-genome shotgun assembly and analysis of the genome of *Fugu rubripes*. *Science* **297**: 1301–1310.

Arabidopsis Genome Initiative (2000) Analysis of the genome sequence of the flowering plant *Arabidopsis thaliana*. *Nature* **408**: 796–815.

Arantes-Oliveira, N., Apfeld, J., Dillin, A., and Kenyon, C. (2002) Regulation of life-span by germ-line stem cells in *Caenorhabditis elegans*. *Science* **295**: 502–505.

Aravind, L., Tatusov, R.L., Wolf, Y.I., Walker, D.R., and Koonin, E.V. (1998) Evidence for massive gene exchange between archaeal and bacterial hyperthermophiles. *Trends in Genetics* **14**: 442–444.

Arbeitman, M.N., Furlong, E.E.M., Imam, F., Johnson, E., Null, B.H., Baker, B.S., Krasnow, M.A., Scott, M.P., Davis, R.W., and White, K.P. (2002) Gene expression during the life cycle of *Drosophila melanogaster*. *Science* **297**: 2270–2275.

Armbrust, E.V., Berges, J.A., Bowler, C., Green, B.R., Martinez, D., Putnam, N.H., Zhou, S., Allen, A.E., Apt, K.E., Bechner, M. *et al.* (2004) The genome of the diatom *Thalassiosira pseudonana*: ecology, evolution, and metabolism. *Science* **306**: 79–86.

Armstrong, M.R., Blok, V.C., and Phillips, M.S. (2000) A multipartite mitochondrial genome in the potato cyst nematode *Globodera pallida*. *Genetics* **154**: 181–192.

Attwood, T.K. and Parry-Smith, D.J. (1999) *Introduction to Bioinformatics*. Prentice Hall, Harlow.

Ausín, I., Alonso-Blanco, C., Jarillo, J.A., Ruiz-Garcia, L., and Martínez-Zapater, J.M. (2004) Regulation of flowering time by FVE, a retinoblastome-associated protein. *Nature Genetics* **36**: 162–166.

Azumi, K., Fujie, M., Usami, T., Miki, Y., and Satoh, N. (2004) A cDNA microarray technique applied for analysis of global gene expression profiles in tributyltin-exposed ascidians. *Marine Environmental Research* **58**: 543–546.

Baker, B.J., Lutz, M.A., Dawson, S.C., Bond, P.L., and Banfield, J.F. (2004) Metabolically active eukaryotic communities in extremely acidic mine drainage. *Applied and Environmental Microbiology* **70**: 6264–6271.

Baldwin, I.T. (2001) An ecologically motivated analysis of plant-herbivore interactions in native tobacco. *Plant Physiology* **127**: 1449–1458.

Ball, K.D. and Trevors, J.T. (2002) Bacterial genomics: the use of DNA microarrays and bacterial artificial chromosomes. *Journal of Microbiological Methods* **49**: 275–284.

Barnes, A.I. and Partridge, L. (2003) Costing reproduction. *Animal Behaviour* **66**: 199–204.

Barnett, M.J., Fisher, R.F., Jones, T., Komp, C., Abola, A.P., Barloy-Hubler, F., Bowser, L., Capela, D., Galibert, F., Gouzy, J. *et al.* (2001) Nucleotide sequence and predicted functions of the entire *Sinorhizobium meliloti* megaplasmid. *Proceedings of the National Academy of Sciences USA* **98**: 9883–9888.

Bartels, D. and Sunkar, R. (2005) Drought and salt tolerance in plants. *Critical Reviews in Plant Sciences* **24**: 23–58.

Bartke, A., Wright, J.C., Mattison, J.A., Ingram, D.K., Miller, S.R., and Roth, G.S. (2001) Extending the lifespan of long-lived mice. *Nature* **414**: 412.

Bastow, R. and Dean, C. (2003) Deciding when to flower. *Science* **302**: 1695–1697.

Bastow, R., Mylne, J.S., Lister, C., Lippman, Z., Martienssen, R.A., and Dean, C. (2004) Vernalization requires epigenetic silencing of *FLC* by histone methylation. *Nature* **427**: 164–167.

Bavykin, S.G., Akowksi, J.P., Zakhariev, V.M., Barsky, V.E., Perov, A.N., and Mirzabekov, A.D. (2001) Portable system for microbial sample preparation and oligonucleotide microarray analysis. *Applied and Environmental Microbiology* **67**: 922–928.

Bayne, B.L. (1989) Measuring the biological effects of pollution: the mussel watch approach. *Water Science and Technology* **21**: 1089–1100.

Becker, A. and Theißen, G. (2003) The major clades of MADS-box genes and their role in the development and evolution of flowering plants. *Molecular Phylogenetics and Evolution* **29**: 464–489.

Beebee, T. and Rowe, G. (2004) *Introduction to Molecular Ecology*. Oxford University Press, Oxford.

Béjà, O., Aravind, L., Koonin, E.V., Suzuki, M.T., Hadd, A., Nguyen, L.P., Jovanovich, S.B., Gates, C.M., Feldman, R.A., Spudich, J.L. *et al.* (2000a) Bacterial rhodopsin: evidence for a new type of phototrophy in the sea. *Science* **289**: 1902–1906.

Béjà, O., Suzuki, M.T., Koonin, E.V., Aravind, L., Hadd, A., Nguyen, L.P., Villacorta, R., Amjadi, M., Garrigues, C., Jovanovich, S.B. *et al.* (2000b) Construction and analysis of bacterial artificial chromosome libraries from a marine microbial assemblage. *Environmental Microbiology* **2**: 516–529.

Béjà, O., Spudich, E.N., Spudich, J.L., Leclerc, M., and DeLong, E.F. (2001) Proteorhodopsin phototrophy in the ocean. *Nature* **411**: 786–789.

Beldade, P. and Brakefield, P.M. (2002) The genetics and evo-devo of butterfly wing patterns. *Nature Reviews Genetics* **3**: 442–452.

Beldade, P., Brakefield, P.M., and Long, A.D. (2002) Contribution of *Distal-less* to quantitative variation in butterfly eyespots. *Nature* **415**: 315–318.

Benner, S.A., Liberles, D.A., Chamberlin, S.G., Govindarajan, S., and Knecht, L. (2000) Functional inferences from reconstructed evolutionary biology involving rectified databases—An evolutionarily grounded approach to functional genomics. *Research in Microbiology* **151**: 97–106.

Bernardi, G. (2000) Isochores and the evolutionary genomics of vertebrates. *Gene* **241**: 3–17.

Biology Analysis Group and Genome Analysis Group (2004) A draft sequence for the genome of the domesticated silkworm (*Bombyx mori*). *Science* **306**: 1937–1940.

Bishop, W.E., Clarke, D.P., and Travis, C.C. (2001) The genomic revolution: what does it mean for risk assessment? *Risk Analysis* **21**: 983–987.

Blanc, G. and Wolfe, K.H. (2004) Widespread paleopolyploidy in model species inferred from age distributions of duplicate genes. *The Plant Cell* **16**: 1667–1678.

Blattner, F.R., Plunket, III, G., Bloch, C.A., Perna, N.T., Burland, V., Riley, M., Collado-Vides, J., Glasner, J.D., Rode, C.K., Mayhew, G.F. *et al.* (1997) The complete genome sequence of *Escherichia coli* K12. *Science* **277**: 1453–1462.

Blaxter, M.L., De Ley, P., Garey, J.R., Liu, L.X., Scheldeman, P., Vierstraete, A., Vanfleteren, J.R., Mackey, L.Y., Dorris, M., Frisse, L.M. *et al.* (1998) A molecular evolutionary framework for the phylum Nematoda. *Nature* **392**: 71–75.

Blázquez, M.A., Ahn, J.H., and Weigel, D. (2003) A thermosensory pathway controlling flowering time in *Arabidopsis thaliana*. *Nature Genetics* **33**: 168–171.

Blüher, M., Kahn, B.B., and Kahn, C.R. (2003) Extended longevity in mice lacking the insulin receptor in adipose tissue. *Science* **299**: 572–574.

Bochdanovits, Z. and De Jong, G. (2004) Antagonistic pleiotropy for life-history traits at the gene expression level. *Proceedings of the Royal Society of London (supplement)* **271**: S75–S78.

Bochdanovits, Z., Van der Klis, H., and De Jong, G. (2003) Covariation of larval gene expression and adult body size in natural populations of *Drosophila melanogaster*. *Molecular Biology and Evolution* **20**: 1760–1766.

Bodrossy, L., Stralis-Pavese, N., Murrell, J.C., Radajewski, S., Weilharter, A., and Sessitsch, A. (2003) Development and validation of a diagnostic microbial microarray for methanotrophs. *Environmental Microbiology* **5**: 566–582.

Boffeli, D., McAuliffe, J., Ovcharenko, D., Lewis, K.D., Ovcharenko, I., Pachter, L., and Rubin, E.M. (2003) Phylogenetic shadowing of primate sequences to find functional regions of the human genome. *Science* **299**: 1391–1394.

Bohnert, H.J., Ayoubi, P., Borchert, C., Bressan, R.A., Burnap, R.L., Cushman, J.C., Cushman, M.A., Deyholos, M., Fischer, R., Galbraith, D.W. *et al.* (2001) A genomics approach towards salt stress tolerance. *Plant Physiology and Biochemistry* **39**: 295–311.

Bongers, T. (1988) *De Nematoden van Nederland*. KNNV Uitgeverij, Utrecht.

Bongers, T. and Ferris, H. (1999) Nematode community structure as a bioindicator in environmental monitoring. *Trends in Ecology and Evolution* **14**: 224–228.

Borevitz, J.O. and Nordborg, M. (2003) The impact of genomics on the study of natural variation in *Arabidopsis*. *Plant Physiology* **132**: 718–725.

Bowers, J.E., Chapman, B.A., Rong, J., and Paterson, A.H. (2003) Unravelling angiosperm genome evolution by phylogenetic analysis of chromosomal duplication events. *Nature* **422**: 433–438.

Brachat, S., Dietrich, F.S., Voegeli, S., Zhang, Z., Stuart, L., Lerch, A., Gates, K., Gaffney, T.D., and Philippsen, P. (2003) Reinvestigation of the *Saccharomyces cerevisiae* genome annotation by comparison to the genome of a related fungus: *Ashbya gossypii*. *Genome Biology* **4**: R45.

Bradbury, J. (2004) Nature's nanotechnologists: unveiling the secrets of diatoms. *PLoS Biology* **2**: e347–e306.

Bradshaw, Jr, H.D., Ceulemans, R., Davis, J., and Stettler, R. (2000) Emerging model systems in plant biology: poplar (*Populus*) as a model forest tree. *Journal of Plant Growth Regulation* **19**: 306–313.

Braeckman, B.P., Houthoofd, K., and Vanfleteren, J.R. (2001) Insulin-like signaling, metabolism, stress resistance and aging in *Caenorhabditis elegans*. *Mechanisms of Ageing and Development* **122**: 673–693.

Branicky, R., Bénard, C., and Hekimi, S. (2000) clk-1, mitochondria and physiological rates. *BioEssays* **22**: 48–56.

Braun, E.L., Halpern, A.L., Nelson, M.A., and Natvig, D.O. (2000) Large-scale comparison of fungal sequence information: mechanisms of innovation in *Neurospora crassa* and gene loss in *Saccharomyces cerevisiae*. *Genome Research* **10**: 416–430.

Brazma, A. and Voli, J. (2000) Gene expression data analysis. *FEBS Letters* **480**: 17–24.

Brazma, A., Hingamp, P., Quackenbush, J., Sherlock, G., Spellman, P., Stoeckert, C., Aach, J., Ansorge, W., Ball, C.A., Causton, H.C. *et al.* (2001) Minimum information about a microarray experiment (MIAME)—towards standards for microarray data. *Nature Genetics* **29**: 365–371.

Breitbart, M., Salamon, P., Andresen, B., Mahaffy, J.M., Segall, A.M., Mead, D., Azam, F., and Rohwer, F. (2002) Genomic analysis of uncultured marine viral communities. *Proceedings of the National Academy of Sciences USA* **99**: 14250–14255.

Britton, J.S., Lockwood, W.K., Li, L., Cohen, S.M., and Edgar, B.A. (2002) *Drosophila*'s insulin/PI3-kinase pathway coordinates cellular metabolism with nutritional conditions. *Developmental Cell* **2**: 239–249.

Brogiolo, W., Stocker, H., Ikeya, T., Rintelen, F., Fernandez, R., and Hafen, E. (2001) An evolutionary conserved function of the *Drosophila* insulin receptor and insulin-like peptides in growth control. *Current Biology* **11**: 213–221.

Brown, A. (2004) *In the Beginning Was the Worm. Finding the Secrets of Life in a Tiny Hermaphrodite*. Pocket Books, London.

Brunetti, C.R., Selegue, J.E., Monteiro, A., French, V., Brakefield, P.M., and Carroll, S.B. (2001) The generation and diversification of butterfly eyespot color patterns. *Current Biology* **11**: 1578–1585.

Brunner, A.M. and Nilsson, O. (2004) Revisiting tree maturation and floral initiation in the poplar functional genomics area. *New Phytologist* **164**: 43–51.

Brunner, A.M., Busov, V.B., and Strauss, S.H. (2004) Poplar genome sequence: functional genomics in an ecologically dominant plant species. *Trends in Plant Science* **9**: 49–56.

Bult, C.J., White, O., Olsen, G.J., Zhou, L., Fleischmann, R.D., Sutton, G.G., Blake, J.A., Fitzgerald, L.M., Clayton, R.A., Gocayne, J.D. *et al.* (1996) Complete genome sequence of the methanogenic archaeon, *Methanococcus jannaschii*. *Science* **273**: 1058–1073.

Bultelle, F., Panchout, M., Leboulenger, F., and Danger, J.M. (2002) Identification of differentially expressed genes in *Dreissena polymorpha* exposed to contaminants. *Marine Environmental Research* **54**: 385–389.

Burczynski, M.E. (ed.) (2003) *An Introduction to Toxicogenomics*. CRC Press, Boca Raton, FL.

Butte, A. (2002) The use and analysis of microarray data. *Nature Reviews Drug Discovery* **1**: 951–960.

Burnaford, J.L. (2004) Habitat modification and refuge from sublethal stress drive a marine plant-herbivore association. *Ecology* **85**: 2837–2847.

Calow, P. (1989) Proximate and ultimate responses to stress in biological systems. *Biological Journal of the Linnean Society* **37**: 173–181.

Campbell, B.J., Stein, J.L., and Cary, S.C. (2003) Evidence of chemolithoautotrophy in the bacterial community associated with *Alvinella pompejana*, a hydrothermal vent polychaete. *Applied and Environmental Microbiology* **69**: 5070–5078.

Cañestro, C., Bassham, S., and Postlewaith, J.H. (2003) Seeing chordate evolution through the *Ciona* genome sequence. *Genome Biology* **4**: 208.

Cánovas, D., Cases, I., and De Lorenzo, V. (2003) Heavy metal tolerance and metal homeostasis in *Pseudomonas putida* as revealed by complete genome analysis. *Environmental Microbiology* **5**: 1242–1256.

Carroll, R.L. (2000) Towards a new evolutionary synthesis. *Trends in Ecology and Evolution* **15**: 27–32.

Carroll, S.B., Grenier, J.K., and Weatherbee, S.D. (2005) *From DNA to Diversity*. Blackwell Publishing, Malden.

Casal, J.J., Fankhauser, C., Coupland, G., and Blázquez, M.A. (2004) Signalling for developmental plasticity. *Trends in Plant Science* **9**: 309–314.

Causton, H.C., Quackenbush, J., and Brazma, A. (2003) *Microarray Gene Expression Data Analysis. A Beginner's Guide*. Blackwell Science, Malden.

Causton, H.C., Ren, B., Koh, S.S., Harbison, C.T., Kanin, E., Jennings, E.G., Lee, T.I., True, H.L., Lander, E.S., and Young, R.A. (2001) Remodeling of yeast genome expression in response to environmental changes. *Molecular Biology of the Cell* **12**: 323–337.

C. elegans Sequencing Consortium (1998) Genome sequence of the nematode *C. elegans*: a platform for investigating biology. *Science* **282**: 2012–2018.

Chain, P., Lamerdin, J., Larimer, F., Regala, W., Lao, V., Land, M., Hauser, L., Hooper, A., Klotz, M., Norton, J. *et al.* (2003) Complete genome sequence of the ammonia-oxidizing bacterium and obligate chemolithoautotroph *Nitrosomonas europaea*. *Journal of Bacteriology* **185**: 2759–2773.

Chang, L. and Karin, M. (2001) Mammalian MAP kinase signalling cascades. *Nature* **410**: 37–40.

Chapman, S., Schenk, P., Kazan, K., and Manners, J. (2001) Using biplots to interpret gene expression patterns in plants. *Bioinformatics* **18**: 202–204.

Chase, J.M. and Leibold, M.A. (2003) *Ecological Niches. Linking Classical and Contemporary Approaches*. The University of Chicago Press, Chicago.

Chen, M., Chory, J., and Fankhauser, C. (2004) Light signal transduction in higher plants. *Annual Review of Genetics* **38**: 87–117.

Chen, W., Provart, N.J., Glazebrook, J., Katagiri, F., Chang, H.-S., Eulgem, T., Mauch, F., Luan, S., Zou, G., Whitham, S.A. *et al.* (2002) Expression profile matrix of Arabidopsis transcription factor genes suggests their putative functions in response to environmental stresses. *The Plant Cell* **14**: 559–574.

Cheng, C., Pounds, S.B., Boyett, J.M., Pei, D., Kuo, M.-L., and Roussel, M.F. (2004) Statistical significance threshold criteria for analysis of microarray gene expression data. *Statistical Applications in Genetics and Molecular Biology* **3**: 36.

Chin, K.-J., Esteve-Núñez, A., Leang, C., and Lovley, D.R. (2004) Direct correlation between rates of anaerobic respiration and levels of mRNA for key respiratory genes in *Geobacter sulfurreducens*. *Applied and Environmental Microbiology* **70**: 5183–5189.

Cho, J.-C. and Tiedje, J.M. (2001) Bacterial species determination from DNA-DNA hybridization by using genome fragments and DNA microarrays. *Applied and Environmental Microbiology* **67**: 3677–3682.

Cho, J.-C. and Tiedje, J.M. (2002) Quantitative detection of microbial genes by using microarrays. *Applied and Environmental Microbiology* **68**: 1425–1430.

Christoffels, A., Koh, E.G.L., Chia, J.-M., Brenner, S., Aparicio, S., and Venkatesh, B. (2004) Fugu genome analysis provides evidence for a whole-genome duplication early during the evolution of ray-finned fish. *Molecular Biology and Evolution* **21**: 1146–1151.

Christophides, G.K., Zdobnov, E., Barillas-Mury, C., Birney, E., Blandin, S., Blass, C., Brey, P.T., Collins, F.H., Danielli, A., Dimopoulos, G. *et al.* (2002) Immunity-related genes and gene families in *Anopheles gambiae*. *Science* **298**: 159–165.

Clancy, D.J., Gems, D., Harshman, L.G., Oldham, S., Stocker, H., Hafen, E., Leevers, S.J., and Partridge, L. (2001) Extension of life-span by loss of CHICO, a *Drosophila* insulin receptor substrate protein. *Science* **292**: 104–106.

Clancy, D.J., Gems, D., Hafen, E., Leevers, S.J., and Partridge, L. (2002) Dietary restriction in long-lived dwarf flies. *Science* **296**: 319.

Clark, M.S., Crawford, D.L., and Cossins, A. (2003) Meeting review: worldwide genomic resources for non-model fish species. *Comparative and Functional Genomics* **4**: 502–508.

Cliften, P., Sudarsanam, P., Desikan, A., Fulton, L., Fulton, B., Majors, J., Waterston, R., Cohen, B.A., and Johnston, M. (2003) Finding functional features in *Saccharomyces* genomes by phylogenetic footprinting. *Science* **301**: 71–76.

Clontech (2002) *Clontech PCR-select*^{TM} *cDNA Subtraction Kit User Manual*. BD Biosciences Clontech.

Cobbett, C. and Goldsbrough, P. (2002) Phytochelatins and metallothioneins: roles in heavy metal detoxification and homeostasis. *Annual Review of Plant Biology* **53**: 159–182.

Colborn, T., Myers, J.P., and Dumanoski, D. (1996) *Our Stolen Future*. Little, Brown and Company, Boston.

Cole, J.R., Chai, B., Marsh, T.L., Farris, R.J., Wang, Q., Kulam, S.A., Chandra, S., McGarrell, D.M., Schmidt, T.M., Garrity, G.M., and Tiedje, J.M. (2003) The Ribosomal Database Project (RDP-II): previewing a new autoaligner that allows regular updates and the new prokaryotic taxonomy. *Nucleic Acids Research* **31**: 442–443.

Cornish-Bowden, A. and Cárdenas, M.L. (2005) Systems biology may work when we learn to understand the parts in terms of the whole. *Biochemical Society Transactions* **33**: 516–519.

Corona, M., Estrada, E., and Zurita, M. (1999) Differential expression of mitochondrial genes between queens and workers during caste determination in the honeybee *Apis mellifera*. *Journal of Experimental Biology* **202**: 929–938.

Cowan, D.A., Arslanoglu, A., Burton, S.G., Baker, G.C., Cameron, R.A., Smith, J.J., and Meyer, Q. (2004) Metagenomics, gene discovery and the ideal biocatalyst. *Biochemical Society Transactions* **32**: 298–302.

Crawford, D.L. (2001) Functional genomics does not have to be limited to a few select organisms. *Genome Biology* **2**: Interactions/1001.1.

Croal, L.R., Gralnick, J.A., Malasarn, D., and Newman, D.K. (2004) The genetics of geochemistry. *Annual Review of Genetics* **38**: 175–202.

Curtis, T.P. and Sloan, W.T. (2004) Prokaryote diversity and its limits: microbial community structure in nature and implications for microbial ecology. *Current Opinion in Microbiology* **7**: 221–226.

Curtis, T.P., Sloan, W.T., and Scannell, J.W. (2002) Estimating prokaryotic diversity and its limits. *Proceedings of the National Academy of Sciences USA* **99**: 10494–10499.

Cushman, J.C. and Bohnert, H.J. (2000) Genomic approaches to plant stress tolerance. *Current Opinion in Plant Biology* **3**: 117–124.

Dallinger, R., Berger, B., Hunziker, P., and Kägi, J.H.R. (1997) Metallothionein in snail Cd and Cu metabolism. *Nature* **388**: 237–238.

D'Amico, L.J., Davidowitz, G., and Nijhout, H.F. (2001) The developmental and physiological basis of body size evolution in an insect. *Proceedings of the Royal Society of London Series B* **268**: 1589–1593.

Daniel, R. (2004) The soil metagenome—a rich resource for the discovery of novel natural products. *Current Opinion in Biotechnology* **15**: 199–204.

Daniel, R. (2005) The metagenomics of soil. *Nature Reviews Microbiology* **3**: 470–478.

Darwin, C. (1845) *The Voyage of The 'Beagle'*. J.M. Dent & Sons, London.

Darwin, C. (1859) *On the Origin of Species by Means of Natural Selection, or the Preservation of Favoured Races in the Struggle for Life*. Penguin Books, Harmondsworth.

Davidowitz, G., D'Amico, L.J., and Nijhout, H.F. (2003) Critical weight in the development of insect body size. *Evolution & Development* **5**: 188–197.

De Gregorio, E., Spellman, P.T., Rubin, G.M., and Lemaitre, B. (2001) Genome-wide analysis of the *Drosophila* immune response by using oligonucleotide microarrays. *Proceedings of the National Academy of Sciences USA* **98**: 12590–12595.

De Jong, G. (1995) Phenotypic plasticity as a product of selection in a variable environment. *American Naturalist* **145**: 493–512.

De Jong, G. and Bochdanovits, Z. (2003) Latitudinal clines in *Drosophila melanogaster*: body size, allozyme frequencies, inversion frequencies, and the insulin-signalling pathway. *Journal of Genetics* **82**: 207–223.

De la Torre, J.R., Christianson, L.M., Béjà, O., Suzuki, M.T., Karl, D.M., Heidelberg, J., and DeLong, E.F. (2003) Proteorhodopsin genes are distributed among divergent marine bacterial taxa. *Proceedings of the National Academy of Sciences USA* **100**: 12830–12835.

De Meester, L. (1996) Local genetic differentiation and adaptation in freshwater zooplankton populations: Patterns and processes. *Ecoscience* **3**: 385–399.

Dehal, P., Satou, Y., Campbell, R.K., Chapman, J., Degnan, B., De Tomaso, A., Davidson, B., Di Gregorio, A., Gelpke, M., Goodstein, D.M. *et al.* (2002) The draft genome of *Ciona intestinalis*: insights into the chordate and vertebrate origins. *Science* **298**: 2157–2167.

DeLong, E.F. (2001) Microbial seascapes revisited. *Current Opinion in Microbiology* **4**: 290–295.

DeLong, E.F. (2005) Microbial community genomics in the ocean. *Nature Reviews Microbiology* **3**: 459–469.

Demerec, M. and Kaufmann, B.P. (1950) *Drosophila Guide. A Guide to Introductory Studies of the Genetics and Cytology of Drosophila melanogaster. With an Appendix Containing a Series of Experiments to be Conducted by the Beginning Student*. The Lord Baltimore Press, Baltimore.

Denef, V.J., Park, J., Rodrigues, J.L.M., Hashsham, S.A., and Tiedje, J.M. (2003) Validation of a more sensitive method for using spotted oligonucleotide DNA microarrays for functional genomics studies on bacterial communities. *Environmental Microbiology* **5**: 933–943.

Denlinger, D.L. (2002) Regulation of diapause. *Annual Review of Entomology* **47**: 93–122.

Denver, D.R., Morris, K., Lynch, M., and Thomas, W.K. (2004) High mutation rate and predominance of insertions in the *Caenorhabditis elegans* nuclear genome. *Nature* **430**: 679–682.

Deppenheimer, U., Johann, A., Hartsch, T., Merkl, R., Schmitz, R.A., Martinez-Arias, R., Henne, A., Wiezer, A., Bäumer, S., Jacobi, C. *et al.* (2002) The genome of *Methanosarcina mazei*: evidence for lateral gene transfer between Bacteria and Archaea. *Journal of Molecular Microbiology and Biotechnology* **4**: 453–461.

DeRisi, J.L., Iyer, V.R., and Brown, P.O. (1997) Exploring the metabolic and genetic control of gene expression on a genomic scale. *Science* **278**: 680–686.

DeSantis, T.Z., Dubosarskiy, I., Murray, S.R., and Andersen, G.L. (2003) Comprehensive aligned sequence construction for automated design of effective probes (SASCADE-P) using 16S rDNA. *Bioinformatics* **19**: 1461–1468.

De Souza, J.T., Weller, D.M., and Raaijmakers, J.M. (2003) Frequency, diversity, and activity of 2,4-diacetylphloroglucinol-producing fluorescent *Pseudomonas* spp. in Dutch take-all decline soils. *Phytopathology* **93**: 54–63.

Devlin, P.F., Yanovsky, M.J., and Kay, S.A. (2003) A genomic analysis of the shade avoidance response in Arabidopsis. *Plant Physiology* **133**: 1617–1629.

Devonshire, A.L. and Field, L.M. (1991) Gene amplification and insecticide resistance. *Annual Review of Entomology* **36**: 1–23.

Devos, K.M. and Gale, M.D. (2000) Genome relationships: the grass model in current research. *The Plant Cell* **12**: 637–646.

Di Meo, C.A., Wilbur, A.E., Holben, W.E., Feldman, R.A., Vrijenhoek, R.C., and Cary, S.C. (2000) Genetic variation among endosymbionts of widely distributed vestimentiferan tubeworms. *Applied and Environmental Microbiology* **66**: 651–658.

Diatchenko, L., Lau, Y.-F.C., Campbell, A.P., Chenchik, A., Moqadam, F., Huang, B., Lukyanov, S., Lukyanov, K., Gurskaya, N., Sverdlov, E.D., and Siebert, P.D. (1996) Suppression subtractive hybridisation: a method for generating differentially regulated or tissue-specific cDNA probes and libraries. *Proceedings of the National Academy of Sciences USA* **93**: 6025–6030.

Dicke, M., Van Poecke, R.M.P., and De Boer, J.G. (2003) Inducible indirect defence of plants: from mechanisms to ecological functions. *Basic and Applied Ecology* **4**: 27–42.

Dicke, M., Van Loon, J.J.A., and De Jong, P.W. (2004) Ecogenomics benefits community ecology. *Science* **305**: 618–619.

Dietrich, F.S., Voegeli, S., Brachat, S., Lerch, A., Gates, K., Steiner, S., Mohr, C., Pöhlmann, R., Luedi, P., Choi, S. *et al.* (2004) The *Ashbya gossypii* genome as a tool for mapping the ancient *Saccharomyces cerevisiae* genome. *Science* **304**: 304–307.

Dillin, A., Crawford, D.L., and Kenyon, C. (2002) Timing requirements for insulin/IGF-1 signaling in *C. elegans*. *Science* **298**: 830–834.

Dimopoulos, G., Casavant, T.L., Chang, S., Scheetz, T., Roberts, C., Donohue, M., Schultz, J., Benes, V., Bork, P., Ansorge, W. *et al.* (2000) *Anopheles gambiae* pilot gene discovery project: Identification of mosquito innate immunity genes from expressed sequence tags generated from immune-competent cell lines. *Proceedings of the National Academy of Sciences USA* **97**: 6619–6624.

Dionne, M.S. and Schneider, D.S. (2002) Screening the fruitfly immune system. *Genome Biology* **3**: reviews 1010.1–1010.2.

Domsch, K.H. (1984) Effects of pesticides and heavy metals on biological processes in soil. *Plant and Soil* **76**: 367–378.

Doolittle, W.F. (1999) Phylogenetic classification and the universal tree. *Science* **284**: 2124–2128.

Dopazo, H. and Dopazo, J. (2005) Genome-scale evidence of the nematode-arthropod clade. *Genome Biology* **6**: R41.

Dopson, M., Baker-Austin, C., Koppineedi, P.R., and Bond, P.L. (2003) Growth in sulfidic environments: metal resistance mechanisms in acidophilic microorganisms. *Microbiology* **149**: 1959–1970.

Dover, G. (1999) Human evolution: our turbulent genes and why we are not chimps. In *The Human Inheritance. Genes, Language, and Evolution*, B. Sykes (ed.). Oxford University Press, Oxford: 75–92.

Drake, J.W., Charlesworth, B., Charlesworth, D., and Crow, J.F. (1998) Rates of spontaneous mutation. *Genetics* **148**: 1667–1686.

Dujon, B. (1996) The yeast genome project: what did we learn? *Trends in Genetics* **12**: 263–270.

Dujon, B., Sherman, D., Fischer, G., Durrens, P., Casaregola, S., Lafontaine, I., De Montigny, J., Marck, C., Neuvéglise, C., Talla, E. *et al.* (2004) Genome evolution in yeasts. *Nature* **430**: 35–44.

Dumont, M.G. and Murrell, J.C. (2005) Stable isotope probing—linking microbial identity to function. *Nature Reviews Microbiology* **3**: 499–504.

Duret, L. and Hurst, L.D. (2001) The elevated GC content at exonic third sites is not evidence against neutralist models of isochore evolution. *Molecular Biology and Evolution* **18**: 757–762.

Dwight, S.S., Balakrishnan, R., Christie, K.R., Costanzo, M.C., Dolinski, K., Engel, S.R., Feierbach, B., Fisk, D.G., Hirschman, J., Hong, E.L. *et al.* (2004) *Saccharomyces* genome database: underlying principles and organisation. *Briefings in Bioinformatics* **5**: 9–22.

Dykhuizen, D.E., Dean, A.M., and Hartl, D.L. (1987) Metabolic flux and fitness. *Genetics* **115**: 25–31.

Eckardt, N.A. (2002) Alternative splicing and the control of flowering time. *The Plant Cell* **14**: 743–747.

Eddy, S.R. (2005) A model of the statistical power of comparative genome sequence analysis. *PLoS Biology* **3**: e10.

Edwards, R.A. and Rohwer, F. (2005) Viral metagenomics. *Nature Reviews Microbiology* **3**: 504–510.

Egli, D., Selvaraj, A., Yepiskoposyan, H., Zhang, B., Hafen, E., Georgiev, O., and Schaffner, W. (2003) Knockout of 'metal-responsive transcription factor' MTF-1 in *Drosophila* by homologous recombination reveals its central role in heavy metal homeostasis. *EMBO Journal* **22**: 100–108.

Eide, D.J. (2001) Functional genomics and metal metabolism. *Genome Biology* **2**: 1028.1–1028.3.

Eisen, J.A. (2000) Horizontal gene transfer among microbial genomes: new insights from complete genome analysis. *Current Opinion in Genetics & Development* **10**: 606–611.

Eisen, M.B., Spellman, P.T., Brown, P.O., and Botstein, D. (1998) Cluster analysis and display of genome-wide expression patterns. *Proceedings of the National Academy of Sciences USA* **95**: 14863–14868.

El Fantroussi, S., Urakawa, H., Bernhard, A.E., Kelly, J.J., Noble, P.A., Smidt, H., Yershov, G.M., and Stahl, D.A. (2003) Direct profiling of environmental microbial populations by thermal dissociation analysis of native rRNAs hybridized to oligonucleotide arrays. *Applied and Environmental Microbiology* **69**: 2377–2382.

El-Assal, S.E.-D., Alonso-Blanco, C., Peeters, A.J.M., Raz, V., and Koornneef, M. (2001) A QTL for flowering time in *Arabidopsis* reveals a novel allele of *CRY2*. *Nature Genetics* **29**: 435–440.

Elton, C. (1927) *Animal Ecology*. Methuen & Co., London.

Enard, W., Khaitovich, P., Klose, J., Zöllner, S., Heissig, F., Giavalisco, P., Nieselt-Struwe, K., Muchmore, E., Varki, A., Ravid, R. *et al.* (2002) Intra- and interspecific variation in primate gene expression patterns. *Science* **296**: 340–343.

Evans, J.D. and Wheeler, D.E. (1999) Differential gene expression between developing queens and workers in the honey bee, *Apis mellifera*. *Proceedings of the National Academy of Sciences USA* **96**: 5575–5580.

Evans, J.D. and Wheeler, D.E. (2000) Expression profiles during honeybee caste determination. *Genome Biology* **2**: research0001.1–0001.6.

Evans, J.D. and Wheeler, D.E. (2001) Gene expression and the evolution of insect polyphenisms. *BioEssays* **23**: 62–68.

Falkowksi, P.G. and De Vargas, C. (2004) Shotgun sequencing in the sea: a blast from the past? *Science* **304**: 58–60.

Falkowksi, P.G., Katz, M.E., Knoll, A.H., Quigg, A., Raven, J.A., Schofield, O., and Taylor, F.J.R. (2004) The evolution of modern eukaryotic phytoplankton. *Science* **305**: 354–360.

Featherstone, D.E. and Broadie, K. (2002) Wrestling with pleiotropy: genomic and topological analysis of the yeast gene expression network. *BioEssays* **24**: 267–274.

Feder, M.E. and Hofmann, G.E. (1999) Heat shock proteins, molecular chaperones and the stress response: evolutionary and ecological physiology. *Annual Review of Physiology* **61**: 243–282.

Feder, M.E. and Mitchell-Olds, T. (2003) Evolutionary and ecological functional genomics. *Nature Reviews Genetics* **4**: 649–655.

Fell, D.A. (1992) Metabolic Control Analysis: a survey of its theoretical and experimental development. *Biochemical Journal* **286**: 313–330.

Finkel, T. and Holbrook, N.J. (2000) Oxidants, oxidative stress and the biology of ageing. *Nature* **408**: 239–247.

Fitch, D.H.A. and Thomas, W.K. (1997) Evolution. In *C. Elegans II*, D.L. Riddle, T. Blumenthal, B.J. Meyer, and J.R. Priess (eds), Cold Spring Harbor Monograph Series. Cold Spring Harbor Press, Cold Spring Harbor: 815–850.

Fleischmann, R.D., Adams, M.D., White, O., Clayton, R.A., Kirkness, E.F., Kerlavage, A.R., Bult, C.J., Tomb, J.-F., Dougherty, B.A., Merrick, J.M. *et al.* (1995) Whole-genome random sequencing and assembly of *Haemophilus influenzae* Rd. *Science* **269**: 496–512.

Foster, S.A. and Baker, J.A. (2004) Evolution in parallel: new insights from a classic system. *Trends in Ecology and Evolution* **19**: 456–459.

Fox Keller, E. (2000) *The Century of the Gene*. Harvard University Press, Cambridge, MA.

Fraser, A.G. and Marcotte, E.M. (2004) A probabilistic view of gene function. *Nature Genetics* **36**: 559–564.

Fredrickson, H.L., Perkins, E.J., Bridges, T.S., Tonucci, R.J., Fleming, J.K., Nagel, A., Diedrich, K.,

Mendez-Tenorio, A., Doktycz, M.J., and Beattie, K.L. (2001) Towards environmental toxicogenomics—development of a flow-through, high density DNA hybridization array and its application to ecotoxicity assessment. *The Science of the Total Environment* **274**: 137–149.

Friedman, D.B. and Johnson, T.E. (1988) A mutation in the *age-1* gene in *Caenorhabditis elegans* lengthens life and reduces hermaphrodite fertility. *Genetics* **118**: 75–86.

Friedrich, C.G., Rother, D., Bardischewsky, F., Quentmeier, A., and Fischer, J. (2001) Oxidation of reduced inorganic sulfur compounds by bacteria: emergence of a common mechanism? *Applied and Environmental Microbiology* **67**: 2873–2882.

Fütterer, O., Angelov, A., Liesegang, H., Gottschalk, G., Schleper, C., Schepers, B., Dock, C., Antranikian, G., and Liebl, W. (2004) Genome sequence of *Picrophilus torridus* and its implications for life around pH 0. *Proceedings of the National Academy of Sciences USA* **101**: 9091–9096.

Gabor, E.M., Alkema, W.B.L., and Janssen, D.B. (2004) Quantifying the accessibility of the metagenome by random expression cloning techniques. *Environmental Microbiology* **6**: 948–958.

Galagan, J.E., Calvo, S.E., Borkovich, K.A., Selker, E.U., Read, N.D., Jaffe, D., FitzHugh, W., Ma, L.-J., Smirnov, S., Purcell, S. *et al.* (2003) The genome sequence of the filamentous fungus *Neurospora crassa*. *Nature* **422**: 859–868.

Galibert, F., Finan, T.M., Long, S.R., Pühler, A., Abola, P., Ampe, F., Barloy-Hubler, F., Barnett, M.J., Becker, A., Boistard, P. *et al.* (2001) The composite genome of the legume symbiont *Sinorhizobium meliloti*. *Science* **293**: 668–672.

Gallardo, M.H., Bickham, J.W., Honeycutt, R.L., Ojeda, R.A., and Köhler, N. (1999) Discovery of tetraploidy in a mammal. *Nature* **401**: 341.

Gallardo, M.H., Kausel, G., Jiménez, A., Bacquet, C., González, C., Figuerora, J., Köhler, N., and Ojeda, R. (2004) Whole-genome duplications in South American desert rodents (Octodontidae). *Biological Journal of the Linnean Society* **82**: 443–451.

Garbeva, P., Van Veen, J.A., and Van Elsas, J.D. (2004) Microbial diversity in soil: selection of microbial populations by plant and soil type and implications for disease suppressiveness. *Annual Review of Phytopathology* **42**: 243–270.

Gasch, A.P. and Werner-Washburne, M. (2002) The genomics of yeast responses to environmental stress. *Functional and Integrative Genomics* **2**: 181–192.

Gasch, A.P., Spellman, P.T., Kao, C.M., Carmel-Harel, O., Eisen, M.B., Storz, G., Botstein, D., and Brown, P.O. (2000) Genomic expression programs in the response of yeast cells to environmental changes. *Molecular Biology of the Cell* **11**: 4241–4257.

Gems, D. and Partridge, L. (2001) Insulin/IGF signalling and ageing: seeing the bigger picture. *Current Opinion in Genetics & Development* **11**: 287–292.

Gems, D. and McElwee, J.J. (2003) Microarraying mortality. *Nature* **424**: 259–261.

Getz, W.M., Westerhoff, H.V., Hofmeyr, J.-H.S., and Snoep, J.L. (2003) Control analysis of trophic chains. *Ecological Modelling* **168**: 153–171.

Giannakou, M.E., Goss, M., Jünger, M.A., Hafen, E., Leevers, S.J., and Partridge, L. (2004) Long-lived *Drosophila* with over-expressed dFOXO in adult fat body. *Science* **305**: 361.

Gibson, G. (2002) Microarrays in ecology and evolution: a preview. *Molecular Ecology* **11**: 17–24.

Gibson, G. and Mackay, T.F.C. (2002) Enabling population and quantitative genomics. *Genetical Research (Cambridge)* **80**: 1–6.

Gibson, G. and Muse, S.V. (2002) *A Primer of Genome Science*. Sinauer Associates, Sunderland, MA.

Gillespie, D.E., Brady, S.F., Bettermann, A.D., Cianciotto, N.P., Liles, M.R., Rondon, M.R., Clardy, J., Goodman, R.M., and Handelsman, J. (2002) Isolation of antibiotics turbomycin A and B from a metagenomic library of soil microbial DNA. *Applied and Environmental Microbiology* **68**: 4301–4306.

Gillott, C. (1980) *Entomology*. Plenum Press, New York.

Girardot, F., Monnier, V., and Tricoire, H. (2004) Genome wide analysis of common and specific stress responses in adult drosophila melanogaster. *BMC Genomics* **5**: 1471–2164/5/74.

Goff, S.A., Ricke, D., Lan, T.-H., Presting, G., Wang, R., Dunn, M., Glazebrook, J., Sessions, A., Oeller, P., Varma, H. *et al.* (2002) A draft sequence of the rice genome (*Oryza sativa* L. ssp. *japonica*). *Science* **296**: 92–100.

Goffeau, A., Barrell, B.G., Bussey, H., Davis, R.W., Dujon, B., Feldmann, H., Galibert, F., Hoheisel, J.D., Jacq, C., Johnston, M. *et al.* (1996) Life with 6000 genes. *Science* **274**: 546–567.

Golden, T.R. and Melov, S. (2004) Microarray analysis of gene expression with age in individual nematodes. *Aging Cell* **3**: 111–124.

Goldstein, D.B. and Schlötterer, C. (eds) (1999) *Microsatellites. Evolution and Applications*. Oxford University Press, Oxford.

Gómez-Mena, C., Piñeiro, M., Franco-Zorrilla, J.M., Salinas, J., Coupland, G., and Martínez-Zapater, J.M.

(2001) *early bolting in short days*: An Arabidopsis mutation that causes early flowering and partially suppresses the floral phenotype of leafy. *The Plant Cell* **13**: 1011–1024.

Gonzalez, P., Dominique, Y., Massabuau, J.C., Boudou, A., and Bourdineaud, J.P. (2005) Comparative effects of dietary methylmercury on gene expression in liver, skeletal muscle, and brain of the zebrafis (*Danio rerio*). *Environmental Science and Technology* **39**: 3972–3980.

González, V., Bustos, P., Ramírez-Romero, M.A., Medrana-Soto, A., Salgado, H., Hernández-González, I., Hernández-Celis, J.C., Quintero, V., Moreno-Hagelsieb, G., Girard, L. *et al.* (2003) The mosaic structure of the symbiotic plasmid of *Rhizobium etli* CFN42 and its relation to other symbiotic genome compartments. *Genome Biology* **4**: R36.

Goodner, B., Hinkle, G., Gattung, S., Miller, N., Blanchard, M., Qurollo, B., Goldman, B.S., Cao, Y., Askenazi, M., Halling, C. *et al.* (2001) Genome sequence of the plant pathogen and biotechnology agent *Agrobacterium tumefaciens* C58. *Science* **294**: 2323–2328.

Görner, W., Durchschlag, E., Martinez-Pastor, M.T., Estruch, F., Ammerer, G., Hamilton, B., Ruis, H., and Schüller, C. (1998) Nuclear localization of the C_2H_2 zinc finger protein Mns2p is regulated by stress and protein kinase A activity. *Genes & Development* **12**: 586–597.

Goto, S.G. and Denlinger, D.L. (2002) Short-day and long-day expression patterns of genes involved in the flesh fly clock mechanism: *period, timeless, cycle* and *cryptochrome*. *Journal of Insect Physiology* **48**: 803–816.

Govind, S. and Nehm, R.H. (2004) Innate immunity in fruit flies: a textbook example of genomic recycling. *PLoS Biology* **2**: 1065–1068.

Gracey, A.Y. and Cossins, A.R. (2003) Application of microarray technology in environmental and comparative physiology. *Annual Review of Physiology* **65**: 231–259.

Gracey, A.Y., Troll, J.V., and Somero, G.N. (2001) Hypoxia-induced gene expression profiling in the euryoxic fish *Gillichthys mirabilis*. *Proceedings of the National Academy of Sciences USA* **98**: 1993–1998.

Graur, D. and Li, W.-H. (2000) *Fundamentals of Molecular Evolution*. Sinauer Associates, Sunderland, MA.

Gray, M.W., Burger, G., and Lang, B.F. (1999) Mitochondrial evolution. *Science* **283**: 1476–1481.

Gray, N.D. and Head, I.M. (2001) Linking genetic identity and function in communities of uncultured bacteria. *Environmental Microbiology* **3**: 481–492.

Greer, C.W., Whyte, L.G., Lawrence, J.R., Masson, L., and Brousseau, R. (2001) Genomic technologies for environmental science. *Environmental Science and Technology* **35**: 360A–366A.

Gregory, T.R. (ed.) (2005) *The Evolution of the Genome*. Elsevier Academic Press, Amsterdam.

Greilhuber, J., Dolezel, J., Lysák, M.A., and Bennett, M.D. (2005) The origin, evolution and proposed stabilization of the terms 'Genome size' and 'C-value' to describe nuclear DNA contents. *Annals of Botany* **95**: 255–260.

Griffiths, S., Dunford, R.P., Coupland, G., and Laurie, D.A. (2003) The evolution of *CONSTANS*-like gene families in barley, rice, and *Arabidopsis*. *Plant Physiology* **131**: 1855–1867.

Gross, C., Kelleher, M., Iyer, V.R., Brown, P.O., and Winge, D.R. (2000) Identification of the copper regulon in *Saccharomyces cerevisae* by DNA microarrays. *Journal of Biological Chemistry* **275**: 32310–32316.

Grossman, A.R., Harris, E.E., Hauser, C., Lefebvre, P.A., Martinez, D., Rokhsar, D., Shrager, J., Silflow, C.D., Stern, D., Vallon, O., and Zhang, Z. (2003) *Chlamydomonas reinhardtii* at the crossroads of genomics. *Eukaryotic Cell* **2**: 1137–1150.

Guarante, L. and Kenyon, C. (2000) Genetic pathways that regulate ageing in model organisms. *Nature* **408**: 255–261.

Guiliano, D.B., Hall, N., Jones, S.J.M., Clark, L.N., Corton, C.H., Barrell, B.G., and Blaxter, M.L. (2003) Conservation of long-range synteny and microsynteny between the genomes of two distantly related nematodes. *Genome Biology* **3**: research/0057/I–14.

Gutman, B.L. and Niyogi, K.K. (2004) Chlamydomonas and Arabidopsis. A dynamic duo. *Plant Physiology* **135**: 607–610.

Halitschke, R., Ziegler, J., Keinänen, M., and Baldwin, I.T. (2004) Silencing of hydroperoxide lyase and allene oxide synthase reveals substrate and defense signaling crosstalk in *Nicotiana attenuata*. *The Plant Journal* **40**: 35–46.

Hamadeh, H.K., Bushel, P.R., Jayadev, S., Martin, K., DiSorbo, O., Sieber, S., Bennett, L., Tennant, R., Stoll, R., Barrett, J.C. *et al.* (2002) Gene expression analysis reveals chemical-specific profiles. *Toxicological Sciences* **67**: 219–231.

Handelsman, J. (2004) Metagenomics: applications of genomics to uncultured microorganisms. *Microbiology and Molecular Biology Reviews* **68**: 669–685.

Handelsman, J., Liles, M., Mann, D., Riesenfeld, C., and Goodman, R.M. (2002) Cloning the metagenome: culture-independent access to the diversity and functions of the uncultivated microbial world. *Methods in Microbiology* **33**: 241–255.

Haq, F., Mahoney, M., and Koropatnick, J. (2003) Signaling events for metallothionein induction. *Mutation Research* **533**: 211–226.

Harborne, J.B. (1997) *Introduction to Ecological Biochemistry*. Academic Press, London.

Harmer, S.L., Hogenesch, J.B., Straume, M., Chang, H.-S., Han, B., Zhu, T., Wang, X., Kreps, J.A., and Kay, S.A. (2000) Orchestrated transcription of key pathways in *Arabidopsis* by the circadian clock. *Science* **290**: 2110–2113.

Harries, J.E., Sheahan, D.A., Jobling, S., Matthiessen, P., Neall, P., Sumpter, J.P., Tylor, T., and Zaman, N. (1997) Estrogenic activity in five United Kingdom rivers detected by measurement of vitellogenesis in caged male trout. *Environmental Toxicology and Chemistry* **16**: 534–542.

Harris, E.H. (2001) *Chlamydomonas* as a model organism. *Annual Review of Plant Physiology and Plant Molecular Biology* **52**: 363–406.

Harris, T.W., Chen, N., Cunningham, F., Tello-Ruiz, M., Antoshechkin, I., Bastiani, C., Bieri, T., Blasiar, D., Bradnam, K., Chan, J. *et al.* (2004) WormBase: a multi-species resource for nematode biology and genomics. *Nucleic Acids Research* **32**: D411–D417.

Hartl, D.L. and Clark, A.G. (1997) *Principles of Population Genetics*, 3rd edn. Sinauer Associates, Sunderland, MA.

Hayama, R. and Coupland, G. (2004) The molecular basis of diversity in the photoperiodic flowering responses of Arabidopsis and rice. *Plant Physiology* **135**: 677–684.

Hayama, R., Yokoi, S., Tamaki, S., Yano, M., and Shimamoto, K. (2003) Adaptation of photoperiodic control pathways produces short-day flowering in rice. *Nature* **422**: 719–722.

Haymes, K.M., Van de Weg, W.E., Arens, P., Vosman, B., and Den Nijs, A.P.M. (1997) Molecular mapping and construction of SCAR markers of the strawberry *Rpf1* resistance gene to *Phytophthora fragariae* and their use in breeding programmes. *Acta Horticulturae* **439**: 845–851.

He, Y., Michaels, S.D., and Amasino, R.M. (2003) Regulation of flowering time by histone acetylation in *Arabidopsis*. *Science* **302**: 1751–1754.

Heckel, D.G. (2003) Genomics in pure and applied entomology. *Annual Review of Entomology* **48**: 235–260.

Hedges, S.B. (2002) The origin and evolution of model organisms. *Nature Reviews Genetics* **3**: 838–849.

Heemsbergen, D.A., Berg, M.P., Loreau, M., Van Hal, J.R., Faber, J.H., and Verhoef, H.A. (2004) Biodiversity effects on soil processes explained by interspecific functional dissimilarity. *Science* **306**: 1019–1020.

Heidel, A.J. and Baldwin, I.T. (2004) Microarray analysis of salicylic acid- and jasmonic acid-signalling in responses of *Nicotiana attenuata* to attack by insects from multiple feeding guilds. *Plant, Cell and Environment* **27**: 1362–1373.

Heidelberg, J.F., Paulsen, I.T., Nelson, K.E., Gaidos, E.J., Nelson, W.C., Read, T.D., Eisen, J.A., Seshadri, R., Ward, N., Methe, B. *et al.* (2002) Genome sequence of the dissimilatory metal ion-reducing bacterium *Shewanella oneidensis*. *Nature Biotechnology* **20**: 1118–1123.

Heidelberg, J.F., Seshadri, R., Haveman, S.A., Hemme, C.L., Paulsen, I.T., Kolonay, J.F., Eisen, J.A., Ward, N., Methe, B., Brinkac, L.M. *et al.* (2004) The genome sequence of the anaerobic, sulfate-reducing bacterium *Desulfovibrio vulgaris* Hildenborough. *Nature Biotechnology* **22**: 554–559.

Hekimi, S. and Guarante, L. (2003) Genetics and the specificity of the aging process. *Science* **299**: 1351–1354.

Held, M., Gase, K., and Baldwin, I.T. (2004) Microarrays in ecological research: a case study of a cDNA microarray for plant-herbivore interactions. *BMC Ecology* **4**: 13.

Henne, A., Schmitz, R.A., Bömeke, M., Gottschalk, G., and Daniel, R. (2000) Screening of environmental DNA libraries for the presence of genes conferring lipolytic activity on *Escherichia coli*. *Applied and Environmental Microbiology* **66**: 3113–3116.

Hensbergen, P.J., Donker, M.H., Van Velzen, M.J.M., Roelofs, D., Van der Schors, R.C., Hunziker, P.E., and Van Straalen, N.M. (1999) Primary structure of a cadmium-induced metallothionein from the insect *Orchesella cincta* (Collembola). *European Journal of Biochemistry* **259**: 197–203.

Hill, A.A., Hunter, C.P., Tsung, B.T., Tucker-Kellogg, G., and Brown, E.L. (2000) Genomic analysis of gene expression in *C. elegans*. *Science* **290**: 809–812.

Hirotsune, S., Yoshida, N., Chen, A., Garrett, L., Wynshaw-Boris, A., Sugiyama, F., Takahashi, S., Yagami, K.-I., and Yoshiki, A. (2003) An expressed pseudogene regulates the messenger-RNA stability of its homologous coding gene. *Nature* **423**: 91–96.

Hodgkin, J., Plasterk, R.H.A., and Waterston, R.H. (1995) The nematode *Caenorhabditis elegans* and its genome. *Science* **270**: 410–414.

Hoffmann, A.A., and Hercus, M.J. (2000) Environmental stress as an evolutionary force. *BioScience* **50**: 217–226.

Hoffmann, J.A. and Reichhart, J.-M. (2002) *Drosophila* innate immunity: an evolutionary perspective. *Nature Immunology* **3**: 121–126.

Hoffmann, M.H. (2002) Biogeography of *Arabidopsis thaliana* (L.) Heynh. (Brassicaceae). *Journal of Biogeography* **29**: 125–134.

Holt, R.A., Subramanian, G.M., Halpern, A., Sutton, G.G., Charlab, R., Nusskern, D.R., Wincker, P., Clark, A.G., Ribeiro, J.M.C., Wides, R. *et al.* (2002) The genome sequence of the malaria mosquito *Anopheles gambiae*. *Science* **298**: 129–149.

Holt, S.J. and Riddle, D.L. (2003) SAGE surveys *C. elegans* carbohydrate metabolism: evidence for an anaerobic shift in the long-lived dauer larva. *Mechanisms of Ageing and Development* **124**: 770–800.

Holter, N.S., Mitra, M., Maritan, A., Cieplak, M., Banavar, J.R., and Federoff, N.V. (2000) Fundamental patterns underlying gene expression profiles: simplicity from complexity. *Proceedings of the National Academy of Sciences USA* **97**: 8409–8414.

Holzenberger, M., Dupont, J., Ducos, B., Leneuve, P., Géloën, A., Even, P.C., Cervera, P., and Le Bouc, Y. (2003) IGF-1 receptor regulates lifespan and resistance to oxidative stress in mice. *Nature* **421**: 182–187.

Hopkin, S.P. (1989) *Ecophysiology of Metals in Terrestrial Invertebrates*. Elsevier Applied Science, London.

Horner-Devine, M., Carney, K.M., and Bohannan, B.J.M. (2004) An ecological perspective on bacterial biodiversity. *Proceedings of the Royal Society of London, Series B* **271**: 113–122.

Houthoofd, K., Braeckman, B.P., Lenaerts, I., Brys, K., De Vreese, A., Van Eygen, S., and Vanfleteren, J.R. (2002) Axenic growth up-regulates mass-specific metabolic rate, stress resistance, and extends life span in *Caenorhabditis elegans*. *Experimental Gerontology* **37**: 1369–1376.

Houthoofd, K., Braeckman, B.P., Johnson, T.E., and Vanfleteren, J.R. (2003) Life extension via dietary restriction is independent of the Ins/IGF-1 signalling pathway in *Caenorhabditis elegans*. *Experimental Gerontology* **38**: 947–954.

Hoy, M.A. (1994) *Insect Molecular Genetics*. Academic Press, San Diego.

Hsu, A.-L., Murphy, C.T., and Kenyon, C. (2003) Regulation of aging and age-related disease by DAF-16 and heat-shock factor. *Science* **300**: 1142–1145.

Huang, J., Mullapudi, N., Lancto, C.A., Scott, M., Abrahamsen, M.S., and Kissinger, J.C. (2004) Phylogenomic evidence supports past endosymbiosis, intracellular and horizontal gene transfer in *Cryptosporidium parvum*. *Genome Biology* **5**: R88.

Huber, H., Hohn, M.J., Rachel, R., Fuchs, T., Wimmer, V.C., and Stetter, K.O. (2002) A new phylum of Archaea represented by a nanosized hyperthermophilic symbiont. *Nature* **417**: 63–67.

Hughes, A.L. (1999) *Adaptive Evolution of Genes and Genomes*. Oxford University Press, New York.

Hughes, J.B., Hellmann, J.J., Ricketts, T.H., and Bohannan, B.J. (2001) Counting the uncountable: statistical approaches to estimating diversity. *Applied and Environmental Microbiology* **67**: 4399–4406.

Hui, D., Iqbal, J., Lehman, K., Gase, K., Saluz, H.P., and Baldwin, I.T. (2003) Molecular interactions between the specialist herbivore *Manduca sexta* (Lepidoptera, Sphingidae) and its natual host *Nicotiana attenuata*: V. Microarray analysis and further characterization of large-scale changes in herbivore-induced mRNAs. *Plant Physiology* **131**: 1877–1893.

Hulbert, A.J., Clancy, D.J., Mair, W., Braeckman, B.P., Gems, D., and Partridge, L. (2004) Metabolic rate is not reduced by dietary-restriction or by lowered insulin/IGF-1 signalling and is not correlated with individual lifespan in *Drosophila melanogaster*. *Experimental Gerontology* **39**: 1137–1143.

Hurst, L.D. (2002) The Ka/Ks ratio: diagnosing the form of sequence evolution. *Trends in Genetics* **18**: 486–487.

Hurtado, L.A., Lutz, R.A., and Vrijenhoek, R.C. (2004) Distinct patterns of genetic differentiation among annelids of eastern Pacific hydrothermal vents. *Molecular Ecology* **13**: 2603–2615.

Hutchinson, G.E. (1957) Concluding remarks. In *Cold Spring Harbor Symposia on Quantitative Biology. Volume XXII Population Studies: Animal Ecology and Demography*. Cold Spring Harbor Press, New York: 415–427.

Ideker, T., Galitski, T., and Hood, L. (2000) A new approach to decoding life: systems biology. *Annual Review of Genomics and Human Genetics* **2**: 343–372.

International Chicken Genome Sequencing Consortium (2004) Sequence and comparative analysis of the chicken genome provide unique perspectives on vertebrate evolution. *Nature* **432**: 695–716.

Irving, P., Troxler, L., Heuer, T.S., Belvin, M., Kopczynski, C., Reichhart, J.-M., Hoffmann, J.A., and Hetru, C. (2001) A genome-wide analysis of immune responses in *Drosophila*. *Proceedings of the National Academy of Sciences USA* **98**: 15119–15124.

Jablonka, E. and Lamb, M.J. (2002) The changing concept of epigenetics. *Annals of the New York Academy of Sciences* **981**: 82–96.

Jackson, R.B., Linder, C.R., Lynch, M., Purugganan, M.D., Somerville, S., and Thayer, S.S. (2002) Linking molecular insight and ecological research. *Trends in Ecology and Evolution* **19**: 409–414.

Jacob, F. (1977) Evolution and tinkering. *Science* **196**: 1161–1166.

Jacobson, D.J., Powell, A.J., Dettman, J.R., Saenz, G.S., Barton, M., Hiltz, M.D., Dvorachek Jr., W.H., Glass, N.L., Taylor, J.W., and Natvig, D.O. (2004) *Neurospora* in temperate forests of western North America. *Mycologia* **96**: 66–74.

Jaillon, O., Aury, J.-M., Brunet, F., Petit, J.-L., Stange-Thomann, N., Maucell, E., Bouneau, L., Fischer, C., Ozouf-Costaz, C., Bernot, A. *et al.* (2004) Genome duplication in the teleost fish *Tetraodon nigroviridis* reveals the early vertebrate proto-karyotype. *Nature* **431**: 946–957.

Jain, R., Rivera, M.C., and Lake, J.A. (1999) Horizontal gene transfer among genomes: The complexity hypothesis. *Proceedings of the National Academy of Sciences USA* **96**: 3801–3806.

Jaiswal, A.K. (2004) Nrf2 signaling in coordinated activation of antioxidant gene expression. *Free Radical Biology & Medicine* **36**: 1199–1207.

Jansen, R.C. and Stam, P. (1994) High resolution of quantitative traits into multiple loci via interval mapping. *Genetics* **136**: 1447–1455.

Jansen, R.C. and Nap, J.-P. (2001) Genetical genomics: the added value from segregation. *Trends in Genetics* **17**: 388–391.

Jarne, P. and Lagoda, P.J.L. (1996) Microsatellites, from molecules to populations and back. *Trends in Ecology and Evolution* **11**: 424–429.

Jenkins, B.D., Steward, G.F., Short, S.M., Ward, B.B., and Zehr, J.P. (2004) Fingerprinting diazotroph communities in the Chesapeake Bay by using a DNA macroarray. *Applied and Environmental Microbiology* **60**: 1767–1776.

Jensen, P.R., Snoep, J.L., Molenaar, D., Van Heeswijk, W.C., Khodolenko, B.N., Van der Gugten, A.A., and Westerhoff, H.V. (1995) Molecular biology for flux control. *Biochemical Society Transactions* **23**: 367–370.

Jessup, C.M., Kassen, R., Forde, S.E., Kerr, B., Buckling, A., Rainey, P.B., and Bohannan, B.J.M. (2004) Big questions, small worlds: microbial model systems in ecology. *Trends in Ecology and Evolution* **19**: 189–197.

Ji, W., Wright, M.B., Cai, L., Flament, A., and Lindpaintner, K. (2002) Efficacy of SSH PCR in isolating differentially expressed genes. *BMC Genomics* **3**: 12.

Jiang, M., Ryu, J., Kiraly, M., Duke, K., Reinke, V., and Kim, S.K. (2001) Genome-wide analysis of developmental and sex-regulated expression profiles in *Caenorhabditis elegans*. *Proceedings of the National Academy of Sciences USA* **98**: 218–223.

Jin, W., Riley, R.M., Wolfinger, R.D., White, K.P., Passador-Gurgel, G., and Gibson, G. (2001) The contributions of sex, genotype and age to transcriptional variance in *Drosophila melanogaster*. *Nature Genetics* **29**: 389–395.

Jones, S.J.M., Riddle, D.L., Pouzyrev, A.T., Velculescu, V.E., Hillier, L., Eddy, S.R., Stricklin, S.L., Baillie, D.L., Waterston, R., and Marra, M.A. (2001) Changes in gene expression associated with developmental arrest and longevity in *Caenorhabditis elegans*. *Genome Research* **11**: 1346–1352.

Judice, C., Hartfelder, K., and Pereira, G.A.G. (2004) Caste-specific gene expression in the stingless bee *Melipone quadrifasciata*—are there common patterns in highly social bees? *Insectes Sociaux* **51**: 352–358.

Kacser, H. and Burns, J.A. (1995) The control of flux (with additional comments by H. Kacser and D.A. Fell). *Biochemical Society Transactions* **23**: 341–366.

Kaeberlein, T., Lewis, K., and Epstein, S.S. (2002) Isolating 'uncultivable' microorganisms in pure culture in a simulated natural environment. *Science* **296**: 1127–1129.

Kaltz, O. and Bell, G. (2002) The ecology and genetics of fitness in *Chlamydomonas*. XII. Repeated sexual episodes increase rates of adaptation to novel environment. *Evolution* **56**: 1743–1753.

Kanao, T., Fukui, T., Atomi, H., and Imanaka, T. (2001) ATP-citrate lyase from the green sulfur bacterium *Chlorobium limicola* is a heteromeric enzyme composed of two distinct gene products. *European Journal of Biochemistry* **268**: 1670–1678.

Kaneko, T., Sato, S., Kotani, H., Tanaka, A., Asamizu, E., Nakamura, Y., Miyajima, N., Hirosawa, M., Sugiura, M., Sasamoto, S. *et al.* (1996) Sequence analysis of the genome of the unicellular cyanobacterium *Synechocystis* sp. strain PCC6803. II. Sequence determination of the entire genome and assignment of potential protein-coding regions. *DNA Research* **3**: 109–136.

Kassen, R. and Rainey, P.B. (2004) The ecology and genetics of microbial diversity. *Annual Review of Microbiology* **58**: 207–231.

Kasuga, M., Liu, G., Miura, S., Yamaguchi-Shinozaki, K., and Shinozaki, K. (1999) Improving plant drought, salt, and freezing tolerance by gene transfer of a single stress-inducible transcription factor. *Nature Biotechnology* **17**: 287–291.

Katschinski, D.M. and Glueck, S.B. (2003) Hot worms can handle heavy metal. Focus on 'HIF-1 is required for heat acclimation in the nematode *Caenorhabditis elegans*'. *Physiological Genomics* **14**: 1–2.

Kawasaki, S., Borchert, C., Deyholos, M., Wang, H., Brazille, S., Kawai, K., Galbraith, D., and Bohnert, H.J. (2001) Gene expression profiles during the initial phase of salt stress in rice. *The Plant Cell* **13**: 889–905.

Keightley, P.D. and Eyre-Walker, A. (1999) Terumi Mukai and the riddle of deleterious mutation rates. *Genetics* **153**: 515–523.

Keightley, P.D. and Charlesworth, B. (2005) Genetic instability of *C. elegans* comes naturally. *Trends in Genetics* **21**: 67–70.

Kellis, M., Patterson, N., Endrizzi, M., Birren, B., and Lander, E.S. (2003) Sequencing and comparison of yeast species to identify genes and regulatory elements. *Nature* **423**: 241–254.

Kellis, M., Birren, B.W., and Lander, E.S. (2004) Proof and evolutionary analysis of ancient genome duplication in the yeast *Saccharomyces cerevisiae*. *Nature* **428**: 617–624.

Kenyon, C., Chang, J., Gensch, E., Rudner, A., and Tabtiang, R. (1993) A *C. elegans* mutant that lives twice as long as wild type. *Nature* **366**: 461–464.

Kerr, M.K. and Churchill, G.A. (2001) Statistical design and the analysis of gene expression microarray data. *Genetical Research (Cambridge)* **77**: 123–128.

Kersting, K. (1984) Normalized ecosystem strain: a system parameter for the analysis of toxic stress in (micro-)ecosystems. *Ecological Bulletins* **36**: 150–153.

Kessler, A. and Baldwin, I.T. (2002) Plant responses to insect herbivory: the emerging molecular analysis. *Annual Review of Plant Biology* **53**: 299–328.

Kessler, A., Halitschke, R., and Baldwin, I.T. (2004) Silencing the jasmonate cascade: induced plant defenses and insect populations. *Science* **305**: 665–668.

Kim, H.-J., Hyun, Y., Park, J.-Y., Park, M.-J., Park, M.-K., Kim, M.D., Kim, H.-J., Lee, M.H., Moon, J., Lee, I., and Kim, J. (2004) A genetic link between cold responses and flowering time through *FVE* in *Arabidopsis thaliana*. *Nature Genetics* **36**: 167–171.

Kim, K.W., Shin, J.-H., Moon, J., Kim, M., Lee, J., Park, M.-C., and Lee, I. (2003) The function of flowering time gene *AGL20* is conserved in crucifers. *Molecules and Cells* **16**: 136–141.

Kim, P.M. and Tidor, B. (2003) Subsystem identification through dimensionality reduction of large-scale gene expression data. *Genome Research* **13**: 1706–1718.

Kim, S.K., Lund, J., Kiraly, M., Duke, K., Jiang, M., Stuart, J.M., Eizinger, A., Wylie, B.N., and Davidson, G.S. (2001) A gene expression map for *Caenorhabditis elegans*. *Science* **293**: 2087–2092.

Kirkwood, T.B.L. and Austad, S.N. (2000) Why do we age? *Nature* **408**: 233–238.

Kitano, H. (2002) Systems biology: a brief overview. *Science* **295**: 1662–1664.

Knietsch, A., Waschkowitz, T., Bowien, S., Henne, A., and Daniel, R. (2003) Construction and screening of metagenomic libraries derived from enrichment cultures: generation of a gene bank for genes conferring alcohol oxidoreductase activity on *Escherichia coli*. *Applied and Environmental Microbiology* **69**: 1408–1416.

Kobayashi, M. and Yamamoto, M. (2005) Molecular mechanisms activating the Nrf2-Keap1 pathway of antioxidant gene regulation. *Antioxidants & Redox Signaling* **7**: 385–394.

Koeman, J.H. (1996) Toxicology, history and scope of the field. In *Toxicology. Principles and Applications*, R.J.M. Niesink, J. De Vries, and M.A. Hollinger (eds). CRC Press, Boca Raton, FL: 3–14.

Kole, C., Quijada, P., Michaels, S.D., Amasino, R.M., and Osborn, T.C. (2001) Evidence for homology of flowering-time genes *VFR2* from *Brassica rapa* and *FLC* from *Arabidopsis thaliana*. *Theoretical and Applied Genetics* **102**: 425–430.

Kondrashov, F.A., Rogozin, I.B., Wolf, Y.I., and Koonin, E.V. (2002) Selection in the evolution of gene duplications. *Genome Biology* **3**: research0008.1–0008.9.

Kong, A.-N.T., Owuor, E., Yu, R., Hebbar, V., Chen, C., Hu, R., and Mandlekar, S. (2001) Induction of xenobiotic enzymes by the MAP kinase pathway and the antioxidant or electrophile response element (ARE/EpRE). *Drug Metabolism Reviews* **33**: 255–271.

Kooijman, S.A.L.M. (2000) *Dynamic Energy and Mass Budgets in Biological Systems*. Cambridge University Press, Cambridge.

Koonin, E.V., Makarova, K.S., and Aravind, L. (2001) Horizontal gene transfer in prokaryotes: quantification and classification. *Annual Review of Microbiology* **55**: 709–742.

Koornneef, M. (2004) Naturally occurring genetic variation in *Arabidopsis thaliana*. *Annual Review of Plant Biology* **55**: 141–172.

Koornneef, M., Alonso-Blanco, C., Peeters, A.J.M., and Soppe, W. (1998) Genetic control of flowering time in Arabidopsis. *Annual Review of Plant Physiology and Plant Molecular Biology* **49**: 345–370.

Korsloot, A., Van Gestel, C.A.M., and Van Straalen, N.M. (2004) *Environmental Stress and Cellular Response in Arthropods*. CRC Press, Boca Raton, FL.

Korth, K.L. (2003) Profiling the response of plants to herbivorous insects. *Genome Biology* **4**: 221.

Kowalchuk, G.A. and Stephen, J.R. (2001) Ammonia-oxidizing bacteria: a model for molecular microbial ecology. *Annual Review of Microbiology* **55**: 485–529.

Kozlowski, J. (1993) Measuring fitness in life-history studies. *Trends in Ecology and Evolution* **8**: 84–85.

Krebs, C.J. (1999) *Ecological Methodology*. Addison Wesley Longman, Menlo Park, CA.

Kreps, J.A., Wu, Y., Chang, H.-S., Zhu, T., Wang, X., and Harper, J.F. (2002) Transcriptome changes for *Arabidopsis* in response to salt, osmotic en cold stress. *Plant Physiology* **130**: 2129–2141.

Kulaev, I. and Kulakoskaya, T. (2000) Polyphosphate and phosphate pump. *Annual Review of Microbiology* **54**: 709–734.

Lander, E.S. and Waterman, M.S. (1988) Genomic mapping by fingerprinting random clones: a mathematical analysis. *Genomics* **2**: 231–239.

Lang, B.F., Burger, G., O'Kelly, C.J., Cedergren, R., Golding, G.B., Lemieux, C., Sankoff, D., Turmel, M., and Gray, M.W. (1997) An ancestral mitochondrial DNA resembling a eubacterial genome in miniature. *Nature* **387**: 493–497.

Langefors, Å., Lohm, J., Grahn, M., Andersen, Ø., and Von Schantz, T. (2001) Association between major histocompatibility complex class IIB alleles and resistance to *Aeromonas salmonicida* in Atlantic salmon. *Proceedings of the Royal Society of London, Series B* **268**: 479–485.

Larcher, W. (2003) *Physiological Plant Ecology*, 4th edn. *Ecophysiology and Stress Physiology of Functional Groups*. Springer, Berlin.

Larkin, P., Folmar, L.C., Hemmer, M.J., Poston, A.J., Lee, H.S., and Denslow, N.D. (2002) Array technology as a tool to monitor exposure of fish to xenoestrogens. *Marine Environmental Research* **54**: 395–399.

Laverman, A.M., Speksnijder, A.G.C.L., Braster, M., Kowalchuk, G.A., and Verhoef, H.A. (2001) Spatiotemporal stability of an ammonia-oxidizing community in a nitrogen-saturated forest soil. *Microbial Ecology* **42**: 35–45.

Lawton, J.H. (1994) What do species do in ecosystems? *Oikos* **71**: 367–374.

Lederberg, J. and McCray, A.T. (2001) 'Ome sweet' omics—A genealogical treasure of words. *The Scientist* **15**: 8.

Lee, K.A. and Klasing, K.C. (2004) A role of immunology in invasion biology. *Trends in Ecology and Evolution* **19**: 523–529.

Lee, S.S., Kennedy, S., Tolonen, A.C., and Ruvkun, G. (2003) DAF-16 target genes that control *C. elegans* lifespan and metabolism. *Science* **300**: 644–647.

Leemans, R., Egger, B., Loop, T., Kammermeier, L., He, H., Hartman, B., Certa, U., Hirth, F., and Reichert, H. (2000) Quantitative transcript imaging in normal and heat-shocked *Drosophila* embryos by using high-density oligonucleotide arrays. *Proceedings of the National Academy of Sciences USA* **97**: 12138–12143.

Leroi, A.M. (2001) Molecular signals versus the Loi de Balancement. *Trends in Ecology and Evolution* **16**: 24–29.

Leroi, A.M., Bartke, A., De Benedictis, G., Franceschi, C., Gartner, A., Gonos, E., Feder, M.E., Kisivild, T., Lee, S., Kartal-Özer, N. *et al.* (2005) What evidence is there for the existence of individual genes with antagonistic pleiotropic effects? *Mechanisms of Ageing and Development* **126**: 421–429.

Lesk, A.M. (2002) *Introduction to Bioinformatics*. Oxford University Press, Oxford.

Lessels, K. and Colegrave, N. (2001) Molecular signals or the *Loi de Balancement*? *Trends in Ecology and Evolution* **16**: 284–285.

Levine, S.N. (1989) Theoretical and methodological reasons for variability in the responses of aquatic ecosystem processes to chemical stress. In *Ecotoxicology: Problems and Approaches*, S.A. Levin, M.A. Harwell, J.R. Kelly, and K.D. Kimball (eds). Springer Verlag, New York: 145–179.

Li, X., Schuler, M.A., and Berenbaum, M.R. (2002) Jasmonate and salicylate induce expression of herbivore cytochrome P450 genes. *Nature* **419**: 712–715.

Liang, P. and Pardee, A.B. (1992) Differential display of eukaryotic messenger RNA by means of the polymerase chain reaction. *Science* **257**: 967–970.

Liao, V.H.-C., Dong, J., and Freedman, J.H. (2002) Molecular characterization of a novel, cadmium-inducible gene from the nematode *Caenorabditis elegans*. *Journal of Biological Chemistry* **277**: 42049–42059.

Liberles, D.A., Schreiber, D.R., Govindarajan, S., Chamberlin, S.G., and Benner, S.A. (2001) The Adaptive Evolution Database (TAED). *Genome Biology* **2**: research 0028.1–0028.6.

Liles, M.R., Manske, B.F., Bintrim, S.B., Handelsman, J., and Goodman, R.M. (2003) A census of rRNA genes and linked genomic sequences within a soil metagenomic library. *Applied and Environmental Microbiology* **69**: 2684–2691.

Lin, K., Hsin, H., Libina, N., and Kenyon, C. (2001) Regulation of the Caenorhabditis elegans longevity protein DAF-16 by insulin/IGF-1 and germline signaling. *Nature Genetics* **28**: 139–145.

Lipshutz, R., Fodor, S.P.A., Gingeras, T.R., and Lockhart, D.J. (1999) High density synthetic oligonucleotide arrays. *Nature Genetics Supplement* **21**: 20–24.

Lobry, J.R. (1996) Asymmetric substitution pattern in the two DNA strands of bacteria. *Molecular Biology and Evolution* **13**: 660–665.

Lockhart, D.J. and Winzeler, E.A. (2000) Genomics, gene expression and DNA arrays. *Nature* **405**: 827–836.

Lockhart, D.J., Dong, H., Byrne, M.C., Follettie, M.T., Gallo, M.V., Chee, M.S., Mittmann, M., Wang, C., Kobayashi, M., Horton, H., and Brown, E.L. (1996) Expression monitoring by hybridization to high-density oligonucleotide arrays. *Nature Biotechnology* **14**: 1675–1680.

Loreau, M., Naeem, S., Inchausti, P., Bengtsson, J., Grime, J.P., Hector, A., Hooper, D.U., Huston, M.A., Raffaelli, D., Schmid, B. *et al.* (2001) Biodiversity and ecosystem functioning: current knowledge and future challenges. *Science* **294**: 804–808.

Lorenz, P. and Schleper, C. (2002) Metagenome—a challenging source of enzyme discovery. *Journal of Molecular Catalysis B: Enzymatic* **19–20**: 13–19.

Lotka, A.J. (1924) *Elements of Physical Biology*. Dover Publications, New York.

Lovett, R.A. (2000) Toxicologists brace for the genomics revolution. *Science* **289**: 536–537.

Lovley, D.R. (2003) Cleaning up with genomics: applying molecular biology to bioremediation. *Nature Reviews Microbiology* **1**: 35–44.

Lovley, D.R., Holmes, D.E., and Nevin, K.P. (2004) Dissimilatory Fe(III) and Mn(IV) reduction. *Advances in Microbial Physiology* **49**: 219–286.

Loy, A., Lehner, A., Lee, N., Adamczyk, J., Meier, H., Ernst, J., Schleifer, K.-H., and Wagner, M. (2002) Oligonucleotide microarray for 16S rRNA gene-based detection of all recognized lineages of sulfate-reducing prokaryotes in the environment. *Applied and Environmental Microbiology* **68**: 5064–5081.

Loy, A., Horn, M., and Wagner, M. (2003) probeBase: an online resource for rRNA-targeted oligonucleotide probes. *Nucleic Acids Research* **31**: 514–516.

Lumppio, H.L., Shenvi, N.V., Summers, A.O., Voordouw, G., and Kurtz, Jr., D.M. (2001) Rubrerythrin and rubredoxin oxidoreductase in *Desulfovibrio vulgaris*: a novel oxidative stress protection system. *Journal of Bacteriology* **183**: 101–108.

Lund, J., Tedesco, P., Duke, K., Wang, J., Kim, S.K., and Johnson, T.E. (2002) Transcriptional profile of aging in *C. elegans*. *Current Biology* **12**: 1566–1573.

Lynch, M. and Conery, J.S. (2000) The evolutionary fate and consequences of duplicate genes. *Science* **290**: 1151–1155.

Lynch, M. and Conery, J.S. (2003) The origins of genome complexity. *Science* **302**: 1401–1404.

Ma, L., Li, J., Qu, L., Hager, J., Chen, Z., Zhao, H., and Deng, X.W. (2001) Light control of Arabidopsis development entails coordinated regulation of genome expression and cellular pathways. *The Plant Cell* **13**: 2589–2607.

Maas, M.F.P.M., Van Mourik, A., Hoekstra, R.F., and Debets, A.J.M. (2005) Polymorphism for pKALILO based senescence in Hawaiian populations of *Neurospora intermedia* and *Neurospora tetrasperma*. *Fungal Genetics and Biology* **42**: 224–232.

MacDonald, C.C. and McMahon, K.W. (2003) The flowers that bloom in the sping: RNA processing and seasonal flowering. *Cell* **113**: 671–672.

Macknight, R., Duroux, M., Laurie, R., Dijkwel, P., Simpson, G., and Dean, C. (2002) Functional significance of the alternative transcript processing of the Arabidopsis floral promoter *FCA*. *The Plant Cell* **14**: 877–888.

Madigan, M.T., Martinko, J.M., and Parker, J. (2003) *Brock Biology of Microorganisms*. Prentice Hall, Pearson Education, Upper Saddle River, NJ.

Mair, W., Goymer, P., Pletcher, S.D., and Partridge, L. (2003) Demography of dietary restriction and death in *Drosophila*. *Science* **301**: 1731–1733.

Mair, W., Sgrò, C.M., Johnson, A.P., Chapman, T., and Partridge, L. (2004) Lifespan extension by dietary restriction in female *Drosophila melanogaster* is not caused by a reduction in vitellogenesis or ovarian activity. *Experimental Gerontology* **39**: 1011–1019.

Marshall, E. (2004) Getting the noise out of gene arrays. *Science* **306**: 630–631.

Martin, D.E., Demougin, P., Hall, M.N., and Bellis, M. (2004) Rank difference analysis of microarrays (RDAM), a novel approach to statistical analysis of microarray expression profiling data. *BMC Bioinformatics* **5**: 148.

Martin, W., Rujan, T., Richly, E., Hansen, A., Cornelsen, S., Lins, T., Leister, D., Stoebe, B., Hasegawa, M., and Penny, D. (2002) Evolutionary analysis of *Arabidopsis*, cyanobacterial and chloroplast genomes reveals plastid phylogeny and thousands of cyanobacterial genes in the nucleus. *Proceedings of the National Academy of Sciences USA* **99**: 12246–12251.

Martinez, D., Larrondo, L.F., Putnam, N., Sollewijn Gelpke, M.D., Huang, K., Chapman, J., Helfenbein, K.G., Ramaiya, P., Detter, J.C., Larimer, F. *et al.* (2004) Genome sequence of the lignocellulose degrading fungus *Phanerochaete chrysosporium* strain RP78. *Nature Biotechnology* **22**: 695–700; Erratum, *Nature Biotechnology* **22**: 899.

Matsumura, H., Reich, S., Ito, A., Saitoh, H., Kamoun, S., Winter, P., Kahl, G., Reuter, M., Krüger, D.H., and Terauchi, R. (2003) Gene expression analysis of plant host-pathogen interaction by SuperSAGE. *Proceedings of the National Academy of Sciences USA* **100**: 15718–15723.

Matthiessen, P. (2000) Is endocrine disruption a significant ecological issue? *Ecotoxicology* **9**: 21–24.

Matthysse, A.G., Deschet, K., Williams, M., Marry, M., White, A.R., and Smith, W.C. (2004) A functional cellulose synthase from ascidian epidermis. *Proceedings of the National Academy of Sciences USA* **101**: 986–991.

Mayer, G.D., Leach, A., Kling, P., Olsson, P.-E., and Hogstrand, C. (2003) Activation of the rainbow trout metallothionein-A promoter by silver and zinc. *Comparative Biochemistry and Physiology Part B* **134**: 181–188.

McCaig, A.E., Glover, L.A., and Prosser, J.I. (1999) Molecular analysis of bacterial community structure and diversity in unimproved and improved upland grass pastures. *Applied and Environmental Microbiology* **65**: 1721–1730.

McCarroll, S.A., Murphy, C.T., Zou, S., Pletcher, S.D., Chin, C.-S., Jan, Y.N., Kenyon, C., Bargmann, C.I., and Li, H. (2004) Comparing genomic expression patterns across species identifies shared transcriptional profile in aging. *Nature Genetics* **36**: 197–204.

McElwee, J., Bubb, K., and Thomas, J.H. (2003) Transcriptional outputs of the *Caenorhabditis elegans* forkhead protein DAF-16. *Aging Cell* **2**: 111–121.

McElwee, J.J., Schuster, E., Blanc, E., Thomas, J.H., and Gems, D. (2004) Shared transcriptional signature in *Caenorhabditis elegans* dauer larvae and long-lived *daf-2* mutants implicates detoxification system in longevity assurance. *Journal of Biological Chemistry* **279**: 44533–44543.

McKusick, V.A. and Ruddle, F.H. (1987) A new discipline, a new name, a new journal. *Genomics* **1**: 1–2.

Menges, M., Hennig, L., Gruissem, W., and Murray, J.A.H. (2002) Cell cycle-regulated gene expression in *Arabidopsis*. *Journal of Biological Chemistry* **277**: 41987–42002.

Menges, M., Hennig, L., Gruissem, W., and Murray, J.A.H. (2003) Genome-wide gene expression in an *Arabidopsis* cell suspension. *Plant Molecular Biology* **53**: 423–442.

Menzel, R., Bogaert, T., and Achazi, R. (2001) A systematic gene expression screen of *Caenorhabditis elegans* cytochrome P450 gene reveals CYP35 as strongly xenobiotic inducible. *Archives of Biochemistry and Biophysics* **395**: 158–168.

Methé, B.A., Nelson, K.E., Eisen, J.A., Paulsen, I.T., Nelson, W., Heidelberg, J.F., Wu, D., Wu, M., Ward, N., Beanan, M.J. *et al.* (2003) Genome of *Geobacter sulfurreducens*: metal reduction in subsurface environments. *Science* **302**: 1967–1969.

Miller, T.R. and Belas, R. (2004) Dimethylsulfoniopropionate metabolism by *Pfiesteria*-associated *Roseobacter* spp. *Applied and Environmental Microbiology* **70**: 3383–3391.

Mills, L. and Chichester, C. (2005) Review of evidence: are endocrine-disrupting chemicals in the aquatic environment impacting fish populations? *Science of the Total Environment* **343**: 1–34.

Mira, A., Ochman, H., and Moran, N.A. (2001) Deletion bias and the evolution of bacterial genomes. *Trends in Genetics* **17**: 589–596.

Miracle, A.L., Toth, G.P., and Lattier, D.L. (2003) The path from molecular indicators of exposure to describing dynamic biological systems in an aquatic organism:

microarrays and the fathead minnow. *Ecotoxicology* **12**: 457–462.

Miracle, A.L. and Ankley, G.T. (2005) Ecotoxicogenomics: linkages between exposure and effects in assessing risks of aquatic contaminants to fish. *Reproductive Toxicology* **19**: 321–326.

Mitchell-Olds, T. (2001) *Arabidopsis thaliana* and its wild relatives: a model system for ecology and evolution. *Trends in Ecology and Evolution* **16**: 693–700.

Momose, Y. and Iwahashi, H. (2001) Bioassay of cadmium using a DNA microarray: genome-wide expression patterns of *Saccharomyces cerevisiae* response to cadmium. *Environmental Toxicology and Chemistry* **20**: 2353–2360.

Monteiro, A., Prijs, J., Bax, M., Kahhaart, T., and Brakefield, P.M. (2003) Mutants highlight the modular control of butterfly eyespot patterns. *Evolution & Development* **5**: 160–167.

Moran, P.J., Cheng, Y., Cassell, J.L., and Thompson, G.A. (2002) Gene expression profiling of *Arabidopsis thaliana* in compatible plant-aphid interactions. *Archives of Insect Biochemistry and Physiology* **51**: 182–203.

Morimoto, R.I., Tissières, A., and Georgopoulos, C. (eds) (1994) *The Biology of Heat Shock Proteins and Molecular Chaperones*. Cold Spring Harbor Press, New York.

Mouradov, A., Cremer, F., and Coupland, G. (2002) Control of flowering time: interacting pathways as a basis for diversity. *The Plant Cell* (supplement) 2002: S111–S130.

Mousseau, T.A. and Fox, C.H. (1998) The adaptive significance of maternal effects. *Trends in Ecology and Evolution* **13**: 403–407.

Murakami, S. and Johnson, T.E. (2001) The OLD-1 positive regulator of longevity and stress resistance is under DAF-16 regulation in *Caenorhabditis elegans*. *Current Biology* **11**: 1517–1523.

Murphy, C.T., McCarrol, S.A., Bargmann, C.I., Fraser, A., Kamath, R.S., Ahringer, J., Li, H., and Kenyon, C. (2003) Genes that act downstream of DAF-16 to influence the lifespan of *Caenorhabditis elegans*. *Nature* **424**: 277–284.

Murray, A.W. (2000) Whither genomics? *Genome Biology* **1**: comment 003.1–003.6.

Muyzer, G., De Waal, E.C., and Uitterlinden, A.G. (1993) Profiling of complex microbial populations by denaturing gradient gel electrophoresis analysis of polymerase chain reaction-amplified genes coding for 16S rRNA. *Applied and Environmental Microbiology* **59**: 695–700.

Myers, G. (1999) Whole-genome DNA sequencing. *Computing in Science and Engineering* **1**: 33–43.

Naeem, S., Loreau, M., and Inchausti, P. (2002) Biodiversity and ecosystem functioning: the emergence of a synthetic ecological framework. In *Biodiversity and Ecosystem Functioning*, M. Loreau, S. Naeem, and P. Inchausti (eds). Oxford University Press, Oxford: 3–11.

Nakamura, Y., Gojobori, T., and Ikemura, T. (2000) Codon usage tabulated from international DNA sequence databases: status for the year 2000. *Nucleic Acids Research* **28**: 292.

Nardi, F., Spinsanti, G., Boore, J.L., Carapelli, A., Dallai, R., and Frati, F. (2003) Hexapod origins: monophyletic or paraphyletic? *Science* **299**: 1887–1889.

Nebert, D.W., Roe, A.L., Dieter, M.Z., Solis, W.A., Yang, Y., and Dalton, T.P. (2000) Role of the aromatic hydrocarbon receptor and [Ah] gene battery in the oxidative stress response, cell cycle control, and apoptosis. *Biochemical Pharmacology* **59**: 65–85.

Neefs, J.-M., Van der Peer, Y., De Rijk, P., Chapelle, S., and De Wachter, R. (1993) Compilation of small ribosomal subunit RNA structures. *Nucleic Acids Research* **21**: 3025–3049.

Nei, M. and Gojobori, T. (1986) Simple methods for estimating the numbers of synonymous and non-synonymous nucleotide substitutions. *Molecular Biology and Evolution* **3**: 418–426.

Nei, M., Gu, X., and Sitnikova, T. (1997) Evolution by the birth-and-death process in multigene families of the vertebrate immune system. *Proceedings of the National Academy of Sciences USA* **94**: 7799–7806.

Nelson, K.E., Clayton, R.A., Gill, S.R., Gwinn, M.L., Dodson, R.J., Haft, D.H., Hickey, E.K., Peterson, J.D., Nelson, W.C., Ketchum, K.A. *et al.* (1999) Evidence for lateral gene transfer between Archaea and Bacteria from genome sequence of *Thermotoga maritima*. *Nature* **399**: 323–329.

Nesbø, C.L., L'Haridon, S., Stetter, K.O., and Doolittle, W.F. (2001) Phylogenetic analyses of two 'archaeal' genes in *Thermotoga maritima* reveal multiple transfers between Archaea and Bacteria. *Molecular Biology and Evolution* **18**: 362–375.

Neutel, A.-M., Heesterbeek, J.A.P., and De Ruiter, P.C. (2002) Stability in real food webs: weak links in long loops. *Science* **296**: 1120–1123.

Nguyen, T., Sherrat, P.J., and Pickett, C.B. (2003) Regulatory mechanisms controlling gene expression mediated by the antioxidant response element. *Annual Review of Pharmacology and Toxicology* **43**: 233–260.

Nguyen, T., Yang, C.S., and Pickett, C.B. (2004) The pathways and molecular mechanisms regulating Nrf2 activation in response to chemical stress. *Free Radical Biology & Medicine* **37**: 433–441.

Nielsen, C. (1995) *Animal Evolution. Interrelationships of the Living Phyla.* Oxford University Press, Oxford.

Nijhout, H.F. (2003a) Development and evolution of adaptive polyphenisms. *Evolution & Development* **5**: 9–18.

Nijhout, H.F. (2003b) The control of body size in insects. *Developmental Biology* **261**: 1–9.

Nishida, R. (2002) Sequestration of defensive substances from plants by Lepidoptera. *Annual Review of Entomology* **47**: 57–92.

Nordgren, A., Bååth, E., and Söderström, B. (1983) Microfungi and microbial activity along a heavy metal gradient. *Applied and Environmental Microbiology* **45**: 1829–1837.

North, N.N., Dollhopf, S.L., Petrie, L., Istok, J.D., Balkwill, D.L., and Kostka, J.E. (2004) Change in bacterial community structure during *in situ* biostimulation of subsurface sediment cocontaminated with uranium and nitrate. *Applied and Environmental Microbiology* **70**: 4911–4920.

Novina, C.D. and Sharp, P.A. (2004) The RNAi revolution. *Nature* **430**: 161–164.

Nunes, F.M.F., Valente, V., Sousa, J.F., Cunha, M.A.V., Pinheiro, D.G., Maia, R.M., Araujo, D.D., Costa, M.C.R., Martins, W.K., Carvalho, A.F. *et al.* (2004) The use of open reading frame ESTs (ORESTES) for analysis of the honey bee transcriptome. *BMC Genomics* **5**: 84.

Nuwaysir, E.F., Huang, W., Albert, T.J., Singh, J., Nuwaysir, K., Pitas, A., Richmond, T., Gorksi, T., Berg, J.P., Ballin, J. *et al.* (2002) Gene expression analysis using oligonucleotide arrays produced by maskless photolithography. *Genome Research* **12**: 1749–1755.

Odum, E.P. (1953) *Fundamentals of Ecology.* W.B. Saunders Co., Philadelphia.

Odum, E.P., Finn, J.T., and Franz, E.H. (1979) Perturbation theory and the subsidy-stress gradient. *Bioscience* **29**: 349–352.

Ohno, S. (1972) So much 'junk' DNA in our genome. In *Evolution of Genetic Systems*, H.H. Smith, H.J. Price, A.H. Sparrow, F.W. Studier, and J.D. Yourno (eds). Gordon and Breach, New York: 366–370.

Oldham, S. and Hafen, E. (2003) Insulin/IGF and target of rapamycin signaling: a TOR de force in growth control. *Trends in Cell Biology* **13**: 79–85.

Oleksiak, M., Churchill, G.A., and Crawford, D.L. (2002) Variation in gene expression within and among natural populations. *Nature Genetics* **32**: 261–266.

Oliver, J.L., Bernaola-Galván, P., Carpena, P., and Román-Roldán, R. (2001) Isochore maps of eukaryotic genomes. *Gene* **276**: 47–56.

Olsen, A., Sampayo, J.N., and Lithgow, G.L. (2003) Aging in *C. elegans*. In *Aging of Organisms*, H.D. Osiewacz (ed.). Kluwer Academic Publishers, Dordrecht: 163–199.

Osta, M.A., Christophides, G.K., and Kafatos, F.C. (2004) Effect of mosquito genes on *Plasmodium* development. *Science* 303: 2030–2032.

Øvreås, L. (2000) Population and community level approaches for analysing microbial diversity in natural environments. *Ecology Letters* 3: 236–251.

Ozturk, Z.N., Talamé, V., Deyolos, M., Michalowski, C.B., Galbraith, D.W., Gozokirmizi, N., Tuberosa, R., and Bohnert, H.J. (2002) Monitoring large-scale changes in transcript abundance in drought- and salt-stressed barley. *Plant Molecular Biology* 48: 551–573.

Pace, N.R. (1997) A molecular view of microbial diversity and the biosphere. *Science* 276: 734–740.

Palmiter, R.D. (1994) Regulation of metallothionein genes by heavy metals appears to be mediated by a zinc-sensitive inhibitor that interacts with constitutively active transcription factor, MTF-1. *Proceedings of the National Academy of Sciences USA* 91: 1219–1223.

Palsson, B. (2000) The challenges of in silico biology. *Nature Biotechnology* 18: 1147–1150.

Parkinson, J., Mitreva, M., Whitton, C., Thomson, M., Daub, J., Martin, J., Schmid, R., Hall, N., Barrell, B., Waterston, R.H. *et al.* (2004) A transcriptomic analysis of the phylum Nematoda. *Nature Genetics* 36: 1259–1267.

Partridge, L. (2001) Evolutionary theories of ageing applied to long-lived organisms. *Experimental Gerontology* 36: 641–650.

Partridge, L. and Gems, D. (2002a) The evolution of longevity. *Current Biology* 12: R544–R546.

Partridge, L. and Gems, D. (2002b) Mechanisms of ageing: public or private? *Nature Reviews Genetics* 3: 165–175.

Partridge, L. and Pletcher, S.D. (2003) Genetics of aging in *Drosophila*. In: *Aging of Organisms* H.D. Osiewacz (ed.). Kluwer Academic Publishers, Dordrecht: 125–161.

Passarge, E., Horsthemke, B., and Farber, R.A. (1999) Incorrect use of the term synteny. *Nature Genetics* 23: 387.

Pastorian, K., Hawel, III, L., and Byus, C.V. (2000) Optimization of cDNA representational difference analysis for the identification of differentially expressed mRNAs. *Analytical Biochemistry* 283: 89–98.

Patel, N.H. (2004) Time, space and genomes. *Nature* 431: 28–29.

Paul, J.H., Sullivan, M.B., Segall, A.M., and Rohwer, F. (2002) Marine phage genomics. *Comparative Biochemistry and Physiology Part B* 133: 463–476.

Pearson, G., Serrão, E.A., and Cancela, M.L. (2001) Suppression subtractive hybridization for studying gene expression during aerial exposure and desiccation in fucoid algae. *European Journal of Phycology* 36: 359–366.

Peichel, C.L., Nereng, K.S., Ohgi, K.A., Cole, B.L.E., Colosimo, P.F., Buerkle, C.A., Schluter, D., and Kingsley, D.M. (2001) The genetic architecture of divergence between threespine stickleback species. *Nature* 414: 901–905.

Pennie, W.D., Tugwood, J.D., Oliver, G.J.A., and Kimber, I. (2000) The principles and practice of toxicogenomics: applications and opportunities. *Toxicological Sciences* 54: 277–283.

Peplies, J., Glöckner, F.O., and Amann, R. (2003) Optimization strategies for DNA microarray-based detection of bacteria with 16S rRNA-targeting oligonucleotide probes. *Applied and Environmental Microbiology* 69: 1397–1407.

Perkins, E.J. and Lotufo, G.R. (2003) Playing in the mud—using gene expression to assess contaminant effects on sediment dwelling invertebrates. *Ecotoxicology* 12: 453–456.

Petersen, K., Didion, T., Anderson, C.H., and Nielsen, K.K. (2004) MADS-box genes from perennial ryegrass differentially expressed during transition from vegetative to reproductive growth. *Journal of Plant Physiology* 161: 439–447.

Pichersky, E. and Gang, D.R. (2000) Genetics and biochemistry of secondary metabolites in plants: an evolutionary perpsective. *Trends in Plant Science* 5: 439–445.

Piel, J., Hui, D., Fusetani, N., and Matsunaga, S. (2004) Targeting modular polyketide synthases with iteratively acting acyltransferases from metagenomes of uncultured bacterial consortia. *Environmental Microbiology* 6: 921–927.

Pierik, R., Cuppens, M.L.C., Voesenek, L.A.C.J., and Visser, E.J.W. (2004) Interactions between ethylene and gibberellins in phytochrome-mediated shade avoidance responses in tobacco. *Plant Physiology* 136: 2928–2936.

Pigliucci, M. (1996) How organisms respond to environmental changes: from phenotypes to molecules (and vice versa). *Trends in Ecology and Evolution* 11: 168–173.

Piñeiro, M., Gómez-Mena, C., Schaffer, R., Martínez-Zapater, J.M., and Coupland, G. (2003) EARLY BOLTING IN SHORT DAYS is related to chromatin remodeling factors and regulates flowering in Arabidopsis by repressing *FT*. *The Plant Cell* 15: 1552–1562.

Pletcher, S.D., Macdonald, S.J., Marguerie, R., Certa, U., Stearns, S.C., Goldstein, D.B., and Partridge, L. (2002) Genome-wide transcript profiles in aging and calorically restricted *Drosophila melanogaster*. *Current Biology* **12**: 712–723.

Pokarzhevskii, A.D., Van Straalen, N.M., Zaboev, D.P., and Zaitsev, A.S. (2003) Microbial links and element flows in nested detrital food-webs. *Pedobiologia* **47**: 213–224.

Polz, M.F., Bertilsson, S., Acinas, S.G., and Hunt, D. (2003) A(r)ray of hope in analysis of the function and diversity of microbial communities. *Biological Bulletin* **204**: 196–199.

Postma, E. and Van Noordwijk, A.J. (2005) Gene flow maintains a large genetic difference in clutch size at a small spatial scale. *Nature* **433**: 65–68.

Price, P.W., Bouton, C.E., Gross, P., McPheron, B.A., Thompson, J.N., and Weis, A.E. (1980) Interactions among three trophic levels: Influence of plants on interactions between insect herbivores and natural enemies. *Annual Review of Ecology and Systematics* **11**: 41–65.

Prince, V.E. and Pickett, F.B. (2002) Splitting pairs: the diverging fates of duplicated genes. *Nature Reviews Genetics* **3**: 827–837.

Procaccini, G., Pischetola, M., and Di Lauro, R. (2000) Isolation and characterization of microsatellite loci in the ascidian *Ciona intestinalis* (L.). *Molecular Ecology* **9**: 1924–1926.

Pühler, A. and Selbitschka, W. (2003) Genome research on bacteria relevant for agriculture, environment and biotechnology. *Journal of Biotechnology* **106**: 119–120.

Purohit, H.J., Raje, D.V., Kapley, A., Padmanabhan, P., and Singh, R.N. (2003) Genomic tools in environmental impact assessment. *Environmental Science and Technology* **37**: 337A–368A.

Putterrill, J., Laurie, R., and Macknight, R. (2004) It's time to flower: the genetic control of flowering time. *BioEssays* **26**: 363–373.

Quackenbush, J. (2001) Computational analysis of microarray data. *Nature Reviews Genetics* **2**: 410–427.

Quail, P.H. (2002) Phytochrome photosensory signalling networks. *Nature Reviews Molecular Cell Biology* **3**: 85–93.

Quaiser, A., Ochsenreiter, T., Klenk, H.-P., Kleftzin, A., Treusch, A.H., Meurer, G., Eck, J., Sensen, C.W., and Schleper, C. (2002) First insight into the genome of an uncultivated crenarchaeote from soil. *Environmental Microbiology* **4**: 603–611.

Quaiser, A., Ochsenreiter, T., Lanz, C., Schuster, S.C., Treusch, A.H., Eck, J., and Schleper, C. (2003) Acidobacteria form a coherent but highly diverse group within the bacterial domain: evidence from environmental genomics. *Molecular Microbiology* **50**: 563–575.

Quesada, V., Macknight, R., Dean, C., and Simpson, G.G. (2003) Autoregulation of *FCA* pre-mRNA processing controls *Arabidopsis* flowering time. *EMBO Journal* **22**: 3142–3152.

Radajewski, S., McDonald, I.R., and Murrell, J.C. (2003) Stable-isotope probing of nucleic acids: a window to the function of uncultured microorganisms. *Current Opinion in Biotechnology* **14**: 296–302.

Ram, R.J., VerBerkmoes, N.C., Thelen, M.P., Tyson, G.W., Baker, B.J., Blake, II, R.C., Shah, M., Hettich, R.L., and Banfield, J.F. (2005) Community proteomics of a natural microbial biofilm. *Science* **308**: 1915–1920.

Ranson, H., Claudianos, C., Ortelli, F., Abgrall, C., Hemingway, J., Sharakhova, M.V., Unger, M.F., Collins, F.H., and Feyereisen, R. (2002) Evolution of supergene families associated with insecticide resistance. *Science* **298**: 179–181.

Ranz, J.M., Castillo-Davis, C.I., Meiklejohn, C.D., and Hartl, D.L. (2003) Sex-dependent gene expression and evolution of the *Drosophila* transcriptome. *Science* **300**: 1742–1745.

Rasmussen, R., Morrison, R., Herrmann, M., and Wittwer, C. (1998) Quantitative PCR by continuous fluorescence monitoring of a double strand DNA specific binding dye. *Biochimica* **2**: 8–11.

Ratcliffe, O.J. and Riechmann, J.L. (2002) Arabidopsis transcription factors and the regulation of flowering time: a genomic perspective. *Current Issues in Molecular Biology* **4**: 77–91.

Ravasz, E., Somera, A.L., Mongru, D.A., Oltvai, Z.N., and Barabási, A.-L. (2002) Hierarchical organization of modularity in metabolic networks. *Science* **297**: 1551–1555.

Raven, J.A. and Allen, J.F. (2003) Genomics and chloroplast evolution: what did cyanobacteria do for plants? *Genome Biology* **4**: 209.

Rebrikov, D.V., Bulina, M.E., Bogdanova, E., Vagner, L.L., and Lukyanov, S.A. (2002) Complete genome sequence of a novel extrachromosomal virus-like element identified in planarian *Girardia tigrina*. *BMC Genomics* **3**: 15.

Rebrikov, D.V., Desai, S.M., Siebert, P.D., and Lukyanov, S.A. (2004) Suppression subtractive hybridisation. In *Gene Expression Profiling. Methods and Protocols*, R.A. Shimkets (ed.). Humana Press, Totowa, NJ: 107–134.

Reinke, V. and White, K.P. (2002) Developmental genomic approaches in model organisms. *Annual Review of Genomics and Human Genetics* **3**: 153–178.

Rendulic, S., Jagtap, P., Rosinus, A., Eppinger, M., Baar, C., Lanz, C., Keller, H., Lambert, C., Evans, K.J., Goesmann, A. *et al.* (2004) A predator unmasked: life cycle of *Bdellovibrio bacteriovorus* from a genomic perspective. *Science* 303: 689–692.

Reymond, P., Weber, H., Diamond, M., and Farmer, E.E. (2000) Differential gene expression in response to mechanical wounding and insect feeding in *Arabidopsis*. *The Plant Cell* 12: 707–719.

Reysenbach, A.-L. and Shock, E. (2002) Merging genomes with geochemistry in hydrothermal ecosystems. *Science* 296: 1077–1082.

Riddle, D.L. (1988) The dauer larva. In *The Nematode Caenorhabditis elegans*, W.B. Wood and the Community of *C. elegans* Researchers (eds). Cold Spring Harbor Press, Cold Spring Harbor: 393–412.

Riesenfeld, C.S., Schloss, P.D., and Handelsman, J. (2004a) Metagenomics: genome analysis of microbial communities. *Annual Review of Genetics* 38: 525–552.

Riesenfeld, C.S., Goodman, R.M., and Handelsman, J. (2004b) Uncultured soil bacteria are a reservoir of new antibiotic resistance genes. *Environmental Microbiology* 6: 981–989.

Rizhsky, L., Liang, H., Shuman, J., Shulaev, V., Davletova, S., and Mittler, R. (2004) When defense pathways collide. The response of Arabidopsis to a combination of drought and heat stress. *Plant Physiology* 134: 1683–1696.

Robbins, A.H., McRee, D.E., Williamson, M., Collet, S.A., Xuong, N.H., Furey, W.F., Wang, B.C., and Stout, C.D. (1991) Refined crystal structure of Cd, Zn metallothionein at 2.0 Å resolution. *Journal of Molecular Biology* 221: 1269–1293.

Roda, A., Halitschke, R., Steppuhn, A., and Baldwin, I.T. (2004) Individual variability in herbivore-specific elicitors from the plant's perspective. *Molecular Ecology* 13: 2421–2433.

Rodrigues-Pousada, C.A., Nevitt, T., Menezes, R., Azevedo, D., Pereira, J., and Amaral, C. (2004) Yeast activator proteins and stress response: an overview. *FEBS Letters* 567: 80–85.

Roelofs, D., Overhein, L., De Boer, M.E., Janssens, T.K.S., and Van Straalen, N.M. (2006) Additive genetic variation of transcriptional regulation: metallothionein expression in the soil insect *Orchesella cincta*. *Heredity* 96: 85–92.

Roesijadi, G. (1996) Metallothionein and its role in toxic metal regulation. *Comparative Biochemistry and Physiology* 113C: 117–123.

Roff, D.A. (2002) *Life History Evolution*. Sinauer Associates, Sunderland, MA.

Rogina, B., Helfand, S.L., and Frankel, S. (2002) Longevity regulation by *Drosophila* Rpd3 deacetylase and caloric restriction. *Science* 298: 1745.

Röling, W.F.M., Van Breukelen, B., Braster, M., Lin, B., and Van Verseveld, H.W. (2001) Relationships between microbial community structure and hydrochemistry in a landfill leachate-polluted aquifer. *Applied and Environmental Microbiology* 67: 4619–4629.

Röling, W.F.M., Milner, M.G., Jones, D.M., Fratepietro, F., Swannell, R.P.J., Daniel, F., and Head, I.M. (2004) Bacterial community dynamics and hydrocarbon degradation during a field-scale evaluation of bioremediation on a mudflat beach contaminated with buried oil. *Applied and Environmental Microbiology* 70: 2603–2613.

Rondon, M.R., Goodman, R.M., and Handelsman, J. (1999) The Earth's bounty: assessing and accessing soil microbial diversity. *Trends in Biotechnology* 17: 403–409.

Rondon, M.R., August, P.R., Bettermann, A.D., Brady, S.F., Grossman, T.H., Liles, M.R., Loiacono, K.A., Lynch, B.A., MacNeil, I.A., Minor, C. *et al.* (2000) Cloning the soil metagenome: a strategy for accessing the genetic and functional diversity of uncultured microorganisms. *Applied and Environmental Microbiology* 66: 2541–2547.

Rosenberg, S.M. and Hastings, P.J. (2004) Worming into genetic instability. *Nature* 430: 625–626.

Roskam, J.C. and Brakefield, P.M. (1999) Seasonal polyphenism in *Bicyclus* (Lepidoptera: Satyridae) butterflies: different climates need different cues. *Biological Journal of the Linnean Society* 66: 345–356.

Rubin, G.M. and Lewis, E.B. (2000) A brief history of Drosophila's contributions to genome research. *Science* 287: 2216–2218.

Rubin, G.M., Yandell, M.D., Wortman, J.R., Gabor Miklos, G.L., Nelson, C.R., Hariharan, I.K., Fortini, M.E., Li, P.W., Apweiler, R., Fleischmann, W. *et al.* (2000) Comparative genomics of the eukaryotes. *Science* 287: 2204–2215.

Russell, P.J. (2002) *iGenetics*. Pearson Education/Benjamin Cummings, San Fransisco.

Rutledge, R.G. and Côté, C. (2003) Mathematics of quantitative kinetic PCR and the application of standard curves. *Nucleic Acids Research* 31: e93.

Saccone, C. and Pesole, G. (2003) *Handbook of Comparative Genomics*. John Wiley & Sons, Hoboken, NJ.

Saint André, A.V., Blackwell, N.M., Hall, L.R., Hoerauf, A., Brattig, N.W., Volkmann, L., Taylor, M.J., Ford, L., Hise, A.G., Lass, J.H. *et al.* (2002) The role of endosymbiotic Wolbachia bacteria in the pathogenesis of river blindness. *Science* 295: 1892–1895.

Sanger, F., Air, G.M., Barrell, B.G., Brown, N.L., Coulson, A.R., Fiddes, J.C., Hutchison III, C.A., Slocombe, P.M., and Smith, M. (1977a) Nucleotide sequence of bacteriophage ΦX174 DNA. *Nature* **265**: 687–695.

Sanger, F., Nicklen, S., and Coulson, A.R. (1977b) DNA sequencing with chain-terminating inhibitors. *Proceedings of the National Academy of Sciences USA* **74**: 5463–5467.

Sansone, S.A., Morrisson, N., Rocca-Serra, P., and Fostel, J.M. (2004) Standardization initiatives in the (eco)toxicogenomics domain: a review. *Comparative and Functional Genomics* **5**: 633–641.

Santoro, M.G. (2000) Heat shock factors and the control of the stress response. *Biochemical Pharmacology* **59**: 55–63.

Schaffer, R., Landgraf, J., Accerbi, M., Simon, V., Larson, M., and Wisman, E. (2001) Microarray analysis of diurnal and circadian-regulated genes in Arabidopsis. *The Plant Cell* **13**: 113–123.

Schena, M., Shalon, D., Davis, R.W., and Brown, P.O. (1995) Quantitative monitoring of gene expression patterns with a complementary DNA microarray. *Science* **270**: 467–470.

Schepens, I., Duek, P., and Fankhauser, C. (2004) Phytochrome-mediated light signalling in Arabidopsis. *Current Opinion in Plant Biology* **7**: 564–569.

Schindler, D.W. (1987) Detecting ecosystem responses to anthropogenic stress. *Canadian Journal of Fisheries and Aquatic Sciences* **44**: 6–25.

Schink, B. and Friedrich, M. (2000) Phosphite oxidation by sulphate reduction. *Nature* **406**: 37.

Schittko, U., Hermsmeier, D., and Baldwin, I.T. (2001) Molecular interactions between the specialist herbivore *Manduca sexta* (Lepidoptera, Sphingidae) and its natural host *Nicotiana attenuata*. II. Accumulation of plant mRNAs in response to insect-derived cues. *Plant Physiology* **125**: 701–710.

Schleper, C., Jurgens, G., and Jonuscheit, M. (2005) Genomic studies of uncultivated Archaea. *Nature Reviews Microbiology* **3**: 479–488.

Schlichting, C.D. and Smith, H. (2002) Phenotypic plasticity: linking molecular mechanisms with evolutionary outcomes. *Evolutionary Ecology* **16**: 189–211.

Schloss, P.D. and Handelsman, J. (2003) Biotechnological prospects from metagenomics. *Current Opinion in Biotechnology* **14**: 303–310.

Schmid, M., Uhlenhaut, N.H., Godard, F., Demar, M., Bressan, R., Weigel, D., and Lohmann, J.U. (2003) Dissection of floral induction pathways using global expression analysis. *Development* **130**: 6001–6012.

Schmidt-Nielsen, K. (1997) *Animal Physiology. Adaptation and Environment*, 5th edn. Cambridge University Press, Cambridge.

Schulte, P.M. (2004) Changes in gene expression as biochemical adaptations to environmental change: a tribute to Peter Hochachka. *Comparative Biochemistry and Physiology Part B* **139**: 519–529.

Schulze, A. and Downward, J. (2001) Navigating gene expression using microarrays—a technology review. *Nature Cell Biology* **3**: E190–E195.

Schulze, E.-D. and Mooney, H.A. (eds) (1993) *Biodiversity and Ecosystem Function*. Springer-Verlag, Berlin.

Schweitzer, J.A., Balley, J.K., Rehill, B.J., Martinsen, G.D., Hart, S.C., Lindroth, R.L., Keim, P., and Whitham, T.G. (2004) Genetically based trait in a dominant tree affects ecosystem processes. *Ecology Letters* **7**: 127–134.

Sebat, J.L., Colwell, F.S., and Crawford, R.L. (2003) Metagenomic profiling: microarray analysis of an environmental genomic library. *Applied and Environmental Microbiology* **69**: 4927–4934.

Segal, E., Shapira, M., Regev, A., Pe'er, D., Botstein, D., Koller, D., and Friedman, N. (2003) Module networks: identifying regulatory modules and their condition-specific regulators from gene expression data. *Nature Genetics* **34**: 166–176.

Seki, M., Narusaka, M., Abe, H., Kasuga, M., Yamaguchi-Shinozaki, K., Carninci, P., Hayashizaki, Y., and Shonozaki, K. (2001) Monitoring expression pattern of 1300 *Arabidopsis* genes under drought and cold stresses by using a full-length cDNA microarray. *The Plant Cell* **13**: 61–72.

Seki, M., Narusaka, M., Ishida, J., Nanjo, T., Fujita, M., Ouno, Y., Kamiya, A., Nakajima, M., Enju, A., Sakurai, T. *et al.* (2002) Monitoring the expression profiles of 7000 *Arabidopsis* genes under drought, cold and high-salinity stresses using a full-length cDNA microarray. *The Plant Journal* **31**: 279–292.

Sharbel, T.F., Huabold, B., and Mitchell-Olds, T. (2000) Genetic isolation by distance in *Arabidopsis* biogeography and postglacial colonization of Europe. *Molecular Ecology* **9**: 2109–2118.

She, X., Jiang, Z., Clark, R.A., Liu, G., Cheng, Z., Tuzun, E., Church, D.M., Sutton, G., Halpern, A.L., and Eichler, E.E. (2004) Shotgun sequence assembly and recent segmental duplications within the human genome. *Nature* **431**: 927–930.

Shimkets, R.A. (ed.) (2004) *Gene Expression Profiling. Methods and Protocols*. Humana Press, Totowa, NJ.

Short, S.M. and Suttle, C.A. (2002) Sequence analysis of marine virus communities reveals that groups of related algal viruses are widely distributed in nature. *Applied and Environmental Microbiology* **68**: 1290–1296.

Shrader, E.A., Henry, T.R., Greeley, Jr., M.S., and Bradley, B.P. (2003) Proteomics in zebrafish exposed to endocrine disrupting chemicals. *Ecotoxicology* **12**: 485–488.

Simpson, G.C. and Dean, C. (2002) Arabidopsis, the Rosetta stone of flowering time? *Science* **296**: 285–289.

Simpson, G.G., Dijkwel, P.P., Quesada, V., Henderson, I., and Dean, C. (2003) FY is an RNA 3′ end-processing factor that interacts with FCA to control the *Arabidopsis* floral transition. *Cell* **113**: 777–787.

Slack, J.M.W., Holland, P.W.H., and Graham, C.F. (1993) The zootype and the phylotypic stage. *Science* **361**: 490–492.

Small, J., Call, D.R., Brockman, F.J., Straub, T.M., and Chandler, D.P. (2001) Direct detection of 16S rRNA in soil extracts by using oligonucleotide microarrays. *Applied and Environmental Microbiology* **67**: 4708–4716.

Smith, L.M., Sanders, J.Z., Kaiser, R.J., Hughes, P., Dodd, C., Connell, C.R., Heiner, C., Kent, S.B.H., and Hood, L.E. (1986) Fluorescence detection in automated DNA sequence analysis. *Nature* **321**: 674–679.

Smyth, G.K., Yang, Y.H., and Speed, T. (2003) Statistical issues in cDNA microarray analysis. In *Functional Genomics: Methods and Protocols*, M.J. Brownstein and A.B. Khodurksy (eds). Humana Press, Totowa, NJ: 111–136.

Snape, J.R., Maund, S.J., Pickford, D.B., and Hutchinson, T.H. (2004) Ecotoxicogenomics: the challenge of integrating genomics into aquatic and terrestrial ecotoxicology. *Aquatic Toxicology* **67**: 143–154.

Snel, B., Bork, P., and Huynen, M. (1999) Genome phylogeny based on gene content. *Nature Genetics* **21**: 108–110.

Snell, T.W., Brogdon, S.E., and Morgan, M.B. (2003) Gene expression profiling in ecotoxicology. *Ecotoxicology* **12**: 475–483.

Snyder, M. and Gerstein, M. (2003) Defining genes in the genomics era. *Science* **300**: 258–260.

Sokal, R.R. and Rohlf, F.J. (1995) *Biometry. The Principles and Practice of Statistics in Biological Research*, 3rd edn. W.H. Freeman and Company, San Fransisco.

Sorrells, M.E., La Rota, M., Bermudez-Kandianis, C.E., Greene, R.A., Kantety, R., Munkvold, J.D., Miftahudin, Mahmoud, A., Ma, X., Gustafson, P.J. et al. (2003) Comparative DNA sequence analysis of wheat and rice genomes. *Genome Research* **13**: 1818–1827.

Soto, A.M., Justicia, H., Wray, J.W., and Sonnenschein, C. (1991) p-Nonyl-phenol: an estrogenic xenobiotic released from 'modified' polystyrene. *Environmental Health Perspectives* **92**: 167–173.

Spellman, P.T. and Rubin, G.M. (2002) Evidence for large domains of similarly expressed genes in the Drosophila genome. *Journal of Biology* **1**: 5.

Spencer, J.F.T. and Spencer, D.M. (1997a) Ecology: where yeasts live. In *Yeasts in Natural and Artificial Habitats*, J.F.T. Spencer and D.M. Spencer (eds). Springer-Verlag, Berlin: 33–58.

Spencer, J.F.T. and Spencer, D.M. (1997b) Taxonomy: the names of the yeasts. In *Yeasts in Natural and Artificial Habitats*, J.F.T. Spencer and D.M. Spencer (eds). Springer-Verlag, Berlin: 11–32.

Sprague, J., Doerry, E., Douglas, S., Westerfield, M., and the ZFIN Group (2001) The Zebrafish Information Network (ZFIN): a resource for genetic, genomic and developmental research. *Nucleic Acids Research* **29**: 87–90.

Spring, J. (1997) Vertebrate evolution by interspecific hybridisation—are we polyploid? *FEBS Letters* **400**: 2–8.

Stearns, S.C. (1992) *The Evolution of Life Histories*. Oxford University Press, Oxford.

Stearns, S.C. and Magwene, P. (2003) The naturalist in a world of genomics. *American Naturalist* **161**: 171–180.

Stein, L.D. (2004) End of the beginning. *Nature* **431**: 915–916.

Sterenborg, I. and Roelofs, D. (2003) Field-selected cadmium tolerance in the springtail *Orchesella cincta* is correlated with increased metallothionein mRNA expression. *Insect Biochemistry and Molecular Biology* **33**: 741–747.

Steward, G.F., Jenkins, B.D., Ward, B.B., and Zehr, J.P. (2004) Development and testing of a DNA macroarray to assess nitrogenase (nifH) gene diversity. *Applied and Environmental Microbiology* **70**: 1455–1465.

Stinchcombe, J.R., Weinig, C., Ungerer, M., Olsen, K.M., Mays, C., Halldorsdottir, S.S., Purugganan, M.D., and Schmitt, J. (2004) A latitudinal cline in flowering time in *Arabidopsis thaliana* modulated by the flowering time gene *FRIGIDA*. *Proceedings of the National Academy of Sciences USA* **101**: 4712–4717.

Stolc, V., Gauhar, Z., Mason, C., Halasz, G., Van Batenburg, M.F., Rifkin, S.A., Hua, S., Herreman, T., Tongprasit, W., Barbano, P.E. et al. (2004) A gene expression map for the euchromatic genome of *Drosophila melanogaster*. *Science* **306**: 655–660.

Strange, K. (2005) The end of 'naïve reductionism': rise of systems biology or renaissance of physiology? *American Journal of Physiology Cell Physiology* **288**: 968–974.

Straub, P.F., Higham, M.L., Tanguy, A., Landau, B.J., Phoel, W.C., Hales, Jr., L.S., and Thwing, T.K.M. (2004) Suppression subtractive hybridization cDNA libraries to identify differentially expressed genes from contrasting fish habitats. *Marine Biotechnology* **6**: 386–399.

Strauss, S.H. and Martin, F.M. (2004) Poplar genomics comes of age. *New Phytologist* **164**: 1–4.

Stürzenbaum, S.R. and Kille, P. (2001) Control genes in quantitative molecular biological techniques: the variability of invariance. *Comparative Biochemistry and Physiology B* **130**: 281–289.

Stürzenbaum, S.R., Winters, C., Galay, M., Morgan, A.J., and Kille, P. (2001) Metal ion trafficking in earthworms. Identification of a cadmium-specific metallothionein. *Journal of Biological Chemistry* **276**: 34013–34018.

Sultan, A., Abelson, A., Bresler, V., Fishelson, L., and Mokady, O. (2000) Biomonitoring marine environmental quality at the level of gene expression—testing the feasibility of a new approach. *Water Science and Technology* **42**: 269–274.

Sumpter, J.P. (2005) Endocrine disrupters in the aquatic environment: an overview. *Acta Hydrochimica et Hydrobiologica* **33**: 9–16.

Sung, S. and Amasino, R.M. (2004) Vernalization in *Arabidopsis thaliana* is mediated by the PHD finger protein VIN3. *Nature* **427**: 159–164.

Sunnucks, P. (2000) Efficient genetic markers for population biology. *Trends in Ecology and Evolution* **15**: 199–203.

Talbert, P.B., Bryson, T.D., and Henikoff, S. (2004) Adaptive evolution of centromere proteins in plants and animals. *Journal of Biology* **3**: 1–17.

Tamaoki, M., Nakajima, N., Kubo, A., Aono, M., Matsuyama, T., and Saji, H. (2003) Transcriptome analysis of O_3-exposed *Arabidopsis* reveals that multiple signal pathways act mutually antagonistically to induce gene expression. *Plant Molecular Biology* **53**: 443–456.

Tamayo, P., Slonim, D., Mesirov, J., Zhu, Q., Kitareewan, S., Dmitrovsky, E., Lander, E.S., and Golub, T.R. (1999) Interpreting patterns of gene expression with self-organizing maps: methods and application to hematopoietic differentiation. *Proceedings of the National Academy of Sciences USA* **96**: 2907–2912.

Tamura, S., Hanada, M., Ohnishi, M., Katsura, K., Sasaki, M., and Kobayashi, T. (2002) Regulation of stress-activated protein kinase signaling pathways by protein phosphatases. *European Journal of Biochemistry* **269**: 1060–1066.

Tan, P.K., Downey, T.J., Spitznagel, E.L., Xu, P., Fu, D., Dimitrov, D.S., Lempicki, R.A., Raaka, B.M., and Cam, M.C. (2003) Evaluation of gene expression measurements from commercial microarray platforms. *Nucleic Acids Research* **31**: 5676–5684.

Tanguy, A., Boutet, I., Laroche, J., and Moraga, D. (2005) Molecular identification and expression study of differentially regulated genes in the Pacific oyster *Crassostrea gigas* in response to pesticide exposure. *FEBS Journal* **272**: 390–403.

Tanksley, S.D. (1993) Mapping polygenes. *Annual Review of Genetics* **27**: 205–233.

Taroncher-Oldenburg, G., Griner, E.M., Francis, C.A., and Ward, B.B. (2003) Oligonucleotide microarray for the study of functional gene diversity in the nitrogen cycle in the environment. *Applied and Environmental Microbiology* **69**: 1159–1171.

Tatar, M., Bartke, A., and Antebi, A. (2003) The endocrine regulation of aging by insulin-like signals. *Science* **299**: 1346–1351.

Tatar, M., Kopelman, A., Epstein, D., Tu, M.-P., Yin, C.-M., and Garofalo, R.S. (2001) A mutant *Drosophila* insulin receptor homolog that extends life-span and impairs neuroendocrine function. *Science* **292**: 107–1110.

Taylor, G. (2002) *Populus*: Arabidopsis for forestry. Do we need a model tree? *Annals of Botany* **90**: 681–689.

Teske, A., Dhillon, A., and Sogin, M.L. (2003) Genomic markers of ancient anaerobic microbial pathways: sulfate reduction, methanogenesis, and methane oxidation. *Biological Bulletin* **204**: 186–191.

Thiele, D.J. (1992) Metal-regulated transcription in eukaryotes. *Nucleic Acids Research* **20**: 1183–1191.

Thomas, M.A. and Klaper, R. (2004) Genomics for the ecological toolbox. *Trends in Ecology and Evolution* **19**: 439–445.

Thompson, D.A.W. (1917) *On Growth and Form*. Cambridge University Press, Cambridge.

Tiedje, J.M. and Zhou, J. (2004) Future perspectives: genomics beyond single cells. In *Microbial Functional Genomics*, J. Zhou, D.K. Thompson, Y. Xu, and J.M. Tiedje (eds). John Wiley & Sons, Hoboken, NJ: 477–486.

Tillier, E.R.M. and Collins, R.A. (2000) The contribution of replication orientation, gene direction, and signal sequences to base-composition asymmetries in bacterial genomes. *Journal of Molecular Evolution* **50**: 249–257.

Tiquia, S.M., Wu, L., Chong, S.C., Passovets, S., Xu, D., Xu, Y., and Zhou, J. (2004) Evaluation of 50-mer oligonucleotide arrays for detecting microbial populations in environmental samples. *BioTechniques* **36**: 664–675.

Tissenbaum, H.A. and Guarante, L. (2001) Increased dosage of a *sir-2* gene extends lifespan in *Caenorhabditis elegans*. *Nature* **410**: 227–230.

Tong, A.H.Y., Lesage, G., Bader, G.D., Ding, H., Xu, H., Xin, X., Young, J., Berriz, G.F., Brost, R.L., Chang, M. *et al*. (2004) Global mapping of the yeast genetic interaction network. *Science* **303**: 808–813.

Tonsor, S.J., Alonso-Blanco, C., and Koornneef, M. (2005) Gene function beyond the single trait: natural variation, gene effects and evolutionary ecology in *Arabidopsis thaliana*. *Plant, Cell and Environment* **28**: 2–20.

Treinin, M., Shliar, J., Jiang, H., Powell-Coffman, J.A., Bromberg, Z., and Horowitz, M. (2003) HIF-1 is required for heat acclimation in the nematode *Caenorhabditis elegans*. *Physiological Genomics* **14**: 17–24.

Treusch, A.H., Kletzin, A., Raddatz, G., Ochsenreiter, T., Quaiser, A., Meurer, G., Schuster, S.C., and Schleper, C. (2004) Characterization of large-insert DNA libraries from soil for ennvironmental genomic studies of Archaea. *Environmental Microbiology* **6**: 970–980.

Tringe, S.G., Von Mering, C., Kobayashi, A., Salamov, A.A., Chen, K., Chang, H.W., Podar, M., Short, J.M., Mathur, E.J., Detter, J.C. *et al.* (2005) Comparative metagenomics of microbial communities. *Science* **308**: 554–557.

Trivedi, S., Ueki, T., Yamaguchi, N., and Michibata, H. (2003) Novel vanadium-binding proteins (vanabins) identified in cDNA libraries and the genome of the ascidian *Ciona intestinalis*. *Biochimica et Biophysica Acta* **1630**: 64–70.

Tudge, C. (2000) *The Variety of Life*. Oxford University Press, Oxford.

Tusher, V.G., Tibshrinai, R., and Chu, G. (2001) Significance analysis of microarrays applied to the ionizing radiation response. *Proceedings of the National Academy of Sciences USA* **98**: 5116–5121.

Tyson, G.W., Chapman, J., Hugenholtz, P., Allen, E.A., Ram, R.J., Richardson, P.M., Solovyev, V.V., Rubin, E.M., Rokhsar, D.S., and Banfield, J.F. (2004) Community structure and metabolism through reconstruction of microbial genomes from the environment. *Nature* **428**: 37–43.

Uchiyama, T., Abe, T., Ikemura, T., and Watanabe, K. (2005) Substrate-induced gene-expression screening of environmental metagenomic libraries for isolation of catabolic genes. *Nature Biotechnology* **23**: 88–93.

Ueki, T., Adachi, T., Kawano, S., Aoshima, M., Yamaguchi, N., Kanamori, K., and Michibata, H. (2003) Vanadium-binding proteins (vanabins) from a vanadium-rich ascidian *Ascidia sydneiensis samea*. *Biochimica et Biophysica Acta* **1626**: 43–50.

Urakawa, H., Noble, P.A., El Fantroussi, S., Kelly, J.J., and Stahl, D.A. (2002) Single-base-pair discrimination of terminal mismatches by using oligonucleotide microaarrays and neural network analyses. *Applied and Environmental Microbiology* **68**: 235–244.

Valinsky, L., Della Vedova, G., Scupham, A., Alvey, S., Figueroa, A., Yin, B., Hartin, R.J., Chrobak, M., Crowley, D.E., Jiang, T., and Borneman, J. (2002) Analysis of bacterial community composition by oligonucleotide fingerprinting of rRNA genes. *Applied and Environmental Microbiology* **68**: 3243–3250.

Valls, M., Bofill, R., Romero-Isart, N., Gonzàlez-Duarte, R., Abián, J., Carrascal, M., Gonzàlez-Duarte, P., Capdevila, M., and Atrian, S. (2000) *Drosophila* MTN: a metazoan copper-thionein related to fungal forms. *FEBS Letters* **467**: 189–194.

Valverde, F., Mouradov, A., Soppe, W., Ravenscroft, D., Samach, A., and Coupland, G. (2004) Photoreceptor regulation of CONSTANS protein in photoperiodic flowering. *Science* **303**: 1003–1006.

Van der Wielen, P.W.J.J., Bolhuis, H., Borin, S., Daffonchio, D., Corselli, C., Giuliano, L., D'Auria, G., De Lange, G.J., Huebner, A., Varnavas, S.P. *et al.*, and the BioDeep Scientific Party (2005) The enigma of prokaryotic life in deep hypersaline anoxic basins. *Science* **307**: 121–123.

Van der Wurff, A.W.G., Chan, Y.L., Van Straalen, N.M., and Schouten, J. (2000) TE-AFLP: combining rapidity and robustness in DNA fingerprinting. *Nucleic Acids Research* **28**: e105.

Van Elsas, J.D., Garbeva, P., and Salles, J. (2002) Effects of agronomical measures on the microbial diversity of soils as related to the suppression of soil-borne plant pathogens. *Biodegradation* **13**: 29–40.

Van Noordwijk, A.J. and De Jong, G. (1986) Acquisition and allocation of resources: their influence on variation in life history tactics. *American Naturalist* **128**: 137–142.

Van Regenmortel, M.H.V. (2004) Reductionism and complexity in molecular biology. *EMBO Reports* **5**: 1016–1020.

Van Spanning, R.J.M., Delgado, M.J., and Richardson, D.J. (2005) The nitrogen cycle: denitrification and relationship to N$_2$ fixation. In *Nitrogen Fixation in Agriculture, Forestry, Ecology and the Environment*, D. Werner and E. Newton (eds). Kluwer Academic Publishers, Dordrecht: 277–342.

Van Straalen, N.M. (1985) Comparative demography of forest floor Collembola populations. *Oikos* **45**: 253–265.

Van Straalen, N.M. (2002) Assessment of soil contamination—a functional perspective. *Biodegradation* **13**: 41–52.

Van Straalen, N.M. (2003) Ecotoxicology becomes stress ecology. *Environmental Science and Technology* **37**: 324A–330A.

Van Straalen, N.M. and Hoffmann, A.A. (2000) Review of experimental evidence for physiological costs of tolerance to toxicants. In *Demography in Ecotoxicology*, J.E. Kammenga and R. Laskowski (eds). John Wiley and Sons, Chichester: 147–161.

Van Straalen, N.M., Hensbergen, P.J., Sterenborg, I., Janssens, T.K.S., and Roelofs, D. (2006) The role of metallothionein in adaptation to heavy metals. In *Environmental Risk Assessment of Metals: New Concepts and Applications*, C.R. Janssen and H.E. Allen (eds). CRC Press, Boca Raton, FL.

Van Valen, L. (1973) A new evolutionary law. *Evolutionary Theory* **1**: 1–30.

Velculescu, V., Zhang, L., Vogelstein, B., and Kinzler, K.W. (1995) Serial analysis of gene expression. *Science* **270**: 484–487.

Velculescu, V., Vogelstein, B., and Kinzler, K.W. (2000) Analysing unchartered transcriptomes with SAGE. *Trends in Genetics* **16**: 423–425.

Venkatesh, B. (2003) Evolution and diversity of fish genomes. *Current Opinion in Genetics & Development* **13**: 588–592.

Venter, J.C., Adams, M.D., Sutton, G.G., Kerlavage, A.R., Smith, H.O., and Hunkapiller, M. (1998) Shotgun sequencing of the human genome. *Science* **280**: 1540–1542.

Venter, J.C., Remington, K., Heidelberg, J.F., Halpern, A.L., Rusch, D., Eisen, J.A., Wu, D., Paulsen, I., Nelson, K.E., Nelson, W. *et al.* (2004) Environmental genome shotgun sequencing of the Sargasso Sea. *Science* **304**: 66–74.

Vido, K., Spector, D., Lagniel, G., Lopez, S., Toledano, M.B., and Labarre, J. (2001) A proteome analysis of the cadmium response in *Saccharomyces cerevisae*. *Journal of Biological Chemistry* **276**: 8469–8474.

Vinogradov, A.E. (2004) Testing genome complexity. *Science* **304**: 389–390.

Voelckel, C. and Baldwin, I.T. (2004) Generalist and specialist lepidopteran larvae elicit different transcriptional responses in *Nicotiana attenuata*, which correlate with larval FAC profiles. *Ecology Letters* **7**: 770–775.

Voelckel, C., Weisser, W., and Baldwin, I.T. (2004) An analysis of plant-aphid interactions by different microarray hybridization techniques. *Molecular Ecology* **13**: 3187–3195.

Voget, S., Leggewie, C., Uesbeck, A., Raasch, C., Jaeger, K.-E., and Streit, W.R. (2003) Prospecting for novel biocatalysts in a soil metagenome. *Applied and Environmental Microbiology* **69**: 6235–6242.

Voordouw, G., Voordouw, J.K., Karkhoff-Schweizer, R.R., Fedorak, P.M., and Westlake, D.W.S. (1991) Reverse sample genome probing, a new technique for identification of bacteria in environmental samples by DNA hybridization, and its application to the identification of sulfate-reducing bacteria in oil field samples. *Applied and Environmental Microbiology* **57**: 3070–3078.

Vos, P., Hogers, R., Bleeker, M., Reijans, M., Van de Lee, T., Hornes, M., Frijters, A., Pot, J., Peleman, J., Kuijper, M., and Zabeau, M. (1995) AFLP: a new technique for DNA fingerprinting. *Nucleic Acids Research* **23**: 4407–4414.

Wahlund, T.M., Hadaegh, A.R., Clark, R., Nguyen, B., Fanelli, M., and Read, B. (2004) Analysis of expressed sequence tags from calcifying cells of marine coccolithophorid (*Emiliania huxleyi*). *Marine Biotechnology* **6**: 278–290.

Walbot, V. (2000) A green chapter in the book of life. *Nature* **408**: 794–795.

Walker, B., Kinzig, A., and Langridge, J. (1999) Plant attribute diversity, resilience, and ecosystem function: the nature and significance of dominant and minor species. *Ecosystems* **2**: 95–113.

Walker, D.W., McGoll, G., Jenkins, N.L., Harris, E.E., and Lithgow, G.L. (2000) Evolution of lifespan in *C. elegans*. *Nature* **405**: 296–297.

Walker, C.H., Hopkin, S.P., Sibly, R.M., and Peakall, D. (2001) *Principles of Ecotoxicology*, 2nd edn. Taylor & Francis, London.

Walker, J.J., Spear, J.R., and Pace, N.R. (2005) Geobiology of a microbial endolithic community in the Yellowstone geothermal environment. *Nature* **434**: 1011–1014.

Walsh, B. (2001) Quantitative genetics in the age of genomics. *Theoretical Population Biology* **59**: 175–184.

Wang, J. and Kim, S.K. (2003) Global analysis of dauer gene expression in *Caenorhabditis elegans*. *Development* **130**: 1621–1634.

Wang, W., Cherry, M., Botstein, D., and Li, H. (2002) A systematic approach to reconstructing transcription networks in *Saccharomyces cerevisiae*. *Proceedings of the National Academy of Sciences USA* **99**: 16893–16898.

Ward, B.B. (2002) How many species of prokaryotes are there? *Proceedings of the National Academy of Sciences USA* **99**: 10234–10236.

Waters, M.D. and Fostel, J.M. (2004) Toxicogenomics and systems toxicology: aims and prospects. *Nature Reviews Genetics* **5**: 936–948.

Wayne, L.G., Brenner, D.J., Colwell, R.R., Grimont, P.A.D., Kandler, O., Krichevsky, M.I., Moore, L.H., Moore, W.E.C., Murray, R.G.E., Stackebrandt, E. *et al.* (1987) Report of the *ad hoc* committee on reconciliation of approaches to bacterial systematics. *International Journal of Systematic Bacteriology* **37**: 463–464.

Weber, J.L. and Myers, E.W. (1997) Human whole genome shotgun sequencing. *Genome Research* **7**: 401–409.

Weinbauer, M.G. and Rassoulzadegan, F. (2004) Are viruses driving microbial diversification and diversity? *Environmental Microbiology* **6**: 1–11.

Weinig, C., Ungerer, M.C., Dorn, L.A., Kane, N.C., Toyonaga, Y., Halldorsdottir, S.S., Mackay, T.F.C., Purugganan, M.D., and Schmitt, J. (2002) Novel loci control variation in reproductive timing in *Arabidopsis thaliana* in natural environments. *Genetics* **162**: 1875–1884.

Weitzman, J.B. (2002) Transcriptional territories in the genome. *Journal of Biology* **1**: 2.

Weller, D.M., Raaijmakers, J.M., McSpadden Gardener, B.B., and Thomashow, L.S. (2002) Microbial populations responsible for specific soil suppressiveness to plant pathogens. *Annual Review of Phytopathology* **40**: 309–348.

Wellington, E.M.H., Berry, A., and Krsek, M. (2003) Resolving functional diversity in relation to microbial community structure in soil: exploiting genomics and stable isotope probing. *Current Opinion in Microbiology* **6**: 295–301.

Wenger, R.H. (2002) Cellular adaptation to hypoxia: O_2-sensing protein hydroxylases, hypoxia-inducible transcription factors and O_2-regulated gene expression. *FASEB Journal* **16**: 1151–1162.

Werck-Reichhart, D. and Feyereisen, R. (2000) Cytochromes P450: a success story. *Genome Biology* **1**: reviews 3003.1–3003.9.

Werck-Reichhart, D., Bak, S., and Paquette, S. (2002) Cytochromes P450. In *The Arabidopsis Book*, C.R. Somerville and E.M. Meyerowitz (eds). American Society of Plant Biologists, Rockville, IL: 10.119/tab.0028.

Westerhoff, H.V. and Palsson, B.O. (2004) The evolution of molecular biology into systems biology. *Nature Biotechnology* **22**: 1249–1252.

Westerhoff, H.V., Hofmeyr, J.-H.S., and Khodolenko, B.N. (1994) Getting into the inside of cells using metabolic control analysis. *Biophysical Chemistry* **50**: 273–283.

Whitaker, R.J., Grogan, D.W., and Taylor, J.W. (2003) Geographic barriers isolate endemic populations of hyperthermophilic Archaea. *Science* **301**: 976–978.

Whitfield, C.W., Band, M.R., Bonaldo, M.F., Kumar, C.G., Liu, L., Pardinas, J.R., Robertson, H.M., Soares, M.B., and Robinson, G.E. (2002) Annotated expressed sequence tags and cDNA microarrays for studies of brain and behavior in the honey bee. *Genome Research* **12**: 555–566.

Whitaker, R.J., Grogan, D.W. and Taylor, J.W. (2003) Geographic barriers isolate endemic populations of hyperthermophilic Archaea. *Science* 301: 976–978.

Williams, J.G.K., Kubelik, A.R., Livak, K.J., Rafalski, J.A., and Tingey, S.V. (1990) DNA polymorphisms amplified by arbitrary primers are useful as genetic markers. *Nucleic Acids Research* **18**: 6531–6535.

Wilson, K.H., Wilson, W.J., Radosevich, J.L., DeSantis, T.Z., Viswanathan, V.S., Kuczmarski, T.A., and Andersen, G.L. (2002) High-density microarray of small-subunit ribosomal DNA probes. *Applied and Environmental Microbiology* **68**: 2535–2541.

Wimp, G.M., Young, W.P., Woolbright, S.A., Keim, P., and Whitham, T.G. (2004) Conserving plant genetic diversity for dependent animal communities. *Ecology Letters* **7**: 776–780.

Wittstock, U. and Gershenzon, J. (2002) Constitutive plant toxins and their role in defense against herbivores and pathogens. *Current Opinion in Plant Biology* **5**: 300–307.

Wolf, Y.I., Rogozin, I.B., Grishin, N.V., and Koonin, E.V. (2002) Genome trees and the tree of life. *Trends in Genetics* **18**: 472–479.

Wood, W.B., Hecht, R., Carr, S., Vanderslice, R., Wolf, N., and Hirsh, D. (1980) Parental effects and phenotypic characterization of mutations that affect early development in *Caenorhabditis elegans*. *Developmental Biology* **74**: 446–469.

Wood, D.W., Setubal, J.C., Kaul, R., Monks, D.E., Kitajima, J.P., Okura, V.K., Zhou, Y., Chen, L., Wood, G.E., Almeida, Jr, N.F. *et al.* (2001) The genome of the natural genetic engineer *Agrobacterium tumefaciens* C58. *Science* **294**: 2317–2323.

Woods, I.G., Kelly, P.D., Chu, F., Ngo–Hazelett, P., Yan, Y.-L., Huang, H., Postlewaith, J.H. and Talbot, W.S. (2000) A comparative map of the zebrafish genome. *Genome Research* **10**: 1903–1914.

Wray, G.A., Hahan, M.W., Abouheif, E., Balhoff, J.P., Pizer, M., Rockman, M.V., and Romano, L.A. (2003) The evolution of transcriptional regulation in eukaryotes. *Molecular Biology and Evolution* **20**: 1377–1419.

Wu, L., Thompson, D.K., Li, G., Hurt, R.A., Tiedje, J.M., and Zhou, J. (2001) Development and evaluation of functional gene arrays for detection of selected genes in the environment. *Applied and Environmental Microbiology* **67**: 5780–5790.

Wu, M., Sun, L.V., Vamathevan, J., Riegler, M., Deboy, R., Brownlie, J.C., McGraw, E.A., Martin, W., Esser, C., Ahmadinejad, N. *et al.* (2004) Phylogenomics of the reproductive parasite *Wolbachia pipientis* wMel: a streamlined genome overrun by mobile elements. *PLoS Biology* **2**: 0327–0341.

Wullschleger, S.D., Jansson, S., and Taylor, G. (2002) Genomics and forest biology: *Populus* emerges as the perennial favourite. *The Plant Cell* **14**: 2651–2655.

Wynne-Edwards, K.E. (2001) Evolutionary biology of plant defenses against herbivory and their predictive implications for endocrine disruptor susceptibility in vertebrates. *Environmental Health Perspectives* **109**: 443–448.

Yang, Z. and Bielawski, J.P. (2000) Statistical methods for detecting molecular adaptation. *Trends in Ecology and Evolution* **15**: 496–503.

Yanovsky, M.J. and Kay, S.A. (2003) Living by the calendar: how plants know when to flower. *Nature Reviews Molecular Cell Biology* **4**: 265–275.

Yarzábal, A., Appia-Ayme, C., Ratouchniak, J., and Bonnefoy, V. (2004) Regulation of the expression of the *Acidithiobacillus ferrooxidans rus* operon encoding two cytochromes c, a cytochrome oxidase and rusticyanin. *Microbiology* **150**: 2113–2123.

Ye, R.W. and Thomas, S.M. (2001) Microbial nitrogen cycles: physiology, genomics and applications. *Current Opinion in Microbiology* **4**: 307–312.

Ye, R.W., Wang, T., Bedzyk, L., and Croker, K.M. (2001) Applications of DNA microarrays in microbial systems. *Journal of Microbiological Methods* **47**: 257–272.

Yoch, D.C. (2002) Dimethylsulfonopropionate: its sources, role in the marine food web, and biological degradation to dimethylsulfide. *Applied and Environmental Microbiology* **68**: 5804–5815.

Yu, C.-W., Chen, J.-H., and Lin, L.-Y. (1997) Metal-induced metallothionein gene expression can be inactivated by protein kinase C inhibitor. *FEBS Letters* **420**: 69–73.

Yu, J., Hu, S., Wang, J., Wong, G.K.-S., Li, S., Liu, B., Deng, Y., Dai, L., Zhou, Y., Zhang, X. *et al.* (2002) A draft sequence of the rice genome (*Oryza sativa* L. ssp. *indica*). *Science* **296**: 79–92.

Zavala, J.A., Patankar, A.G., Gase, K., and Baldwin, I.T. (2004) Constitutive and inducible trypsin proteinase inhibitor production incurs large fitness costs in *Nicotiana attenuata*. *Proceedings of the National Academy of Sciences USA* **101**: 1607–1612.

Zdobnov, E.M., Von Mering, C., Letunic, I., Torrents, D., Suyama, M., Copley, R.R., Christophides, G.K., Thomasova, D., Holt, R.A., Subramanian, G.M. *et al.* (2002) Comparative genome and proteome analysis of *Anopheles gambiae* and *Drosophila melanogaster*. *Science* **298**: 149–159.

Zera, A.J. and Harshman, L.G. (2001) The physiology of life history trade-offs in animals. *Annual Review of Ecology and Systematics* **32**: 95–126.

Zhang, B., Egli, D., Georgiev, O., and Schaffner, W. (2001) The *Drosophila* homolog of mammalian zinc finger factor MTF-1 activates transcription in response to heavy metals. *Molecular and Cellular Biology* **21**: 4505–4514.

Zhang, L.V., King, O.D., Wong, S.L., Goldberg, D.S., Tong, A.H.Y., Lesage, G., Andrews, B., Bussey, H., Boone, C., and Roth, F.P. (2005) Motifs, themes and thematic maps of an integrated *Saccharomyces cerevisiae* interaction network. *Journal of Biology* **4**: 6.

Zhou, J. (2003) Microarrays for bacterial detection and microbial community analysis. *Current Opinion in Microbiology* **6**: 288–294.

Zhou, J. and Thompson, D.K. (2004) Application of microarray-based genomic technology to mutation analysis and microbial detection. In *Microbial Functional Genomics*, J. Zhou, D.K. Thompson, Y. Xu, and J.M. Tiedje (eds). John Wiley & Sons, Hoboken, NJ: 451–476.

Zhou, J., Thompson, D.K., and Tiedje, J.M. (2004) Genomics: toward a genome-level understanding of the structure, functions, and evolution of biological systems. In *Microbial Functional Genomics*, J. Zhou, D.K. Thompson, Y. Xu, and J.M. Tiedje (eds). John Wiley & Sons, Hoboken, NJ: 1–19.

Zhu-Salzman, K., Salzman, R.A., Ahn, J.-E., and Koiwa, H. (2004) Transcriptional regulation of sorghum defense determinants against a phloem-feeding aphid. *Plant Physiology* **134**: 420–431.

Zimmerman, P., Hirsch-Hoffmann, M., Hennig, L., and Gruissem, W. (2004) GENEVESTIGATOR. Arabidopsis microarray database and analysis toolbox. *Plant Physiology* **136**: 2621–2632.

Zou, S., Meadows, S., Sharp, L., Jan, L.Y., and Jan, Y.N. (2000) Genome-wide study of aging and oxidative stress response in *Drosophila melanogaster*. *Proceedings of the National Academy of Sciences USA* **97**: 13726–13731.

Index

vanabins 108
vanadium 108
V-ATPase 131, 141
Venn diagram 13
Venter, J.C. 5
vernalization 187, 190, 191
Vestimentifera 159
viral communities 149, 150
vital rates 161
vitamin E 221
vitellogenin 180, 253
volcano plot 50

wFleaBase 95
WGS 31, 129, 147, 148

white rot 84, 133
whole genome shotgun
    sequencing 31, 129, 147, 148
wildfire 86, 242
Wolbachia 81
worker caste 198, 199
WormBase 92
WW domain 193

Xenbase 112
xenobiotic responsive element 213,
    228, 229
xenobiotics 170, 174, 252
Xenopus laevis 112
XRE 213, 228, 229

YAC 27, 28
Yap 231
Yarrowia lipolytica 88
yeast see *Saccharomyces*
yeast activator protein 231
yeast artificial chromosome 27, 28
yolk protein 172

Zea mays 59, 99
zebrafish 6, 13, 110
Zebrafish Information
    Network 110
zinc 209, 223, 224
zootype 69